THINKING SPACE

Social and cultural theory has recently taken a spatial turn – using geographical concepts and metaphors to think about the currently complex and differentiated world. *Thinking Space* looks at a range of social theorists and asks what role space plays in their work, what difference (if any) it makes to their concepts, and what difference such an appreciation makes to the way we might think about space. It thus looks to a two-way exchange between the appropriation of geographical ideas and the work that those spatial sensitivities perform in various theories.

Contributions from a range of geographical writers each take the work of one thinker, ranging from early this century to contemporary writers and from a wide range of disciplines. They draw out how these theorists use spatial ideas, what role these ideas play in their thinking and what this may mean for how we think, not only about theory, but also about space itself. This is done by introducing the work of the key thinkers, then taking the ideas forward and examining their potential and pitfalls. Each of the chapters takes on one approach and sees where it will go, following the implications of works for both thinking theory through a spatial lens and thinking about space.

Few other books have addressed this range of thinkers, have focused on the role space plays in their thought or what the implications are for thinking about space. For this reason, it will be of use to those looking to learn about the 'spatial turn' in theory and for those looking to see what difference space makes.

Mike Crang is Lecturer in Geography at the University of Durham and **Nigel Thrift** is Professor of Geography at the University of Bristol.

CRITICAL GEOGRAPHIES

Edited by

Tracey Skelton, *Lecturer in International Studies, Nottingham Trent University,*
and **Gill Valentine**, *Professor of Geography, The University of Sheffield.*

This series offers cutting-edge research organised into four themes: concepts, scale, transformations and work. It is aimed at upper-level undergraduates, research students and academics, and will facilitate inter-disciplinary engagement between geography and other social sciences. It provides a forum for the innovative and vibrant debates which span the broad spectrum of this discipline.

THINKING SPACE

*Edited by Mike Crang
and Nigel Thrift*

London and New York

First published 2000
by Routledge
11 New Fetter Lane, London EC4P 4EE

Simultaneously published in the USA and Canada
by Routledge
29 West 35th Street, New York, NY 10001

Routledge is an imprint of the Taylor and Francis Group

Typeset in Baskerville by Florence Production Ltd, Stoodleigh, Devon
Printed and bound in Great Britain by
St Edmundsbury Press, Bury St Edmunds, Suffolk.

British Library Cataloguing in Publication Data
A catalogue record for this book is available from the British Library

Library of Congress Cataloging in Publication Data
Thinking space / edited by Mike Crang and Nigel Thrift.
p. cm. — (Critical geographies)
Includes bibliographical references and index.
ISBN 0–415–16415–4 (hardback: alk. paper)
— ISBN 0–415–16016–2 (pbk.: alk. paper)
1. Social sciences – Philosophy. 2. Scaling (Social sciences).
3. Space in economics. 4. Geography.
I. Crang, Mike. II. Thrift, N. J. III. Series.
H61 .T473 2000

300—dc21   99–052190

ISBN 0–415–16015–4 (hbk)
ISBN 0–415–16016–2 (pbk)

CONTENTS

CONTENTS

FIGURES AND TABLES

Figures

Tables

CONTRIBUTORS

John Allen, Faculty of Social Sciences, The Open University, Milton Keynes.

Nick Bingham, Faculty of Social Sciences, The Open University, Milton Keynes.

Virginia Blum, Department of English, University of Kentucky, Lexington.

Alastair Bonnett, Department of Geography, University of Newcastle-upon-Tyne.

Mike Crang, Department of Geography, University of Durham.

Michael R. Curry, Department of Geography, University of California, Los Angeles.

Marcus A. Doel, Department of Geography, University of Loughborough.

Derek Gregory, Department of Geography, University of British Columbia.

Julian Holloway, Department of Geography, Manchester Metropolitan University.

James Kneale, Department of Geography, University of Exeter.

Tim Luke, Department of Political Science, Virginia Polytechnic Institute and State University.

Andy Merrifield, Department of Geography, Clark University, Mass.

Heidi Nast, International Studies Program, De Paul University, Chicago.

Joe Painter, Department of Geography, University of Durham.

Chris Philo, School of Geography and Topographical Science, University of Glasgow.

Steve Pile, Faculty of Social Sciences, The Open University, Milton Keynes.

Mike Savage, Department of Sociology, University of Manchester.

Pam Shurmer-Smith, Department of Geography, University of Portsmouth.

Nigel Thrift, School of Geographical Sciences, Bristol University.

Gearóid Ó Tuathail, Department of Geography, Virginia Polytechnic Institute and State University.

PREFACE

While scanning the shelves in the basement of San Francisco's famous City Lights Books recently, Charlie noticed a new section across from 'Commodity Aesthetics' called 'Topographies'. Although it was still being filled, it was clear that it would include books not only from the traditional 'terrain' of cultural geography, but also from less established fields for which thinking about spaces and places is a primary concern, from gender studies to the study of virtual reality. What the books there all have in common is a desire to 'map' some aspect of contemporary life, whether literally or metaphorically. As many people have been saying, 'Space is hot'.

(Bertsch and Sterne 1994)

This is a book about the relationships between space and theory, inspired by developments within and beyond the discipline of geography. Within the discipline there has been a burgeoning interest in social thought that has both extended and pluralised the influences drawn upon by geographers. Beyond the discipline social thought appeared to be increasingly smitten with a geographical idiom of margins, spaces and borders. However, this spatial turn was not a cause for a disciplinary triumphalism that others were turning to geography since much of it seemed resolutely ignorant of geographers and geography as a discipline. Indeed, it seemed at various times to show both deliberate ignorance of geography while – lest anyone might become chauvinistic or proprietary over the claims of the discipline – also displaying how limited much geographical thought had been.

But we were still worried that much of the geographical sensitivity to spatial terms was missing in social thought. Sometimes, a spatialised vocabulary seemed a way of drawing in a natural grounding to sustain and enable various theoretical manoeuvres. Yet, deploying the theoretical sensitivities of contemporary social thought within the discipline had problematised and denaturalised many of these terms and groundings.

And it seemed that sometimes geography functioned through a perceived Cinderella status, as an atheoretic discipline of maps and place names, which could vouch for the power of theory and provide a new realm of exploration. Yet at the same time the best geographical writing – not all of it necessarily within the discipline – had been uncovering the latent and repressed theoretical premises of projects of empirical (and imperial) description (e.g. Matless 1999, Carter 1987, Naylor and Jones 1997, Ryan 1994, Pratt 1992), unpacking the rhetorical claims of scientific geography (e.g. Livingstone 1992, Barnes 1996, Gibson-Graham 1996), setting out the workings of spatial imaginaries in contemporary issues of identity (e.g. Keith and Pile 1991, Pile and Thrift 1997, Carter, Donald and Squires 1993) and, increasingly, paying attention to nonrepresentational issues like the push of embodied practice (e.g. Thrift 1996).

This collection, then, wants to unpack the effectivity space has in social theory. It does not just appropriate theory that appears to be of a conveniently spatial nature for geographers, but also asks geographers to consider the role that space enacts in particular schools of thought. We have asked contributors who have engaged with particular writers and thinkers to unpack the way their approaches utilise spaces rather than appropriate them unproblematically. They have been asked to go travelling with the ideas.

We must then say a word about the choice of writers and theories to be discussed. It is inevitably partial and doomed to exclude many whose ideas could no doubt be profitably discussed. We have attempted to bring together a range of thinkers, from those whose work has become almost ubiquitous, to those who are less well known. Not every one of the chosen writers has been drawn upon to the same extent in geography, but they all have something interesting to say about geography. We think they relate and offer points of purchase on what we might call, following Perec (1996), species of spaces.

ACKNOWLEDGEMENTS

This book has had a long gestation and many people have contributed beyond writing parts. We would also like to thank the contributors for patience, for suggestions and developing the ideas from their bare bones. Indeed we would like to thank all those whose labour went into writing this collection, especially Anna Paskowicz. We would thank the people at Routledge, both past and present, for support and encouragement – particularly Tristan Palmer, Sarah Lloyd and Sarah Carty. The editors would like to thank Pion Ltd., for permissions to reproduce the chapters by Chris Philo, Derek Gregory and Mike Savage from *Society and Space*.

INTRODUCTION

Mike Crang and Nigel Thrift

Space is the everywhere of modern thought. It is the flesh that flatters the bones of theory. It is an all-purpose nostrum to be applied whenever things look sticky. It is an invocation which suggests that the writer is right on without her having to give too much away. It is flexibility as explanation: a term ready and waiting in the wings to perform that song-and-dance act one more time.

The problem is not so much that space means very different things – what concepts do not – but that it is used with such abandon that its meanings run into each other before they have been properly interrogated. For example, in the literature it is common to mix up what is going on in the 'real' world – for instance, changes in the space of communication which mean that certain kinds of geographical distance are compromised – spaces in theory – for example, the assumption of mobility in all its forms – and actual space – say, cities like Paris or Berlin or Naples, to name but three cities which now stand as idiolects. Then again, different disciplines do space differently. For example, in literary theory, space is often a kind of textual operator, used to shift registers. In anthropology, it is a means of questioning how communities are constituted in an increasingly cosmopolitan world. In media theory it tends to signify an aesthetic shift away from narrative – and temporal – modes of structuring primarily visual media. In geography and sociology, it is a means of questioning materiality; for example, space can be used to move closer to 'experience'. And so on. And in all disciplines, space is a representational strategy.

Then there is one more problem – space is exceedingly difficult to write about shorn of its relation to time. Though part of the reason for the turn to space in many disciplines has been a drive to move away from the tyrannies of historicism and developmentalism, the fact remains that space without time is as improbable as time without space. Thus Foucault's celebrated announcement that the era of space was succeeding that of time needs to be taken with a pinch of salt. Henri Bergson's warning about 'spatialising time' – and what we thus assume about space and time – has continued pertinence, emphasising the caution we need to muster

1

in applying the metaphor of mapping – with all the particular assumptions and practices that cartography involves – to a current era of apparent urban dissolution and reghettoisation (Jameson 1992: 2–3). Similar caution needs to be applied to claims of pluritemporalism and juxtaposition as the dominant motifs of global cultures, converting the world into a museum of contemporaneously available styles (Roberts 1988). The world can no doubt be represented as a 'teleimagistic global collage, forever in movement, . . . composed of fragments ripped from their contexts, their serrated boundaries advancing and receding in an unending deadly dance with their neighbors, their imbricated times violently clashing, diverging – only to collide again' (Burgin 1996: 185). But this representation itself requires the invention of particular theoretical spaces and times if it is to resonate.

How, then, to make some sense of this Babel of conflicting interpretations, without either producing a bland and domesticated common ground or an abandoned battlefield with only crows for company? In this book, we have tried to produce a field in which space can be indexed within constraints sufficient to say something meaningful. These constraints are the writings of certain modern philosophers and social theorists. Some of these writers are explicitly spatial in orientation. That is, they make space for space in their account. In some of their work (e.g. Lefebvre), space is indeed central. In other accounts by these writers, space is an implicit operator which needs to be teased out. But what we believe is that, through the work of these different writers, we can start to produce a more nuanced account of how and why space may be important.

What is very clear is that space is not considered by any of these writers to be outside of the realm of social practice. Equally, the ecology of thought is no longer seen as somehow standing outside of the spatial. Geography has taken the same path, moving away from a sense of space as a practico-inert container of action towards space as a socially produced set of manifolds. The former position has been engaged with and problematised through two manoeuvres. First, the long-running criticism of abstract models and thought in geography has shown over and over again that theoretical models based on reducing the world to a spaceless abstraction are of very limited utility. Not only that but they often bury within them a quest for purity and abstract reason that simulates some of the worst aspects of Enlightenment thinking (Sibley 1998). The space of theory is a purified space, defined by the purging of real spatiality and the creation of a space of thought where processes appear to be able to operate without geographical location or extent. In this vision, geography became a contaminant, threatening the pristine realm of theory.[1] Second, critiques of the abstracted space of observation as a methodological and epistemological practice illustrate that there is no Cyclops eye of theory that can stand apart from the world (Hetherington 1998) and that knowledge is always emplaced and localised (Harding 1991, Haraway 1989). Equally, the practices of knowledge are bound into a messy entanglement of the knowing and the known (Cook and Crang 1995).[2] Theory can no longer (openly at least) claim that the author stands outside what is depicted and that the position of authorship is both exterior and superior – standing not only outside space but also time (Curry 1996: 179–83).

If space then, is not a neutral medium that stands outside the way it is conceived, we can trace, and dispute, various shifts in the organisation of space alongside different forms of knowledge and social institutions. For instance, Lefebvre (1991) looks to the advent of open-ended and quantifiable space as sustaining the processes of imperial expansion and capital accumulation, while variants of Heidegger's arguments link modern subjectivity to a detachment from the world, and the relegation of the world to a pictorial object. These kinds of schematic histories can be inflected to provide all manner of histories of techniques of vision which show that far from being a given, space has a history that is bound up in ways of knowing and creates different objects of knowledge (Burgin 1996, Crary 1990, Curry 1996).

In this book, we therefore want to point to the spatiality of theory in a number of senses. Certainly, in the sense that no social process exists without geographical extent and historical duration, we need to consider the embeddedness of action in the world. In the sense that every theoretical endeavour is also geographically, historically and institutionally located, we can indeed work to unpack the travels and travails of theory as it evolves and circulates about the globe, as it is translated, transformed, channelled and reproduced. However, further than that, we also want to suggest that the role of space in the construction of theory is itself important, not only in the ways that theory might apply to a spatially distributed world, but in the spatialities that allow thought to develop particular effectivities and intensities.

1. Species of spaces

In the sections that follow, we will try to draw out some of the ways in which space figures in the strata of current philosophical and social theoretical writing. Our intention is not to be comprehensive. That would be an impossible task. But it is meant to be at least indicative of the main passage points in current writing on space, all of which in one sense or the other move away from the Kantian perspective on space – as an absolute category – towards *space as process* and *in process* (that is space and time combined in becoming).

So what species of spaces have we decided to fix on? We will begin with two of the spaces through which so much contemporary theoretical work has proceeded: spaces of language and spaces of self and other. We

3

will then move to a consideration of two of the more 'concrete' thinking spaces of 'modernity'; spaces of place and spaces of agitation. Then we will end with two spaces which, enlivened by the toil of these concrete spaces, both echo and extend the careers of the preceding spaces; namely, spaces of experience and spaces of writing.

2.1 Spaces of language

Thinking about space occurs through the medium of language. Just as there is no pristine 'thought' about the world that does not require the mediation of language, and conversely no world that is not already spoken and written, just as texts are worldly and worlds textual, so we also need to consider the relationship of space and language. When the 'textual' metaphor is applied, when the model of language has become so prominent in interpretation, it seems we need to pay more attention to the relationship between space–time and language. All too often language appears pre-ontological, prior to the worldly categories of action in space and time. And yet, the models of language that have become so prominent are actually founded upon rather particular models of space and time within language. In the end not only is space seen as linguistic but language is seen as spatial.

We can begin by taking the enormously influential work of Ferdinand de Saussure and his development of a structural linguistics. Saussure's work emerged in reaction to, and in competition with, historicist models of language which took evolutionary and developmental approaches, tracing the mutation and shifts of languages through time and cultural space. In contrast, structural linguistics discarded the diachronic, developmental view of language and created an analytical space for a synchronic pattern to be discerned. Thus, it created a linguistic space defined through its atemporality, a systemic space of 'la langue', where the only events remaining are the acts of speaking – 'la parole'. The separation of the two remains at the heart of some theories (see chapter 6) and is reworked in others (see chapter 3). This systemic space is the semiotic grid through which elements of language relate to each other, outside of time. The relationship of signifiers to each other (as opposed to signified and referents) structures language through a series of oppositions, binaries and absences. In the systemic realm, signifiers relate to each other in chains of mutual absence. The static analytic space both allows these mutually referring chains to complete circles and the space of language to cut itself free from reference, thereby creating the closed prison house that has been the cause of so much argument. Language becomes a series of synchronous spatial relationships that work to defer meaning not in time but in space, a mental space that overwhelms interpretation by not reinserting signification in lived spatiality (Lefebvre 1991: 133). As later chapters in the book will argue, this is a very specific and limited sense of what space might mean. Moreover, it produces a static model of language operating as a closed system rather than as an evolving or emergent system. The whole, in the sense of the ordering and spatial structure, determines the work of the parts (cf. De Landa 1998).

This model has ramifications beyond formal linguistics. In Derrida's deconstruction of structuralist models his main technique has been to destabilise the spatial structure (or show how unstable it was anyway). An illustration might be the critique of the law of the genre (Derrida 1980), unravelling closed semantic structures through their constitutive outside – that foundational law that is not included within the system. The deferral of meaning, the undecidability of language, seems to come not from its temporal development or deployment but from its spatial constitution.

It is just this limit to the explanatory universe offered by a language that fascinated Lacan (see chapter 9). His work looked at the creation of this structural universe and concluded that it required the introduction of a master signifier – that is a law-making and ordering principle (in his post-Freudian vision, the Law of the Name of the Father). This master signifier guaranteed the interchangeability of the other elements, and the limit of its applicability defined the edge of that realm of coherence and meaning. Reading Lacan sympathetically with Slavoj Zizek (1989, 1991a, 1991b), it is this sense of both boundedness and perforation that forms the kernel of his concerns with the limits of language. But Lacan also suggested that, although the symbolic system would be internally relational, it could not use its internal logic to explain itself or support itself. There had to be an irrational (or at least incommensurable) start point. In this model, then, there is a leap of faith, an irrational move that, once taken, enables the other rules to apply perfectly logically – offering internal coherence but also an edge to the system. After entering the symbolic realm the ladder is, as it were, kicked away.

However, we might wish to think of this moment the other way round – as retroactive. Thus we might see this limit not as an edge – the circular coherence of a symbolic realm means we never reach an edge – but rather as a hole or a tear (Gasché 1986). Or, as Zizek (1991b) would have it, like a cartoon where Wiley Coyote runs over the edge of the cliff, but stays up in the air till he looks down and sees that there is nothing beneath him. Similarly the Symbolic realm functions through believing it is grounded – and trauma is finding where it is not. In this sense then the relationship of the Real to language is as an inarticulable and traumatic exteriority that cannot be fitted within the symbolic universe (Gregory 1996). Lacan thus moved to offering an idea of language as an involuted, knotted space with complex relationships between different registers of Symbolic, Imagined and Real.

MIKE CRANG AND NIGEL THRIFT

All of which still tends to leave a sense of time and space in language where the formal relationships are a static pattern of places. Temporality, in the sense of activity, then becomes the recombination, activation and mobilisation of these points. At the level of social narrative, this view of language echoes the Russian formalist work of Vladimir Propp on folk tales. His comparative studies suggested that there were only a very few basic stories or plot lines behind most folk tales. At a general level these stories involved particular characters (hero, villain, companions) alongside particular actions (destabilising, redemptive, leaving home) and particular locales (especially home and wilderness). These basic parameters provided the scripts which action then followed. The limited number of scripts, the way events were emplotted into them, and how empirical people get assigned roles make this a useful tool to analyse the discursive shaping of social action (Shotter and Gergen 1989). However, yet again action is orchestrated around points of fixity.

It therefore seems important to address action and practice and the possibilities that then arise for rethinking the space of language (Threadgold 1997). For example, the work of Deleuze (see chapter 5) has attempted to move away from notions of representation to see language as a performativity or practice (Curry 1996: 190, see chapter 4). This is not performance as enacting a script but as creating effects. It might be the sort of transcendent empiricism (Boundas 1996) in Deleuze's model of language – one where the idea of structure lying behind the 'units' and emplaced parts of language is abandoned (cf. Taylor and Cameron 1987). As Donald (1997: 183) puts it, 'Space is less the already existing setting for such stories, than the production of space through that taking place, through the act of narration'. Here Donald draws upon the literary critic J. Hillis Miller's (1995: 7) sense of the 'atopical', wherein space is an eventful and unique happening. Like Heidegger's boundary, space is less a limit than a creation of what it encircles, more to do with doing than knowing, less a matter of 'how accurate is this?' than of 'what happens if I do it?'.

This eventful space may be the grounding of communication, but the relationship between the enunciated and the place of enunciation remains a thorny issue. One of the more sustained attempts to think through enunciation is Bakhtin's notion of the utterance (see chapter 3). His theorisation grounds language in both space and time through the chronotope as the historical realm of dominance of particular forms of language. Moreover the notion of heteroglossia, of languages always functioning in the plural and between people and places, acts to underline the diversity in languages that allows them to evolve (cf. De Landa 1998). Bakhtin embeds all language in the context of the utterance through the three terms of the speaker, the addressee and the relationship between them. He connects the incompleteness of symbolic systems to the action of the social – where two speakers do not hold the same view, where the speaker and audience see the same conversational practice from different positions. These asymmetries of knowledge and position suggest language is a dialogical process. Seen in this way, language does not stand outside time as a spatial system, but is bound into the times and spaces of action.

At one level, then, this collection asks us to engage with the role that the concept of space plays in structuring thought and language. When we are facing some of the difficulties and imponderables in different theories relating theory and practice, or when we are tempted by the spatialised vocabulary of linguistic theory, it is important to think what spaces are deployed and with what effects. This is never more important than when considering spaces of self and other, and the way that the spatial categories of interior and exterior have structured socio-spatial thought.

2.2 Spaces of self and other, interiority and exteriority

The dichotomy between, and the ethical imperatives of self and other, is often interpreted through a language of spatial containers. Instead of the notion of a qualified presence between here and there, we have the binaries of inside and outside and present and absent operating at a range of scales and historical–geographical configurations. At the level of the person, a divide between consciousness and being has long been a feature of Western thought, and has come under sustained critique. The division is evident in a variety of registers from the commonsensical notion of self being inside a fleshy container, so when someone suffers neurological damage, very often friends and relatives end up discussing whether their loved one is 'still in there', through to popular (and scientific) nineteenth-century accounts that attempted to read character from appearance, visually classifying people by features purportedly linked to sexual, criminal or racialised behaviours (Gilman 1985).

Partly in reaction to such crude socio-biology, and later determinist claims of the dominance of biological evolutionary patterns, social sciences have engaged with constructivist approaches that open out the fictive nature of many claims for biological facts. However, one side effect has been an implicit tendency to emphasise the autonomy of mind from body. The inadequacies of this inversion become apparent when we are confronted with models of self, psyche and body which try to relate organism and social persona without simply subordinating one to the other, or indeed with models which try to avoid the idea of two realms full stop. We have, then, a renewed interest in 'vitalist' or biological philosophies (De Landa 1998: 67; Deleuze 1991), in neuro-psychiatry (e.g. Sacks 1985) and in the work of Luria (e.g. Luria 1968). These are all approaches which reassert the embodiedness of the mind. More than just embodied orientation

7

(Lakoff and Johnson 1980) these approaches push us to consider the functioning of minds as part of embodiment. However, at the same time there is also a resocialisation of the ways we know the biological body (Moore 1994). Accordingly, Elizabeth Grosz (1995: 103) has argued that the body is a socio-cultural artefact, that 'corporeality is itself psychically, socially and sexually, and representationally reproduced'.

Here we have the Foucauldian sense of a body that is a site of unco-ordinated possibilities until it is trained, administered and taught how to be a body. (Indeed, Freud left a legacy of a model of 'progress' that is the movement from open living matter exposed to all forms of external excitation into a fortress of psychical defence systems (Mandarini 1998: 94)). Bodies are prime sites of communication through practical action – whether that be through purity taboos making the body a corporeal marker and performance of ethnic identity or the enactment of variable feminine and masculine roles by Peruvian prostitutes through sexual positions (Palmer 1998). From a different angle, through its historicisation, the naturalness and givenness of biology can be contested to suggest that 'culture constructs a biological order in its own image' (Grosz 1995: 104). The biological order is itself culturally known, through the changing and evolving disciplines of sciences which tell us what biology means (Moore 1994). Our analysis of a space of self must indeed become complex, when we can no longer see the cultural sat atop of biology, no longer see the body as a container, no longer see a Manichean inside–outside division. As Grosz (1995: 103) argues, we can now map the psyche onto the outside, and indeed the surface of the body onto the psyche.

A careful historical geography of a spatialised selfhood can make clear that the relationship of body and individual has indeed become complex and socially distinctive. Thus, the emergence of the modern individual is linked to the textualisation of the self and especially to the rise of diaries and autobiographies, as well as legal personae, in the early modern period (Stone 1991). Meanwhile there was the historical development of spaces for the self through the evolution of privacy in the home (Ariès and Duby 1988). The person is thereby reshaped in time and space, defined as an individual through particular spatialities of existence. These technologies, coupled with the idea of bodies as containers, produced a self defined through disciplining boundaries, and a process of mastery and self-control. Thus van den Abbeele (1991) tracks the philosophical body's travels, tracking the different spatialities of thought, and practices of thinking, in the self-mastery of the walking philosopher – be that the regular strolls of Kant or the autonomous wanderings of Rousseau – and the bridling and taming of power in Montesqieu's rides through the landscape. Technologies allowed the self to become at one level more predictable and more self-contained, and at another more extended and dispersed. For example, through the expansion of textual media conversations occurred over greater distances, linking people in disparate times and places. In other words, the modern self was created through mediating technologies and thus through not being self-present (Kittler 1999). The self could therefore be shaped through textual forms – into the narrative form of a life-like story which produced a very particular sense of agency (Somers 1992, 1994).

In a different register, the world has become full of things, objects of all sorts that can be taken up and used to create senses of the self. For example, bound together as (in most cases) shifting and incomplete projects, collections of objects offer ways of connecting to other times and places, to shape a sense of ourselves. These personal material maps, these 'auto-topographies' (Gonzalez 1995), bind the self into the world. Selves do not occur preformed, nor do they even 'interact' with the world as though self and world were pre-existing entities rubbing at the edges. Rather selves are created through, as Heidegger would have it, being-in-the-world. Boundaries are not the limits of the self but rather they create that sense of self.

We might then look at the evolution of a modern spatial self through these lenses of practice and spatialised selfhood. The unification and frag-mentation of that self through new transport, communications and media and technologies, into the *bricoleur* of urban experience (Bouchet 1998) that Simmel portrays (chapter 2), or the distracted wanderer to be found in Benjamin or Kracaeur (Crary 1992), poses different issues of thinking about the spatial self. We might distinguish a mode of experience whereby detachment and enclosure from the world reshape the nature of engage-ment – attention might turn not only to the velocity of information bombarding individuals, but also to the shift between individuals moving through informational space and information moving through individual space. Thus Virilio (1997) points to what he regards as a symptomatic shift from the modern metropolis shaped by mass movement of people and a postmodern environment of couch potatoes to which the world comes ready made (see chapter 16). As others have remarked, this is an infor-mational world where increasingly our self is linked to the world (or divided from it) through the screen – the glass pane of a car windscreen, the computer terminal or the television set (Virilio 1997, Friedberg 1993).

Yet these stories of the spatial self also imply other functions and scales. At one level, there are issues about freedom of motion – and command of space. The expansiveness of imaginative space has again and again been shown to be inflected by gender, class and historical circumstance (hooks 1991). We have only to reflect briefly on the sense of agency offered to white boys through imperialist fiction (e.g. Phillips 1996) to remind ourselves of the way that senses of self are both positioned and enabled through different configurations of global forces. These processes of spatial selfhood have been amplified through dualist models of 'self' and 'other' which support and are reinforced by a series of territorial imaginaries of inside and outside. We might typify these imaginaries as geographical fantasies, sustaining ideas of a territory of self-identity set against a radical and exoticised alterity.

This spatiality has been a repressed element of much social thought. Thus while claims of universalist rights and theory could be sustained within the west, they relied on the non-west as an arena of material and symbolic support. Claims of Enlightenment projects bringing emancipation were at their inception cast in the shadows of imperial expansion that brought domination to most. It is not simply the case that through the unleashing of a particular instrumental reason the tools of liberation eventually became forms of domination (*pace* the Frankfurt School), but that they functioned in a system founded on an exteriority that could be used as foil and counterpoint – allowing a model of Progress set up in relationship to peoples portrayed as locked in cyclical time, a model of history and agency set against peoples without history (Chakrabarty 1992, Kalpagam 1999).

Our purpose here is not to review modernisation theories, or the dynamism of capital or the refracted notions of traditional societies required by these models. It is rather to note that western theory required for itself a space of identity and homogeneity. In the words of Cornelius Castoriadis (1987), the gaze of theoria could only read what was written in terms of the same – it did not admit of spaces of alterity. We might then examine the resultant imposition of a grid of, on the one hand, western categories making non-western societies legible, ordered and controllable (Edney 1997) and, on the other, the creation of a radically unknowable alterity. Equally though, we need to allow for the internal marking of categories of western thought by the repressed exterior. The difficult positions this creates for contemporary theory, and the not always entirely successful responses, are explored in chapters 12 and 15. Notably, these chapters draw on the growing influence of postcolonial thought which has moved from attempting to decolonise the self-identities of those formerly ruled by western powers, to trying to unpick the colonial legacy in categories of western thought. If the Frankfurt school's attempt to discern the dark shadows of the enlightenment as internal and necessary parts of modernisation suggests the seeds of the holocaust lie within administered modernity (Bauman 1989), post-colonial critiques have made clear that these very ideas of modernity relate to the colonial circumstances of their creation. It is surely more than coincidental that the privileged, rational Master subject of so much theory evolved in contradistinction to an abject colonised subject. The categories and ideas of 'modern' western thought therefore need to be located geographically – their claims to universality need circumscribing and locating.

Through this hinge we might then turn to approaches which have criticised the model of the transcendent knowing subject – as formed by taking colonial practices into theoretical orbit (Bondi and Domosh 1992) and becoming locked into a logic of Mastery. It is no longer novel, but still needs reiterating, that the subject of geography was bound to these imperial knowledges. From the cult of the explorer (see Riffenburgh 1993, Driver 1991), embodying geography through a sublimated masculinity, to the creation of administered territories (Edney 1997), the practices and vision of geography have to be interrogated to disinter the assumptions of these colonising knowledges (Avery 1995, Ó Tuathail 1997).

This critique has had the effect of dislocating theory, or better, re-placing theoretical space. As Grosz notes (1995: 97) 'there is an historical correlation between the ways in which space (and to a lesser extent, time) is represented, and the ways in which subjectivity represents itself.' The striking shifts in cartographic practice, so elegantly outlined by Conley (1996), offer one example, where we can follow the shifts from cordiform maps based on the humanist analogy of the body and world, to the '*isolario*' descriptions of new lands, that came with the post-Columbian fragmentation of the coherent and known classical world, where the world becomes a never-ending series of pieces to be assembled. Thus 'the view shifts from one of the microcosmic self as mirror of the macrocosmic world to one in which both the reader and the characters discover that every figure counts as an insular entity among thousands of others' (Conley 1996: 177). Conley thereby draws attention not only to shifting orders of knowledge but also to the type of space created by theory. Thus the Cartesian foundational fantasy of self-possession depends on an alliance with a strongly marked geographical consciousness (p14). Indeed, the epistemic significance of Descartes' *Dioptrics* is that it was 'a complex technique of power, it was a means of legislating for the observer what constituted perceptual truth' (Crary 1988: 31). Considering what space is made for thinking we might take an example from the advent of the camera obscura, the darkened room wherein a secluded observer could behold a projection (or introjection) of the world outside. The camera obscura functioned in part as a spatial figure at the heart of Descartes' conception of the subject (Bailey 1989, Conley 1996, Ihde 1995: 150), where the mind functions as an inner space, and perception and thought are understood as quasi-observational activities. 'It is a figure for the observer who is nominally a free sovereign individual but who is also a privatized isolated subject enclosed in quasi-domestic space separated from a public exterior world' (Crary 1988: 33).

Following Crary's (1990, 1988: 47) exhortation to consider the practices of observation rather than the objects of the theoretical gaze forces us to think about the spaces of thought. From Vermeer's painting of the geographer, hunched over a map inside a darkened room, illuminated by light pouring in through the window, this sense of interiority and detachment in thinking, along with the consequent disembedding of the products of

11

that thought, has marked geographical endeavour. We might caricature the model here as a dichotomous pattern still repeated in critiques of vision and society that offer a pattern of 'spectacle' and 'receptacle' (Jenks 1995). This location for theorising or, to gloss Michel de Certeau (see chapter 6), this geographical operation, makes a place where facts become truths (1984: 11). It creates an observational model with an attendant representational form of knowledge and correspondence theory of truth, the form of analytical space still perpetuated in the semiotics of Barthes, where knowledge follows a model of visual representation objectifying the world with the viewer at the apex of a cone of vision (Burgin 1996: 39). This form of visual space has gone on to form a powerful model of the psychical space of mastery. But the critiques of the effect of producing an abstract world – or indeed the world as a picture – risk adopting that very subject position (*ibid.*: 47), a position we suggest can be better problematised through different ways of thinking about the spaces of knowledge and desire.

We would therefore contrast this space of interpretation with the growing concern for both non-representational modes of knowing (Thrift 1996, 1999a) as found in actor-network theory (see chapter 13), theories of practice (chapter 11), performative knowledges (see chapter 4) and the spaces of theory in dialogue (chapter 3). In particular, thinkers have been concerned with the role of the other in shaping an ethical basis for theory and this inevitably means attempting to refashion space. Thus, there are the ethical concerns of Emmanuel Levinas about the absolute Other, which rework the dialogical operations of Bakhtin, and feed into modes of engagement with the other suggested by de Certeau (cf. Godzich 1986). And yet we would do well to be reminded that Levinas develops an ethical principle of knowledge around conversation with an Other that has been purged of any sexual specificity – a rather different manoeuvre compared with the approach to unsayable difference in the psychoanalytically 'inspired' Irigaray or Cixous for example (cf. Grosz 1995: 74–5).

However, the concept of dialogic interaction points to the subject's embeddedness in places, with a shift from the centred space of the classical subject as modern visual technologies displace and deprivilege the observer – making the observer into a part of the field of vision (Crary 1990), just as the modern metropolis assimilates the subject into the space of the city (Grosz 1995: 90). This sense then produces what Guattari calls the subject as a 'specific enunciative consistency' (1992: 34) composed of a 'machinic heterogenesis', that brings parts of different orders (e.g. cognitive, affective, material, social) into contact through action. There is then

> no univocal subjectivity based on cut, lack or suture, but there are
> ontologically heterogenous modes of subjectivity, constellations
> of incorporeal universes of reference which take the position of

partial enunciators in mutiple domains of alterity, or more precisely
domains of alterification

(Ibid.: 45)

The space of knowledge therefore is not self-maintaining, but generative
of difference. These entanglements of different orders get worked out in
concrete spaces and it is to the concrete spaces of 'modernity' that we
now turn.

2.3 Metonymic spaces

There is a spectre haunting social theory and that spectre is nineteenth-
century Paris. Much of the social theory rediscovered from the first part of
this century revolves around, returns to, and is orchestrated by arguments
grounded in the history of the Parisian metropolis. And contemporary
urban theory holds a number of debts to that now lost city. At the simplest
level, we can find the roots of much urban thinking in the work of those
like Benjamin (see chapter 1) who wrote about urban life from a Parisian
context. If modernity meant the urbanisation of the mind (Schlör 1998: 16)
it often implied a specifically urban experience, whereby Paris came to be
a metonym for both urban life, urbanity, and modernity. As social theory
has become urbanised, that urban space has become generalised. For example, some theorists have taken Paris as a metonym for modernity and
Los Angeles as a metonym for post-modernity (e.g. Soja 1989, 1996). This
sort of archetypal selection and epochal mapping seems to us to miss the
point. Often it seems social theory hovers anxiously over a range of spaces
of sociality that actually work to sustain the models put forward. For
instance, the communicative reason of Habermas seems at once abstract
and cut off from the historical geography of the city and yet deeply allusive of particular forms of urbane sociality in cafes and salons (Howell
1993). The general theories of the commodity produced by Benjamin and
Marcuse are in turn marked by the retail space of Paris and the suburban
consumption of California respectively. In other words, social theory often
relies upon the 'dark matter' of a hidden city to animate its concerns.

Yet Paris does seem to hold a privileged point, as the city of imagination and theory – or at least as a theoretical imagination. In part this may
be traced to the rise of Francophone thought coupled with a centralised
and metropolitan French intellectual culture (Bourdieu 1988). Paris is
certainly implicated in the historical-geography of social thought, when
we consider how many influential theorists have located themselves and
much of their work there – in this collection, Lacan, Deleuze, Lefebvre,
Benjamin, de Certeau, Bourdieu, and Virilio. But we think it goes further,
feeding in as naturalised (or, better, urbanised) assumptions. Whilst not
reducing thought to the place where it occurs, a recurring theme in the

collection is thinking through the places, and the imagination of places, that produces theory. Thus we can then draw out three motifs that link Paris as lived and concrete space with social theory. The first is the position of Paris within the field of artistic production and thence more widely across other cultural forms. The second is the autopoietic cycling of Parisian mythology that makes the city itself a permanent intertextual field. The third is the sense of urbanity produced in a particular inflection of the urban experience.

In the first motif, Paris is the capital of art (Millan, Rigby and Forbes 1995: 15), not simply because of the random emergence of different schools but as part of a sustained effort to become a technological and leisure metropolis (Herbert 1988). Recalling Benjamin's celebrated acclamation of Paris as the capital of the nineteenth century, it is the home to a range of artistic movements that supposedly offer the ur-texts of modern life. Indeed, it would be hard to imagine some of these movements without the city. But when theory has turned to literature, with Hugo and Zola, or poetry, with Baudelaire or Rimbaud, or art, with Manet, Seurat, Rodin, or Augustus John, to grasp the urban experience, it has also turned to Paris (e.g. Harvey 1985, Clark 1985, Buck-Morss 1989, Ross 1988). This artistic pre-eminence is not without its own specificities as a field of cultural activity (Bourdieu 1995). We might note, in particular, the linkage of artistic experience with a particular intersection of sexuality and urban space. Paris also figures as one of the cities around which debates over the sexualisation of the public sphere have revolved. Thus, the issues of sexual desire in art and its relationship to the streets are often inflected by a specifically Parisian experience (Wolff 1985, Pollock 1988, Millan, Rigby and Forbes 1995: 44). Part of this orientation has been bound up with the figure of the flâneur as a trope for artistic, intellectual and urban practices (Tester 1994) as in Baudelaire's sexualised and ambivalent 'A une passante' (Wilson 1992) – with its legacy of a masculine gaze, visual consumption, commodified (and, in the sexualised public sphere, feminised) objects, which is still being taken up and reworked (Buck-Morss 1986, Shields 1989, Friedberg 1993, Wilson 1997: 136). In this sense the debates of high theory are bound up with and sustained by the very particular histories of sexual regulation of not just the urban sphere, but Paris. Studies of the regulation of 'nightlife' indicate both congruences but also antinomies between the great nineteenth-century European cities – where the concerns of sexualised space differ notably between Berlin, London and Paris (Schlör 1998).

The second motif we pick up from the reworking of these artistic and aesthetic practices is the problematic of Paris as a textualised city. As Prendergast (1992: 22, 205) notes, the mapping of paysages to pages is both appealing and problematic because of 'the great tentacular myth-making machine of "Paris"' itself. So much has been written about the city that the city has become an intertextual field where the instability and over-determination of sites seems at least in part related to the volume of past theoretical and cultural investment. And this investment in itself forms part of a practice of dislocation – when we think not just of travelling theory but of emigré writers who have particular biographies which are also geographies. Added to which, cities are already inscribed in a relational field with other cities. Thus the relationship between Glasgow and Edinburgh has been likened to that between Sparta and Athens (McArthur 1997). Second-Empire Paris set out to define itself as cultural capital of Europe against Rome. Paris is also often contrasted to Berlin as where Kracauer contrasted the latter as a city whose streets had no memory (Wilson 1997: 128). For de Certeau, Paris was a means of contrasting spaces of 'fortuitous creation' with the more obviously readable 'consciously formed' spaces (Prendergast 1992: 210). For Benjamin, Paris's 'streets served as a mnemonic system, bringing images of the past into the present' (Friedberg 1993: 73), both in relation to the ruins of modernity and to his childhood in Berlin. In other words; Paris is one of those 'cities whose greatness emerges from the interstices of their own ruins' (Olalquiaga 1992: xxi; see also Réda 1996).

Out of this semiotic swirl comes a third motif, that of a particular urbanity, an urbanity reflected in the practices of cultural life, and in the way that these are deterritorialised as universal and reflexively reterritorialised as the good city. Even in the rarefied theory of Habermas, we find echoes of the city that boasted 600 cafes by 1716, twice that by the revolution (Hetherington 1997: 15) and, by the end of the nineteenth century, a quite staggering 24, 000 (Millan, Rigby and Forbes 1995: 15). Through the nineteenth century there is also the shift from a nocturnal city composed of a labyrinth of routes to a nocturnal city composed of located places associated with particular practices (Schlör 1998). However, these places of city life form an even more privileged core of theory – in the way they function in relation to the outskirts of the city. From the nineteenth-century zones of abjection (where the chiffoniers colonised the ring of forts) to the circling of the city by the *péripherique*, Paris is a centred city whose story can be told through a history of concentric boundaries (Forbes 1995: 254). The suburbs, beyond the arrondissements within the wall of 1859, are expelled despite attempts to reincorporate them with the RER line. Indeed, it is the deprived 'banlieues' that form an unreadable alternate Paris that escapes the well-worn myths of the city. Here is a purgatory ringing the paradise of the city (Maspero 1994: 16). It is along the RER line that Maspero (1994) told his ethnographic travelogue trying to stitch the fragments of Parisian life together. Maspero's and Augé's (1986) peripatetic ethnographies offer a different take on suburbs which now seem like liners on long motionless journeys, leaving everyone in transit (Maspero 1994: 37, Forbes 1995). These marginal sites, of motion held in place, have become increasingly emblematic in both theory and art.

15

And yet the very idea of the city as unreadable seems bound up with Parisian life. '*Paris-inconnu*' forms a founding moment for anthropological excursions old and new into the city, spawning a whole series of 'Parisian-ismes' (Prendergast 1992: 3), archaeological clichés of the city (Rifkin 1993: 24). Indeed the problem of representing the incoherence of the popula-tion at the margins of the city – what Privat d'Anglemont called the 'faubourg impossible' (Prendergast 1992: 85) – is one of the driving forces behind the problematisation of representation *tout court*. For example, Baudelaire engages in a 'defiguration', that is a 'move from the unifying power of poetic metaphor to a language of heterogeneous metonymies vainly gesturing towards a whole in a context where it is precisely the sense of "wholeness" that is lacking' (*Ibid*: 130, cf. Cappetti 1993: 35). One might follow up the connections between the city and the exoticisation of the Other within by mapping this defiguration into the anthropologies of the faubourgs and those of the Chicago School as an attempt to create 'immobile landscapes' that offer a topographic legibility (Cappetti 1993: 54). Instead we would prefer to 'work from the primacy of the 'under-side', not as a mystery to be revealed but as the substantive detail of the dream' (Rifkin 1993: 9).The result though is that aesthetic forms struggle to cope with the very multiplicity of knowledges in the city, what de Certeau calls a 'heterology' (Sheringham 1996).

Paris has thus been used as emblematic of modernity, and modern life is read through an account of speed as a shattering of a spatial orienta-tion that registers in a range of aesthetic practices (Kern 1982, Lefebvre 1991). Certainly the relationship of Paris to a planning and politics of circulation and light is dense but it is more nuanced than a simple motif of flow conquering place (see Evenson 1999, Schlör 1998).[3] Paris becomes a metonym for modernity as a 'frenzy of the visible, social multiplication of images, not just circulation but extension of a geographical field of the visible, whole world' (Friedberg 1993: 22). A range of technologies, from exhibitionary practices like dioramas and photography, intersect with new transportation technologies, like the train and the automobile, to open up the city progressively to new forms of knowing. An argument can be made that these technologies open up new sights in ways which render the gaze both increasingly mobile and increasingly virtual. Thus the city rolls by the window of the seated yet moving observer or is echoed in new media like the cinema where the immobilised spectator witnesses the city in motion (Friedberg 1993).

2.4 Agitated spaces

Let us move then, to the figure of *agitated space* (Latour 1997), a figure of modernity which has become increasingly commonplace in the literature but usually as a trope radically reduced in dimension to the identification

16

of only a few key narratives, narratives which then provide all the action. One of these narratives is particularly relevant here. That is the narrative of time–space compression, as propagated by authors like Bauman, Harvey, Jameson, and Virilio. In a sense, it argues two mutually exclusive things: space becomes more important exactly as it becomes less important. This epochal story dates from the eighteenth century when in most countries in Europe comment begins to be passed on the gathering speed of travel and communication, and the simple fact that places therefore start to come closer together in time. As this process of time–space compression bites (see especially, Kern 1982; Harvey 1989; Studeny 1995) the spatialities of traditional societies and their limited incorporations are gradually replaced by a new world full of intermediary machines which enable bodies to travel and communicate more swiftly, thus rewriting the horizons of experience, including notions of space. In turn, space becomes a playground for new modes of organisation, most especially that of the state which, through the powers granted to it by these intermediaries (and the 'facts' they make possible) is able to parcel out and govern territory in ways heretofore undreamed of.

But then, as this process continues, it reaches on to a new millennial phase, especially attractive to those from the 'apocalypse now' school of social theory. The process of speed-up, boosted especially by new electronic communications media, reaches a new plane where travel is increasingly a by-product of all but instantaneous communication, rather than vice versa. This may only produce a restless 'space of flows' as in the work of Castells (1997) or it may produce something like a total dissolution of space, space as an isochronic plane, space degree zero, space as the 'lost dimension' (Virilio 1991). Over the old territorial space looms a new cybernetic spacing which is

> devoid of spatial dimensions, but inscribed in the singular temporality of an instantaneous decision. From here on, people can't be separated by physical obstacles or by temporal distances. With the interfacing of computer terminals and video-monitors, distinctions of *here* and *there* no longer mean anything.
>
> (Virilio 1991: 13)

What is remarkable is how very little criticism depictions like this – and other similar readings around notions like cyberspace – have received. Indeed in much academic literature, they are simply taken as a given, as a faithful rendition of now – or soon now. This is even though the account offered is chock full of cardinal errors: riven by a technological determinism that constantly transposes the characteristics of machines on to human subjects (Thrift 1995, 1996); by a humanism that posits a sacred human entity being invaded by machinic imperatives and transplantation;

17

indifferent to the constant backup work that is needed by mediaries and intermediaries to keep telecommunications instantaneous, especially embodied work (from sitting with backache at a terminal, to repairing a system), and; generally unable to see that mechanisms are elements of projected communities, not something set apart – 'humanity in another "state", the way that water, vapour and ice are different states of the same substance' (Latour and Powers 1998: 188). Most serious of all, such accounts fail to sense the continual process of slow adjustment in practices (including accounts) which have typified speed-up, of the addition of new cultural layers which negate the idea of a simple transmission from technology onto space.

Yet, at the same time, the notion that we live in a speeded-up world – a 'faster', 'more mobile' world – has become a resource for western cultures, a means of both making identity and making new metaphors (Heise 1997). In an age when even a non-determinist realist like Latour (1993) can liken new modes of reason to a cable-television network, this is clear. Seen in this way, writers like Virilio are not reporting back from reality but are actively producing new senses of space which, in certain senses, are the tropes of modernity powered up, renewing their cultural grip and changing our spatial sensibility in the process. In a sense, early modernist movements like futurism have, through notions like cyberspace, become a part of the everyday vocabulary the West uses to understand itself and others.

In turn, the narrative of speed-up feeds through to, and off, another one, that the world is in the midst of a phase of 'globalisation'. Again, there is the same sense of a process in which western nations are living on the leading edge of time, of a process with an historical inevitability, of a process which must produce endless spatial copies of itself. However, the narrative of globalisation seems to be one around which it is easier to gather counter-memories and minoritarian themes, and for four reasons. First, globalisation produces a much greater propensity to play on and with difference. Since borders are crossed so often in this world, issues like identity become more rather than less important – and the imaginative power that has to go into sustaining them. Increasingly 'the most powerful feature of contemporary life is cultural variety of societies, rather than variety of cultures in society: acceptance or rejection of a cultural form is no more (if it ever was) a package deal' (Bauman 1999: xliii). Second, it is much easier to see that globalisation is not a total geographical makeover. It is a process passed on networks which only go so far and so fast (Thrift 1995). Cultures stop, meet, mix, eddy, set off again. Globalisation is a space of cultural wanderings amongst cultures increasingly likely to wander. Third, globalisation is leading to the increasing questioning of fixed 'national' cultures (e.g. Beck 1999). It is possible to make entirely too much of this point as any glance at a whole series of rather nasty nationalistic wars makes clear, but, at the same time, it is increasingly clear (from both historical and contemporary evidence) that cultures are not hermetic. Especially in a world of global media cultural

> identities retain their distinct identities only in so far as they go on ingesting and divesting cultural matter seldom of their own making. Identities do not rest on the impressiveness of their traits, but consist increasingly in distinct ways of selecting/recycling/ rearranging the cultural matter which is common to all. It is the movement and capacity to change, not the ability to cling to once established form and contexts, that secures their continuity.
>
> (Bauman 1999: xiv)

It is no surprise then, that, fourth, globalisation has produced a host of spatial metaphors. These metaphors of longing and belonging tend to be 'open', based on 'points of encounter', 'contact zones', 'borderlands' and 'hybridity', and are most common in disciplines like anthropology which are precisely trying to move away from the old managed territory concept of culture with its humanistic dialectic of strange and familiar or orient and occident towards something both more and less exotic where the practices of cultural mixing can be tracked in practice, in ethnography, and in theory.

2.5 Spaces of experience

This notion of reaching out and touching things brings us, in a different incarnation, to another means of thinking space, through the concept of experience, with its implications of self-presence. Nowadays such a stream of work, represented especially but not only by the phenomenological tradition, may seem problematic as the sense of a centre or a ground or a self is undermined; many communities seem less and less local, the ground of experience is no longer necessarily ground and certainly moves and changes, the self may no longer be seen as just the body, and so on.[4]

Thus, the notion of experience as a self-evident 'thisness' clearly has to change to something more distributed. In modern philosophy and social theory, a number of streams of thought have been produced which, added together, constitute a determined assault on 'thisness', all of which, interestingly, relate in some way or another to issues of *mobility*. One such stream is the move from notions of *the body*, especially notions of the body as a privileged centre of perception, to *embodiment*, in which carnality becomes a field which only ever has a partial grip on the world and which constantly interacts with other fields, mimetically and otherwise. The second is the increasing attention to the object world. In traditions like actor–network theory, for example, thought itself always comes heavily

19

equipped, surrounded by a vast apparatus of devices and metrics which are not incidental but through a series of mediated shifts produce their own object. The third is the attention to travel. By and large, thought has often been associated with stillness, but writing from experience is increasingly considered to involve travel, both as a means of providing experience and as a means of thinking it. And, fourth, experience increasingly involves the model of writing, as the mode of inscription best able to express, through Derridean notions of the trace and deferral, the illusion of self-presence, the here and there of travel, and the need to produce models which can do something other.

In a sense, each of these literatures puts more emphasis on *practice* (Thrift 1996) but a distributed and distracted practice galvanised into action by connection in spaces which are therefore depicted as a swarm of movements and counter-movements.[5] Three main writers have attempted to produce models of these kinds of spaces, spaces which are movement and which are the sum of movements.

One is, of course, Jacques Derrida, who has produced a generalised model of writing, one in which inscription becomes a property of nature itself: hence Derrida insists that even 'the most elementary processes within the living cell' should be considered as a writing (cited in Kirby 1997: 63). In turn, this 'writing in the general sense'

> articulates a *differential* of space/time, an inseparability between representation and substance that rewrites causality. It is as if the very tissue of substance, the ground of Being, is the mutual inter-text – a 'writing' that both circumscribes and exceeds the conventional divisions of nature and culture
>
> (Kirby 1997: 61)

Yet Derrida's model of writing, performed in writing, often seems to be writing air. Though Derrida has tried to perform such qualities, his writing often seems to lack a sense of the sticky viscosity of life, of the friction of movement, as well as movement itself.[6]

For this sense we need to turn to two other authors. One is Gilles Deleuze. For Deleuze, life is an impersonal non-organic power that goes beyond any lived experience. Operating at a number of different levels, life has an overflowing transformative quality which, through encounters, constantly opens up new possibilities, and his overall aim is simply to multiply forms of life.[7]

Thus, for Deleuze, space is a crucial dimension – his 'geophilosphy' is about creating new conceptual spaces but he also wants to acknowledge other territories: of new perceptual and affective spaces (artists and actors), new image spaces (painters, filmmakers), new sound spaces (musicians), and so on. And his sensibility is inherently geographical in the sense of being surficial rather than imaginary and accurate. Thus writing of the unconscious, Deleuze (1997: 63–4) notes that

> Maps . . . are superimposed in such a way that each map finds itself modified in the following map, rather than finding its origin in the preceding one: from one map to the next, it is not a matter of searching for an origin, but of evolutionary displacements. Every map is a redistribution of impasses and breakthroughs, of thresholds and enclosures, which necessarily go from bottom to top. There is not only a renewal of directions, but also a difference in nature: the unconscious no longer deals with persons and objects but with *trajectory* and *becoming*: it is no longer an unconscious of commemoration but one of mobilisation, an unconscious whose objects take flight rather than remaining buried in the ground. [But] . . . Maps should not be understood only in extension, in relation to a space constructed by trajectories, there are also maps of intensity, that are concerned with what fills space, what subtends the trajectory. . . . A list or constellation of affects, an intensive map, is a becoming.

This kind of mobile sense of space is paralleled in certain ways by that of Michel Serres. For Serres, and his legate Bruno Latour[8], experience is also a mobile quality in which time and space are briefly patched together (brought into being) by the work of communicative operators. Time and space therefore are a 'multiple foldable diversity. If you think about it for two minutes, this intuition is clearer than one that imposes a constant distance between moving objects, and it explains more' (Serres and Latour 1995: 59). One might be able to make a basic grammar of modes of passing between these folds but

> one must be wary of the spatial image. Networks, even if you add the idea of virtual modes of tracing, leave an image in space that is almost too stable. But, if you immerse it in time, the network itself is going to fluctuate, become very unstable and bifurcate endlessly . . .
>
> This is why I use examples of turbulences in fluid, liquid or air.
>
> (Serres and Latour, 1995: 109)

Yet clearly Serres (Serres and Latour, 1995: 111–12) yearns for a synthesis and he is willing to use a spatial image, the map:

> When you are working on relationships that are in process, you're like a man who takes a plane from Toulouse to Madrid, travels

21

by car from Geneva to Lausanne, goes on foot from Paris towards the Chevreuse valley, or from Cervina to the top of the Matterhorn (with spikes on his shoes, a rope and an ice axe), who goes by boat from Le Havre to New York, who swims from Calais to Dover, who travels by rocket towards the moon, travels by semaphore, telephone or fax, by diaries from childhood to old age, by monuments from antiquity to the present, by lightning bolts when in love. One may well ask 'What in the world is this man doing?'

There are dilemmas in the mode of travelling, the reasons for the trip, the point of departure and the destination, in the places through which one will pass; the speed, the means, the vehicle, the obstacles to be overcome, make that space active. And, since I have used diverse methods, the coherence of my project is suspect. In fact, I have always analysed the mode of travel in my movements from place to place. Admittedly, the differentiation of gestures and operations can only make things difficult but, in fact, it was always a matter of establishing a relation, constructing it, fine-tuning it. And once established, thousands of relations, here, there, everywhere – after a while, when you step back and look, a picture emerges. Or at least a map. You see a general theory of relations, without any point focalising the construction or solidifying it, like a pyramid. The turbulences keep moving. The flows keep dancing.

Perhaps, in the end, this ambition for a partial, mobile synthesis was best brought together by Lefebvre as he sought to convey a quality of experience which arises out of the conglomeration of different spaces and times, sometimes in harmony, sometimes in discord, but always mobile – encountering – alive, to be found in modern societies. Lefebvre believed that this quality of experience varied and could be named through a method of 'rhythmanalysis'. Whether he was right or wrong, he perhaps came closest to producing a sense of an embodied, inhuman, travelling means of inscription. It was never very close, yet, in a sense, the value was – as in so much of this work – in the journey, rather than the destination.

2.6 Spaces of writing

Perhaps the problem may be with writing itself. This may explain the turn to 'performance' across the social sciences and humanities today (Thrift 1999a). Performances can register all the senses, they can work more closely on affect, they can communicate 'the now' and, in so far as they are not written down, in an age of writing, they can appear mysterious, even when banal. Or perhaps what is needed is more attention to the spaces of writing themselves. It is no coincidence, we suspect, that there is currently such a concentration of authors in the social sciences and humanities interested in 'performative writing'. Much has been made of the phrase 'performative writing' in recent years, an impetus to naming that has three main sources. The most well known of these is probably the work of Jacques Derrida. His numerous 'writing performances' (Derrida and Wolfreys 1998) based on boosting the 'play' of semiosis through the communicative power of the intertext, writing together traces and a search after productive entanglements in general, have challenged the whole spatio-temporal contest of language. His infamous erasures, parentheticals, ellipses and other word play, inspired in part by Barthes, may produce a breathless parataxis or a spectral delicacy – or a dense and unforgiving academicism. Then there is Judith Butler's work on discursive performativity: Butler is intent on mobilising discourse as a

> play of substitution, enabled by a founding absence that the sign attempts to fill. In other words, a sign is a 'sign of' or a 'substitute for' something other than itself. . . . Butler understands this absence or loss as the originating difference that language is unable to repair. The repeated attempt to surmount or to retrieve the differences is regarded as politically significant by Butler because it can be transformational – an opportunity that performs other possibilities
>
> (Kirby 1997: 109)

The third source is literature more generally. From Laurence Sterne to James Joyce to Samuel Beckett, writers have wanted to make language into a performance. The same impulse is to be found in recent poetry (for example, the American 'language poets' like Robert Creeley, Ron Silliman, Rosemary Waldrop and Lyn Heijinian (see Perloff 1996)), and in certain forms of aphoristic philosophy, including (in their very different ways) Nietzsche and Wittgenstein.

The impetus for this perfomativity is in part quite clearly a desire to think of writing as a space, a space to be travelled and negotiated. In turn, this fixation on writing as a space produces certain consequences.[9] First of all, once writing is seen as a spatial construction then all kinds of parallels with other spatial constructions become clear, with networks of communications and information technology, with travel and transportation, with other inscription devices (like diagrams or screens) which may be mixed liberally with the writing, or with the spaces of science. To take but one example, the spaces of chaos and complexity have become a key mediating device for much modern writing, as interpretative devices, as means of communication and as part of a more general awareness of a problem, a kind of question mark which is both scientific and more generally cultural. They 'oblige us to think of the map of problems as an account of local explorations, of discoveries of possibilities of passage that prove nothing beyond themselves, that authorise neither generalisation nor method' (Latour 1997: 9). Thus Livingston (1997), for example, attempts to produce a study of chaos as a logic at work in the historical and cultural formations of Romanticism and post-modernity by explaining the wandering spaces of writing, as in crossed letters, poetic lists, and the like. And his text, with its boxes, diagrams, and textual plays, makes explicit obeisance to this very impulse. Then second, texts can be seen as a kind of corporeal geography. For example, Genette (1997) provides an anatomy of the modern book[10] concentrating on the paratextual machinery of authorship – the cover, the author's name, the title, the dedication, the epigraph, the preface, footnotes, definitions, glossaries, and the like. As Genette (1997: 4) points out, these paratextual elements necessarily have a *location* in the text. Third, the spaces of the text, as already made clear, are spaces of constant experimentation, which attempt to write beyond current forms of textuality. Writing as a performative practice must involve reaching for new forms of inscription and display which, according to Pollock (1998), usually conform to six principles: evocation (operating metaphorically to render absence present), metonymy (a self-consciously partial or incomplete rendering), subjectivity (as a performed relation among a set of subjects), nervousness (as a general coordination of anxiety and restlessness), citational (in a mix of quotation or re-citation), and consequential (in that it is productively forceful). Thus we arrive at a kind of politics of legibility which is why, of course, 'performative writing' has been of such importance in feminist writings (see, for example, Kristeva, 1986; Sedgwick, 1993), in writings on ethnicity as a practice of 'shape-shifting' (Pollock 1997) and in writings on place (Shephard 1996). Then, fourth, it becomes possible to actually write of literary maps as, well, maps. For example, Moretti (1998: 65) has argued that we should consider mapping the spaces of writing as a method in and of itself:

> What do literary maps allow us to see? Two things basically. First, they highlight the *ortgebunden*, place-bound nature of literary forms, each of them with its peculiar geometry, its boundaries, its spatial taboos and favourite routes. And then, maps bring to light the internal logic of narrative: the semiotic domain around which a plot coalesces and self-organises.

3. Conclusion

No doubt these are just a few of the species of spaces it is possible to think space and thought through. Exotic new hybrids continue to be produced. For example, there is the current emphasis on the transhuman to be found in the transcendences of complexity theory (cf. Thrift 1999b), the productive becomings of ethology (Ansell-Pearson 1997, 1999) and the 'redeeming epidemics' of digital thinking (cf. Plant 1997). What we can say is that the 'where' is now joining the 'who', the 'what' and the 'why' of philosophy and social theory on roughly equal terms and it is providing, in its turn, a willingness 'to live to know and to practice in the complexities of tension' (Law 1999: 12). In other words, distribution may not be all, but all is distribution.

Notes

1 The realist way round a lack of spatial sensitivity through dividing necessary and contingent relations (Sayer 1985) has proved enormously appealing. Yet, as the debates have been played out, the tendency has been to consider social laws as modified by spatially contingent circumstances and geographically embedded histories – which seems to risk leaving space out of theoretical social relations.
2 We might indeed look to the desires and erotics of knowledge buried in desires for abstract and purified knowledge (de Certeau 1984). Alternatively one could study the very concrete and emplaced interests of the academy as an institution, of the labours of knowledge production and the practices through which theory comes into being and circulates (Clifford 1989, Thrift 1999b, Law 1999, Latour 1993). Geography has a seen a host of reworkings of field methodologies and challenges to some of the exoticist legacies of field work, alongside its more chest-beating proponents, but it seems to us that focusing only on the pitfalls of field work acts to reinforce the impression of the apparent transparency and reasonableness of the academy.
3 Thus we might note that in the planning of the city from Haussmann to the conversion of the riverbanks to highways there has been a presumption in favour of circulation, and yet we must also consider what was not done – such as Corbusier's plan for a A City of Three Million that would have hacked out 240 hectares of the city for the sake of uninhibited flow (Evenson 1999).
4 Thus Michel Serres (Serres and Latour 1995: 131–2)

> When I was young I laughed a lot when I read Maurice Merleau-Ponty's *Phenomenology of Perception*. He opens it with these words: 'At the outset of the study of perception, we find in language the notion of sensation'. Isn't this an extraordinary introduction? A collection of examples in the same vein, so austere and meagre, inspire the descriptions that follow. From his window, the author sees some trees, always in bloom, he huddles over his desk, now and again a red blotch appears – it's a quote. What you can decipher in this book is a nice ethnology of city dwellers who are hyper-technicalised, intellectualised, chained to their library chairs, and tragically stripped of any tangible experience. Lots of phenomenology and no sensation – everything is language.

This is actually a little harsh, given Merleau-Ponty's anti-Platonic claims for the corporeal!
5 Ironically, in human geography the force of the non-representational critique which is central to other disciplines still has to register in a subject still too often

MIKE CRANG AND NIGEL THRIFT

caught up in dreams of representation and the anaemic political fantasies that go with them.

6 See, in particular, Irigaray (1999) for a counter.

7 For Deleuze, life is an impersonal and nonorganic power that goes beyond any lived experience – an ontological concept of life that draws on sources as diverse as Nietzsche (life as 'will to power'), Bergson (the elan vital), and modern evolutionary biology (life as 'variation' and 'selection'). (Smith 1997: xiv.)

8 Thus in Latour the packing together of time and space is accomplished by networks and the communicative operators are immutable mobiles. Note that, on the whole, Serres takes a longer-term view than Latour.

9 There is also a large literature to be found on writing as a means of producing space and time, as in the work of Innis and Goody and others.

10 Similarly, a geographical history of the practices of the modern book has now emerged in the writings of Chartier and others.

References

Abbeele, G. van den (1991) *Travel as Metaphor: from Montaigne to Rousseau*. Minneapolis, University of Minnesota Press.
Ansell-Pearson, K. (1997) *Viroid Life: Perspectives on Nietzsche and the Transhuman Condition*. London, Routledge.
Ansell-Pearson, K. (1999) *Germinal Life. Deleuze*. London, Routledge.
Ariès P. and Duby, G. [general eds] (1988) *A History of Private Life*. Cambridge, Mass., Belknap.
Augé, M. (1986) *Un Ethnologue dans le Métro*. Paris, Hachette.
Avery, B. (1995) 'The subject of imperial geography', in Brahm, G. and Driscoll, M. (eds) *Prosthetic Territories: Politics and Hypertechnologies*. Boulder, CO, Westview Press: 55–70.
Bailey, L. (1989) 'The skull's darkroom: The camera obscura and subjectivity' in *Philosophy of Technology*. Dordrecht, Kluwer Academic Publishers.
Barnes, T. (1996) *Logics of Dislocation: Models, Metaphors and Meanings of Economic Space*. New York, Guilford Press.
Bauman, Z. (1989) *Modernity and the Holocaust*. Ithaca, NY, Cornell Univ. Press.
Bauman, Z. (1999) *Culture as Praxis*. London, Sage.
Beck, U. (1999) 'A cosmopolitical manifesto'. *Dissent* 46: (1) 53–55 WIN 1999.
Bertsch. C and Sterne J. (1994) 'Personal space'. *Bad Subjects* 17, November.
Bondi, L. and Domosh, M. (1992) 'Other figures in other places: on feminism, postmodernism and geography'. *Society and Space* 10: 199–213.
Bouchet, D. (1998) 'Information technology, the social bond and the city: Georg Simmel updated'. *Built Environment* 24 (2/3): 104–33.
Boundas, C. (1996) 'Deleuze-Bergson: an ontology of the virtual' in Patton, P. (ed.) *Deleuze: a Critical Reader*. Oxford, Blackwell: 81–106.
Bourdieu, P. (1988) *Homo Academicus*. Cambridge, Polity Press.
Bourdieu, P. (1995) *The Field of Cultural Production*. London, Polity Press.
Buck-Morss, S. (1986) 'The Flâneur, the Sandwichman and the Whore: the politics of loitering'. *New German Critique* 39: 99–139.
Buck-Morss, S. (1989) *The Dialectics of Seeing: Walter Benjamin and the Arcades Project*. Mass. MIT Press, Cambridge.
Burgin, V. (1996) *In/Different Spaces: Place and Memory in Visual Culture*. Berkeley, California University Press.
Cappetti, C. (1993) *Writing Chicago: Modernism, Ethnography and the Novel*. NY, Columbia University Press.

26

Carter E., Donald J. and Squires J. (eds) (1993) *Space and Place: Theories of Identity and Location*. London, Lawrence & Wishart.

Carter, P. (1987) *The Road to Botany Bay: An Essay in Spatial History*. London, Faber and Faber.

Castells, M. (1997) *The Network Society*. Oxford, Blackwell.

Castoriadis, C. (1987) *The Imaginary Institution of Society*. Cambridge, Polity Press.

Certeau, M. De (1984) *The Practice of Everyday Life*. Berkeley, California University Press.

Chakrabarty, D. (1992) 'Provincialising Europe: Postcoloniality and the critique of history'. *Cultural Studies* 6 (2): 337–57

Clark, T. J. (1985) *The Painting of Modern Life. Paris in the Art of Manet and His Followers*. London, Thames and Hudson.

Clifford, J. (1989) 'Notes on theory and travel'. *Inscriptions* 5: 177–88.

Conley, T. (1996) *The Self-Made Map: Cartographic Writing in Early Modern France*. Minneapolis, University of Minnesota Press.

Cook, I. and Crang, M. (1995) *Doing Ethnographies*. Catmog 58, Norwich Geobooks.

Crary, J. (1988) 'Modernizing VISION', in Foster, H. (ed.) *Vision and Visuality*. Seattle, Bay Press: 29–49.

Crary, J. (1990) *Techniques of the Observer*. Cambridge, Mass., MIT Press.

Crary, J. (1992) 'Spectacle, attention, counter-memory'. *October* 50: 97–107.

Curry, M. (1996) *The Work in the World: Geographical Practice and the Written Word*. Minneapolis, University of Minnesota Press.

Deleuze, G. (1991) *Bergsonism*. New York, Zone Books.

Deleuze, G. (1997) *Essays Critical and Clinical*. Minneapolis, University of Minnesota Press.

Derrida, J. (1980) 'The law of the genre'. *Critical Inquiry* 7 (1): 55–81.

Derrida, J. (1998) (ed. Wolfreys, J.) *Writing Performances. A Derrida Reader*. Edinburgh, Edinburgh University Press.

Donald, J. (1997) 'This, here, now: Imagining the modern city', in Westwood, S. and Williams, J. (eds) *Imagining Cities: Scripts, Signs, Memory*. London, Routledge: 181 201.

Driver F. (1991) 'Henry Morton Stanley and his critics: Geography, exploration and Empire'. *Past and Present* 133: 134–66.

Edney, M. (1997) *Mapping an Empire: the Geographical Construction of British India, 1765–1843*. Chicago, University of Chicago Press.

Evenson, N. (1999) 'Paris and the automobile', in Goodman, D. (ed.) *The European Cities and Technology Reader: Industrial to Post-Industrial City*. London, Routledge: 175–85.

Forbes, J. (1995) 'Popular culture and cultural politics', in Forbes, J. and Kelly, M. (eds) *French Cultural Studies: an Introduction*. Oxford, Oxford University Press: 232–63.

Friedberg, A. (1993) *Window Shopping: Cinema and the Postmodern*. Berkeley, University of California Press.

Gasché, R. (1986) *The Tain of the Mirror. Derrida and the Philosophy of Reflection* Cambridge, Mass., Harvard University Press.

Genette, G. (1997) *Paratexts. Thresholds of Interpretation*. Cambridge, Cambridge University Press.

Gibson-Graham, J.-K. (1996) *The End of Capitalism (as we knew it)*. Oxford, Blackwell.

Gilman S. (1985) 'Black bodies, white bodies: toward an iconography of female sexuality in late nineteenth century art, medicine and literature', *Critical Inquiry* 12 (1): 223–261.

Godzich, W. (1986) 'The further possibilities of knowledge', Foreword to Certeau, M. De *Heterologies: Discourse on the Other*. Manchester, Manchester University Press.

27

Gonzalez, J. (1995) 'Autotopographies', in Brahm, G. and Driscoll, M. (eds) *Prosthetic Territories: Politics and Hypertechnologies*. Boulder, CO., Westview Press: 133–50.
Gregory, D. (1996) 'Lacan' in Benko, G. and Strohmayer, U. (eds) *Space and Social Theory*. Oxford, Blackwell.
Grosz, E. (1995) *Space, Time and Perversion: Essays on the Politics of Bodies*. London, Routledge.
Guattari, F. (1992) *Chaosmosis: a New Ethico-aesthetic Paradigm*. Sydney, Power Publications.
Haraway, D. (1989) *Primate Visions: Gender, Race and Nature in the World Of Modern Science* New York, Routledge.
Harding, S. (1991) *Whose Science? Whose Knowledge? : Thinking from Women's Lives* Ithaca, NY, Cornell University Press.
Harvey, D. (1985) *Consciousness and the Urban Experience: Studies in the History and Theory of Capitalist Urbanization*. Baltimore, Md., Johns Hopkins University Press.
Harvey, D. (1989) *The Condition of Postmodernity*. Oxford, Blackwell.
Heise, U. K. (1997) *Chronoschisms. The Narrative and Postmodernism*. Cambridge, Cambridge University Press.
Herbert, R. L. (1988) *Impressionism. Art, Leisure and Parisian Society*. New Haven, Yale University Press.
Hetherington, K. (1997) *The Badlands of Modernity. Heterotopia and Social Ordering*. London, Sage.
Hetherington, K. (1998) *Expressions of Identity: Space, Performance, Politics*. London, Sage.
hooks, b. (1991) *Yearning: Race, Gender, and Cultural Politics*. London, Turnaround.
Howell, P. (1993) 'Public space and the public sphere: political theory and the historical geography of modernity'. *Society and Space* 11: 303–22.
Ihde, D. (1995) 'Image technologies and traditional culture', in Feenberg, A. and Hannay, A. (eds) *Technology and the Politics of Knowledge*. Bloomington, Indiana University Press: 147–58.
Irigaray, L. (1999) *The Forgetting of Air*. London, Athlone Press.
Jameson, F. (1992) *The Global Aesthetic: Cinema and Space in the World System*. Bloomington, Indiana University Press.
Jenks, C. (1995) 'The centrality of the eye in western culture', in Jenks, C. (ed.) *Visual Culture*. London, Routledge: 1–25.
Kalpagam, U. (1999) 'Temporalities, histories and routines of rule in colonial India'. *Time and Society* 8 (1): 141–60.
Keith, M. and Pile, S. (eds.) (1991) *Place and the Politics of Identity*. London, Routledge.
Kern, S. (1982) *The Culture of Time and Space 1880–1918*. Cambridge, Mass., Harvard University Press.
Kirby, V. (1997) *Telling Flesh, The Substance of the Corporeal*. London, Routledge.
Kittler, F. (1999) *Gramophone, Film, Typewriter*. Stanford, CA, Stanford University Press.
Kristeva, J. (1986) *The Kristeva Reader*. New York, Columbia University Press.
Lakoff, G. and Johnson, M. (1980) *Metaphors We Live By*. Chicago, University of Chicago Press.
Landa, M. De (1998) 'Virtual environments and the emergence of synthetic reason', in Broadhurst Dixon, J. and Cassidy E. (eds) *Virtual Futures: Cyberotics, Technology and Post-human Pragmatism*. London, Routledge: 65–76.
Latour, B. (1993) *We Have Never Been Modern*. Hassocks, Harvester.
Latour, B. (1997) 'Trains of thought. Piaget, formalism and the fifth dimension'. *Common Knowledge* 6: 170–91.
Latour, B. and Powers, R. (1998) 'Two writers face one Turing test: a dialogue in honour of HAL'. *Common Knowledge* 7: 177–91.
Law, J. (1999) 'After ANT; complexity, naming and topology' in Law, J., Hassard, J. (eds) *Actor Network Theory and After*. Oxford, Blackwell: 1–14.

Lefebvre, H. (1991) *The Production of Space*. Oxford, Blackwell.

Livingston, I. (1997) *Arrow of Chaos. Romanticism and Postmodernity*. Minneapolis, University of Minnesota Press.

Livingstone, D. (1992) *The Geographical Tradition: Episodes in the History of a Contested Enterprise*. Oxford, Blackwell.

Luria, A. (1968) *The Mind of the Mnemonist: a Little Book about a Vast Memory*. Cambridge, Mass., Harvard University Press.

Mandarini, M. (1998) 'From epidermal history to speed politics', in Broadhurst Dixon, J. and Cassidy E. (eds) *Virtual Futures: Cyberotics, Technology and Post-human Pragmatism*. London, Routledge: 88–99.

Maspero, F. (1994) *Roissy Express: a Journey through the Paris Suburbs*. London, Verso.

Matless, D. (1999) 'The uses of cartographic literacy: mapping surveying and citizenship in twentieth-century Britain', in Cosgrove D. (ed.) *Mappings*. London, Reaktion.

McArthur, C. (1997) 'Chinese boxes and Russian dolls: tracking the elusive cinematic city', in Clarke, D. (ed.) *The Cinematic City*. London, Routledge: 19–45.

Millan, G., Rigby, B. and Forbes J. (1995) 'Industrialisation and its discontents (1870–1914)', in Forbes, J. and Kelly, M. (eds) *French Cultural Studies: an Introduction*. Oxford, Oxford University Press: 11–53.

Miller, J. H. (1995) *Topographies*. Stanford, CA, Stanford University Press.

Moore H. (1994) 'Divided we stand: sex, gender and sexual difference'. *Feminist Review* 47: 78–95.

Moretti, F. (1998) *Atlas of the European Novel 1800–1900*. London, Verso.

Naylor S. and Jones G. (1997) 'Writing orderly geographies of distant places: the regional survey movement and Latin America'. *Ecumene* 4 (3): 273–99.

Ó Tuathail, G. (1997) *Critical Geopolitics*. Routledge, London.

Olalquiaga, C. (1992) *Megalopolis: Contemporary Cultural Sensibilities*. Minneapolis, University of Minnesota Press.

Palmer, C. (1998) 'From theory to practice: experiencing the nation in everyday life'. *Journal of Material Culture* 3 (2):175–99.

Perec, G. (1996) *Species of Spaces*. Harmondsworth, Penguin.

Perloff, M. (1996) *Wittgenstein's Ladder. Poetic Language and the Strangeness of the Ordinary*. Chicago, University of Chicago Press.

Phillips, R. (1996) *Masculinity and Adventure Fiction*. New York, Guilford Press.

Pile, S. and Thrift, N. (eds) (1997) *Mapping the Subject*. London, Routledge.

Plant, S. (1997) *Zeros and Ones*. London.

Pollock, D. (ed.) (1997). *Exceptional Spaces. Essays in Performance and History*. Chapel Hill, University of North Carolina Press.

Pollock, D. (1998). 'Performative writing' in Phelan, P., Lane, J. (eds) *The Ends of Performance*. New York, New York University Press: 73–103.

Pollock, G. (1988) *Vision and Difference: Femininity, Feminism and the Histories of Art*. London, Routledge.

Pratt, M. (1992) *Imperial Eyes. Travel Writing and Transculturation*. London, Routledge.

Prendergast, C. (1992) *Paris and the Nineteenth Century*. Oxford, Blackwell.

Réda, J. (1996) *The Ruins of Paris*. London, Reaktion.

Riffenburgh, B. (1993) *The Myth of the Explorer*. Oxford, Oxford University Press.

Rifkin, A. (1993) *Street Noises: Parisian Pleasure, 1900–40*. Manchester, Manchester University Press.

Roberts, D. (1988) 'Beyond progress: The museum and the montage'. *Theory, Culture and Society* 5: 543–57.

Ross, K. (1988) *The Emergence of Social Space: Rimbaud and the Paris Commune*. Minneapolis, University of Minnesota Press.

Ryan, J. (1994) 'Visualising imperial geography'. *Ecumene* 1: 157–76.

Sacks, O. (1985) *The Man who Mistook his Wife for a Hat*. London, Picador.

Sayer, A. (1985) *Method in Social Science*. London, Hutchinson.

Schlör, J. (1998) *Nights in the Big City: Paris, Berlin, London, 1840–1930*. London, Reaktion.

Sedgwick, E. K. (1993) *Tendencies*. Durham, NC, Duke University Press.

Serres, M. and Latour, B. (1995) *Conversations on Science, Culture and Time*. Ann Arbor, University of Michigan Press.

Shephard, K. (1996) Princeton, Princeton University Press.

Sheringham, M. (ed.) (1996) *Parisian Fields*. London, Reaktion.

Shields, R. (1989) 'Social spatialisation and the built environment: the West Edmonton Mall'. *Society and Space* 7: 147–64.

Shotter, J. and Gergen, K. (eds) (1989) *Texts of Identity*. London, Sage.

Sibley, D. (1998) 'Sensations and spatial science: gratification and anxiety in the production of ordered landscapes', *Environment and Planning A*, 30 (2) 235–46.

Smith, D. W. (1997) ' "A life of pure immanence" Deleuze's Critique et Clinique project' in Deleuze, G. *Essays Critical and Clinical*. Minneapolis, University of Minnesota Press: xi-xiii.

Soja, E. (1989) *Postmodern Geographies*. London, Verso.

Soja, E. (1996) *Thirdspace*. Oxford, Blackwell.

Somers, M. (1992) 'Narrative, narrative identity and social action: rethinking English working-class formation'. *Social Science History* 16 (4): 591–630.

Somers, M. (1994) 'The narrative constitution of identity: a relational and network approach'. *Theory and Society* 23: 605–49.

Stone, A. R. (1991) 'Will the real body please stand up?: boundary stories about virtual cultures', in Benedikt, M. (ed.) *Cyberspace: First Steps*. Cambridge, Mass., MIT Press: 81–119.

Studeny, C. (1995) *L'Invention de la Vitesse, France XVIII-XX Siècle*. Paris, Gallimard.

Taylor, T. and Cameron, D. (1987) *Analysing Conversation: Rules and Units in the Structure of Talk*. Oxford, Pergamon Press.

Tester, K. (ed.) (1994) *The Flâneur*. Routledge, London.

Threadgold, T. (1997) *Feminist Poetics. Poiesis, Performance, Histories*. London, Routledge.

Thrift, N. J. (1995) 'A hyperactive world' in Johnston, R. J., Taylor, P. J., Watts, M. (eds). Oxford, Blackwell.

Thrift, N. J. (1996) *Spatial Formations*. London, Sage.

Thrift, N. J. (1999a). 'Afterwords'. *Environment and Planning D. Society and Space*. (forthcoming).

Thrift, N. J. (1999b) 'The place of complexity'. *Theory, Culture and Society* 16: 31–70.

Virilio, P. (1991) *The Lost Dimension*. New York, Semiotext(e).

Virilio, P. (1997) *Open Sky*. London, Verso.

Wilson, E. (1992) The Invisible Flâneur. *New Left Review* 191: 90–110.

Wilson, E. (1997) 'Looking backward: nostalgia and the city', in Westwood, S. and Williams, J. (eds) *Imagining Cities: Scripts, Signs, Memory*. London, Routledge: 127–39.

Wolff, J (1985) 'The Invisible Flâneuse', *Theory Culture and Society* 2 (3): 37–48.

Zizek, S. (1989) *The Sublime Object of Ideology*. London, Verso.

Zizek, S. (1991a) *For They Know Not What They Do: Enjoyment as a Political Factor*. London, Verso.

Zizek, S. (1991b) *Looking Awry: An Introduction to Jacques Lacan through Popular Culture*. Cambridge, Mass., MIT Press.

Part 1

UR-TEXTS AND STARTING POINTS

1

WALTER BENJAMIN'S URBAN THOUGHT

A critical analysis

Mike Savage

1 Introduction

In recent years there has been a major upsurge of interest in the work of the German cultural critic Walter Benjamin (1892–1940). Attention has been directed particularly to the relevance of Benjamin's writings for literary criticism (for example, Eagleton, 1981; Jennings, 1987), philosophy (for example, Benjamin and Osborne, 1993; Roberts, 1982), social theory (for example, Buci-Glucksmann, 1994; Buck-Morss, 1989; Frisby, 1985), and cultural studies (Lury, 1992; McRobbie, 1992). By contrast, the direct significance of Benjamin's ideas to the study of urbanism has attracted relatively little attention (the important exception to this is Szondi, 1988, also, see Frisby, 1985), though there are recent indications that this is changing (Buci-Glucksmann, 1994; Burgin, 1993; Cohen, 1993; Gregory, 1994; Wolff, 1993). My intention in this paper is to examine critically the way Benjamin explored the 'urban' in his work, in order to emphasise the distinctiveness and originality of his writing in this area. I aim not to make an original contribution to the now formidable field of Benjamin scholarship,[1] but rather to position Benjamin's ideas so that they may fruitfully be brought to bear on those working in urban studies and related fields.

The principal issue which I will interrogate with Benjamin's work as my guide concerns the value of 'culturalist' approaches to urbanism, which have undergone a striking revival in recent years. In the late 1970s and early 1980s it was widely believed that concepts of 'urban culture' were unsustainable, as advanced capitalist countries were characterised by the breakdown of a distinction between the city and country, with the result that cities lost any cultural distinctiveness they might once have possessed (see Giddens, 1981; Mellor, 1977; Saunders, 1981; Smith, 1981). Instead, different versions of political economy were championed as being able to explore the various processes which produced specific urban sites (for example, Harvey,

1982). However, ten years later, the situation has been transformed. Largely as a result of poststructuralist influences, there has been a growing interest in reading cultural artifacts as 'texts', and the city has been no exception to this. Keith and Cross (1993: 9) argue that 'the urban narrative has re-emerged triumphantly as a genre in which the city can be read as both emblem and microcosm of society', and there are a number of recent studies which appear to support their contention.[2] The study of representations of cities has become the subject of considerable attention in art history (for example, Clark, 1985; Seed and Wolff, 1988) and literature (for example, Tanner, 1990; Williams, 1973), and there is evidence that the endeavours of cultural geographers to read landscapes is applied increasingly to urban settings (for example, Duncan, 1990; Zukin, 1991). What remains uncertain, however, is the status of the 'urban' in this work. Is the 'urban' simply a discursive term standing in opposition to others (such as 'rural')? Is the 'urban' synonymous with the built environment, in the way that Olsen (1986) suggests in his account of how cities can be seen as 'works of art' in terms of their architectural forms? Or do distinct forms of urban social or cultural relationship exist? Is the 'urban' merely a convenient site in which a variety of social or cultural processes can be explored (as suggested by Savage and Warde, 1993), including, possibly, the various textual processes and narratives which encode the 'urban'? Just as debates about urban studies have never been able to resolve the question of defining what the 'urban' actually is (see Saunders, 1981), so the recent cultural trend in urban studies has tended to dodge this key issue. In this paper I will argue that the work of Walter Benjamin offers a fruitful way of considering the stakes involved in these sorts of issues. Benjamin talks about the urban in very different ways, however, and it is no easy matter to elucidate his views. In what follows I argue that the ideas which he advances in one of his best-known essays, 'On some motifs in Baudelaire' (1968b), where he appears to articulate a fairly conventional view of the city as site of a distinctively modern experience, is not the best starting point. I argue that this essay does not reflect his more complex interest in the 'urban', and suggest that a more interesting angle of approach is to consider how Benjamin examined the relationship between history, experience, memory, and the built environment. This concern was closely related to his fascination with the ways in which cities could (and could not) be represented textually. I argue that attention to this aspect of Benjamin's urban thought illuminates other aspects of his thinking.

I begin in section 2 by offering a brief introductory survey of Benjamin's urban writing in order to emphasise its complexity as well as the central part which it came to play in his later writing. I then offer a thematic discussion of some of the central issues which Benjamin's urban writing provokes. In section 3 I discuss how Benjamin viewed, sociologically, the distinctive nature of urban 'experience' in his later work, emphasising his concern not to describe modern experience but to suggest how it could be redeemed. In section 4 I consider how Benjamin used urban writing critically, as a device for disrupting narrative meaning. Finally, in section 5 I seek to place Benjamin's urban writing in the context of his philosophy of history and relate it to his idea of 'aura' in order to explain why the city was so compelling to him. In the concluding section of the paper I pull out some of the implications of this survey for contemporary approaches to urban culture.

2 Benjamin's urban writings

Benjamin's interest in cities developed only in the course of the 1920s, especially as his work on *The Origins of German Tragic Drama* (1977b) reached its conclusion and he turned to critical issues of more direct political concern. By the late 1920s, it can be argued that Benjamin constantly used cities to frame his inquiries. However, being well aware of the complexities of representational processes, he wrote about cities in a variety of ways, with little consistency of approach, and no single view on urbanism can be discerned in his work. As a starting point it is useful to summarise the different characters of his urban texts. At least five different types of urban writings can be elucidated.

1   Benjamin's city portraits of Naples (1924), Moscow (1927), and Marseilles (1928) (all included in Benjamin, 1979a) are all accessible short essays, written in journalistic, largely descriptive, style, reflecting on urban life and culture in the respective cities. Their interest derives partly from their being the first works where Benjamin's direct interest in cities was manifest.
2   *One Way Street* (1925–26) (in Benjamin, 1979a) is one of Benjamin's best-known works, also written in the mid-1920s, at the time of his developing interest in Marxism and surrealism. Its status as an urban work needs justification. It was written as a series of aphorisms, many of which were titled around typically urban sights such as 'filling station', 'underground works', 'caution: steps', and so forth. He uses urban wandering as a device on which to hang a series of reflections which seem to be triggered by phenomena of the urban built environment.
3   Benjamin's autobiographical sketch, 'A Berlin chronicle' (1932), which incorporates his own childhood memories into an account of Berlin and Paris (in Benjamin, 1979a) and uses a complex narrative form involving 'photographic' recollections of his youth, is distinctive in being one of the relatively few works in which Benjamin discussed directly his own experiences (see Witte, 1991). In the same category might be placed his 'Moscow diary', a recently translated diary of his two-month stay in Moscow in 1927–28 (Benjamin, 1990).

4   The Passagenwerk (or Arcades project) absorbing Benjamin for much of the last decade of his life was a study of Paris, 'Capital of the nineteenth century'. Never completed, it left in published form a number of essays (for example, Benjamin, 1968b; 1968d; 1968e), a series of writings translated as *Charles Baudelaire: A Lyric Poet in the Era of High Capitalism* (1973a), and a series of detailed notes, mostly still untranslated into English (though, see Benjamin, 1985). Benjamin's aim was to use montage techniques to explore the relationship between Baudelaire, urbanism, and the development of capitalism. Much recent Benjamin scholarship has been preoccupied with reconstructing the purpose, methods, and ideas of this work (Buck-Morss, 1989; Tiedmann, 1988), though there has been relatively little attention to its urban dimensions (though see Cohen, 1993).

5   In many of Benjamin's theoretical and philosophical reflections, some of them arising out of the Passagenwerk, there are a number of observations and references to architecture, the built environment, and other urban phenomena. Many of the otherwise disparate essays collected in *Illuminations* (1968a), and *One Way Street* (1979a) contain important asides on cities.

This brief description indicates that after the mid-1920s Benjamin constantly used urban phenomena as devices for exploring the intellectual problems with which he had grappled throughout his life and to which older means of inquiry seemed inappropriate. This does seem to mark an important shift in his thinking. Much critical discussion of Benjamin has examined whether he replaced his early messianic and romantic thought with a more critical (though never orthodox) Marxism in the mid-1920s (see Roberts, 1982) or whether he remained enmeshed in the same sort of intellectual dilemmas throughout his life (see McCole, 1993). In the context of this debate it is interesting to note that at least in methodological terms Benjamin did appear to recast the literary criticism and abstract philosophical–ethical discussion which characterised his early writing (most notably, *The Origins of German Tragic Drama*) into a sort of urban criticism (further, see Cohen, 1993). However, it is evident from the variety of his urban writings that Benjamin wavered and experimented in his textual approach between journalistic narrative, memoir, aphorisms, essays, and montage. Similarly, the urban as 'object' shifted incessantly – from being the general properties of the built environment, to specific buildings, the nature of urban 'experience', accounts of particular cities and their histories, and the ability of certain forms of representation (such as photography) to 'picture' cities. It is the complex and manifold meanings of the urban in Benjamin's work which makes that work of such contemporary interest.

3 Modern urban experience in Benjamin's thought

In many accounts of Benjamin's writing the way he saw the city as being characteristically modern (in Frisby's words, as 'the crucial showpiece of modernity' (1985: 224)) is stressed (for instance, see Turner, 1994: facing 25). Here, Benjamin seems to follow a long, orthodox tradition within social theory. The notion of a distinctly modern urban experience, or 'way of life', has, from the time of Simmel and Wirth, been the central focus of sociological discussion of urban culture (see Savage and Warde, 1993). The transition to modernity is seen as leading to profound changes in the nature and quality of social relationships, and both Tonnies (1988) and Simmel (1950) argued that these changes could be seen most clearly in cities, where modernity was most developed. The city, in opposition to the small-scale community, was the main locale in which new impersonal social relationships, the money economy, and social disorganisation could be observed. This was also the conception which modernist novelists evoked in their writings, conceiving the city as symptomatic of the new and modern (Anderson, 1988; Berman, 1983; Bradbury and MacFarlane, 1976; Williams, 1989).

In the late 1920s and early 1930s Benjamin articulated a theory of history which had many parallels with this account (Honneth, 1993; Roberts, 1982). This theory of history elaborated the replacement of 'experience' (*Erfahrung*) by instrumental reaction (*Erlebnis*). In the former state, found in preindustrial societies, experience is based in habit and repetition of actions, without conscious intention. These experiences are bound to traditions, the socially constructed and legitimated ways of acting, which gain their authority by their uniqueness and specificity. In the latter state, found in modern industrial societies, the mass reproduction of commodities and symbols disperses tradition, so that individuals simply react to the stimuli of the environment and develop instrumental ways of thinking in order to cope in such a changed environment (for a discussion, see Roberts, 1982: 157–95).

However, there is considerable disagreement concerning the status of this theory of history in Benjamin's overall thought. Some commentators (Bauman, 1993) argue that it goes against the persistently nondeterminist, redemptive character of Benjamin's general thought, manifested notably in his 'Theses on the philosophy of history' (see Benjamin, 1968e). Here Benjamin criticised historicist ideas of progress and trend in history so that the possibility of redemptive action could always be held open, a view which appears to contradict his own formulations concerning the replacement of *Erfahrung* by *Erlebnis*. And in line with this view, Benjamin's account of urban experience turns out to be more complex than that of Simmel and Wirth. Consider Benjamin's celebrated account of 'urban experience' in his essay 'On some motifs in Baudelaire' (1968b). Here he seems concerned to establish a relationship between the modern city and the development of the 'shock experience'. Benjamin argued that the daily routine bombardment of people's senses by various shocks forces them to use consciousness as a filter to protect themselves, and he discussed this in relationship to the urban masses as well as to modern factory workers. Benjamin saw Baudelaire as the poet of the 'metropolitan masses', as the first writer who refused to stand apart from the urban mass in order to write detachedly about them but who immersed himself in the experience of the masses.

In this essay he refers mainly to Proust and Freud in developing its arguments, but its formulations appear initially very close to those developed by Simmel in his famous essay 'The metropolis and mental life' (in Wolff, 1950) written thirty years before. And it is a difficult essay to interpret. The published version of 'On some motifs in Baudelaire' was written in response to Adorno's criticisms of an earlier paper, and there are indications that some of Benjamin's formulations (notably his unusual references to Freud) were included only in order to improve its acceptability for publication (see Jennings, 1987). But, even as it stands, the essay does not fully endorse the Simmelian idea of a 'modern urban experience'. References to city dwellers are only important to his argument because they allow the artist (Baudelaire) to come into direct contact with the crowd, and, through Baudelaire's art, allows the redemption of the modern shock experience. Benjamin evokes urban experience not because of its typicality but in order to suggest its redemptive possibility through art. The urban is important in Benjamin's argument not because it is the prime site of modern experience – here the factory served his purposes better – but as the site in which the possibilities for redemption could best be explored (for a modern parallel, see Berman, 1983). This suggests that Benjamin's purpose was rather different from that of Simmel. As he informed Adorno, his point was not to write a sociological account of the mass urban experience: 'I would see the crux of [the essay] in the theory of the flâneur . . . At the core of the text, my critique of the concept of masses, made tangible by the modern metropolis, must be brought out' (in Adorno and Scholem, 1994: 589). Benjamin claims that the metropolis only makes 'tangible' his main concerns rather than embodies them. His prime focus is on the flâneur, the street wanderer who is able to subvert conventional meanings and values and thereby offers a critique of the impersonal notion of the 'mass'. Benjamin's interest in the flâneur, furthermore, is not primarily concerned with delineating it as an actual social type which existed in specific urban historical settings, but as a theoretical, critical, counter to the idea of the mass, as an attempt to indicate the sorts of potential for critique which continued to exist.

Even in this central text, therefore, Benjamin is not primarily concerned with describing modern urban experience. Elsewhere, his interest in this project was even flimsier. In some of his city portraits, such as his accounts of Naples, Marseilles, and Moscow, Benjamin emphasises the way that backwardness, as much as modernity, is evident in city life. In the case of Moscow he points to the extent to which village life is reproduced within Moscow itself (Benjamin 1979a: 202). And indeed in other work Benjamin appears to see the countryside rather than the city as the seat of modernising tendencies, noting that the road systems which come to play an expanding role in modern cities are actually a rural phenomenon (1979a: 59). In other works Benjamin uses the idea of the urban dweller in ways too loose to suggest he placed any weight on the term. An illuminating example occurs in a letter to Scholem in 1938 when Benjamin argued that Kafka's work 'is an ellipse with foci that lie far apart and are determined on the one hand by mystical experience . . . and on the other by the experience of the modern city dweller' (Adorno and Scholem, 1994: 563). He went on (563–4):

> When I speak of the experience of the city dweller I subsume a variety of things under this notion. On the one hand, I speak of the modern citizen who knows he is at the mercy of the vast bureaucratic machinery . . . On the other hand, by modern city dwellers, I am speaking of the contemporary of today's physicist[3] . . . [If] I were to say, as I just did, that there was a tremendous tension between Kafka's experiences that correspond to present day physics and his mystical ones, this would only amount to a half truth.

Although Benjamin begins by appearing to make a great claim for urban experience (in contrast to the mystical experience) he later collapses it into two disparate points about bureaucratic experience and a 'physicist's' frame of mind. Then, in recapping Kafka's thought, Benjamin, presumably inadvertently, substitutes the idea of the physicist for that of the city dweller, so leaving the latter term redundant. This strengthens the argument that Benjamin was not interested in working out an account of modern urban experience, but used the term more loosely. In particular he used it primarily as a foil for the redemptive possibilities inherent in the figure of the flâneur.

Benjamin's account was therefore not of the type to set aside those of Simmel or contemporary urban sociologists such as those of the Chicago School. As Frisby emphasises, unlike Simmel, Benjamin's interest in the city was linked to its role as a 'labyrinth', where all kinds of lost dreams, hopes, and artifacts, swept aside by more recent fashions and developments, resided, yet which the urban explorer might stumble across, allowing her or him to gain access to the 'prehistory of modernity' and so to rupture any naive evolutionary belief that the present marked a state of progress over the past. The urban experience thus reveals modernity not as progress but as the latest episode of the 'ever-same'. Before I develop this point I pass on to consider Benjamin's 'textual' approach to urbanism.

4 Urban textuality as critique

If Benjamin did not evoke the urban in order to describe urban experience, what did he use it for? One alternative possibility is that he saw urban writing as a critical device allowing established and conventional values to be put into question. This might suggest a much more radically 'textual' approach to cities than we have considered so far, in which Benjamin is not interested in describing cities or urban experiences but in using urban writing as a critical device, developing his concern with how 'allegory' can shatter logocentric forms of reason (Buci-Glucksmann, 1994). This is the argument adopted by Gregory (1994) in his claim that Benjamin's interest in cities was related to his 'critique of narrativity'. Conventional narratives promote a linear account of historical progress, and their disruption involves breaking their conventions. Benjamin therefore 'spatialised' time, 'supplanting the narrative encoding of history through a textual practice that disrupted the historiographic chain in which moments were clipped together like magnets' (Gregory, 1994: 234). By invoking the 'city' he allowed himself a way of writing which disconnects and subverts meanings by placing words alongside each other in unconnected ways, in the same way as various urban sites jostle, unconnectedly, against each other.

It is certainly true that much of Benjamin's work saw him avoiding conventional narrative, experimenting with other techniques, such as his use of visual images and diagrammatic devices. Most famously, in his *Passagenwerk* he deliberately eschews narrative in favour of a montage of quotations and aphorisms organised around different, apparently unconnected, headings. Buck-Morss (1989) has argued that, insofar as its overall architecture can be discerned, the *Passagenwerk* is best understood as related to a series of oppositions, which in diagrammatic form contrast the 'waking' to the 'dream', and the 'transitory' to the 'petrified'. This same endeavour to find visual ways of representation explains Benjamin's observation that 'I have long, indeed for years, played with the idea of setting out the sphere of life – bios – graphically on a map' (1979a: 295). It helps explain also Benjamin's fascination with the flâneur, the figure whose aimless wanderings can reveal things hidden to those intent on purposive linear goals.[4]

Benjamin's strategy was therefore to displace, by questioning the boundaries between past and present, the notion of linear historical time which was sustained by narrative form. In his urban writing Benjamin could use a common spatial reference to bring things together in time. Thus the city could be used to disrupt ideas of new and old, with the result that 'antiquity is revealed in modernity, and modernity in antiquity' (Adorno and Scholem, 1994: 557). As he phrased it in his essay 'Central Park' (1985: 34): 'The modern standing opposed to the antique, the new stands in opposition to the always-the-same (the modern: the masses; the antique: the city Paris)'. Burgin (1993) has also noted, focusing especially on Benjamin and Lacis's account of Naples (1979a), how Benjamin saw the urban as problematising typical divisions 'inside' and 'outside', and hence saw the city as dislocating established, conventional, dualisms.

In many respects Benjamin was simply drawing here upon surrealist currents. From the mid-1920s Benjamin became interested in the avant-garde techniques used by the surrealists, especially Breton and Aragon. The surrealists themselves were obsessed by the city, especially Paris, the subject of Aragon's *Payson de Paris*, and Breton's *Nadja* (see Cohen, 1993). Surrealists used the chaos and variety of urban experience to sabotage tradition and order. Benjamin noted that this surrealist approach depended on the city: 'no face is surrealistic in the same degree as the true face of the city' (1979b: 230). Benjamin's own interest in their work sprang from their concern with experience: '[their] writings are concerned literally with experiences, not with theories and still less with phantasms' (1979b: 227). The experiences which the surrealists evoked Benjamin called 'profane illuminations'. The use of disruptive, shocking, artistic techniques put received wisdom and traditions in question, and so developed a critical perspective in which there was a 'substitution of a political for a historical view of the past' (1979b: 230).

It would be wrong to see Benjamin's urban writing simply as echoing surrealism, however. As numerous writers have pointed out (Cohen, 1993: chapter 7; Tiedmann, 1988), Benjamin's attitude to surrealism was not uncritical. He was concerned that the 'profane illuminations' surrealists sought might become an end in themselves rather than a means to a political end. As surrealist interest lay in disrupting 'reality' it could lead towards a symbolism which came to deny any historical sense and which might spill over into a concern with 'art for art's sake'. The surrealist advocacy of avant-garde techniques allowed the disruption of established meaning to slide over into a form of romantic narcissism in which the avant-garde valorised itself (also, see Wolin, 1982). As a result, Tiedmann (1988) argues that, although the *Passagenwerk* in its early years before 1930 was indebted to surrealism, this influence became more nuanced thereafter, partly as a result of Benjamin's renewed interest in Marxism. In his typically evasive way, Benjamin wrote to Scholem in 1929 that his essay on surrealism was 'a screen placed in front of the *Paris Arcades*' (Adorno and Scholem, 1994: 348), a quote which can be taken to indicate both the importance of surrealism and the fact that Benjamin was using it 'as a screen' to come up with something secret behind the facade.[5]

The crucial issue here is that Benjamin was not simply concerned with disrupting meaning, but also with recovering it, and this set him apart from at least some strains within surrealism.

> Language shows clearly that memory is not an instrument for exploring the past but its theatre. It is the medium of past experience, as the ground is the medium in which dead cities lie interred. He who seeks to approach his own buried past must conduct himself like a man digging ... Fruitless searching is as much a part of this as succeeding, and consequently remembrance must not proceed in the manner of a narrative or still less that of a report, but must, in the strictest epic and rhapsodic manner, assay its spade in ever-new places, and in the old ones dig to even deeper levels.
>
> (Benjamin, 1979a: 314)

Benjamin's attempt at recovery was therefore concerned with memory. But his approach to memory was related to his critique of conceptual, theoretical, and narrative knowledge. Benjamin endorsed a view in which truth could not be grasped conceptually by an intentional intellectual process. Instead, he upheld the idea that truth 'is self-representation, and is therefore immanent in it as form' (1977b: 30). People cannot go out and find truth, but truth must reveal itself to people. This much is consistent with his advocacy of surrealism. However, unlike the surrealists, Benjamin was also preoccupied by how the personal could play a part in the recovery of lost meanings. This is where he found Proust's idea of 'involuntary memory' attractive. Proust (1978) argued that 'voluntary memory', where people consciously make an effort to remember a past event, does not have the same quality as those memories which are triggered off by a particular inadvertent stimulus and which seem to envelop the person from their place in the past, so breaking the apparent boundary between past and present and bringing lost hopes and dreams to mind.

For Proust, these sorts of memories are lodged in specific places where people have been. These places continue to bear the traces of past experiences. It is therefore possible that revisiting them may at some time evoke the past and in the same moment unlock past hopes and desires which previously seemed to have been overtaken – and defeated – by the passage of time (Szondi, 1988). This clarifies Benjamin's approach to narrative and history. He argued that Proust's 'true interest is in the passage of time in its most real – that is space-bound – form' (Benjamin, 1968c: 213), and he referred to how Proust's ideas endorsed Daudet's attempt to 'turn his life into a city', in which he attempted to project 'his biography onto the city map' (Benjamin, 1968c: 208). Thus Benjamin's evocation of the city was not simply a textual device but also a practical one, which could be carried out by people in daily life.

In his case this led to reflections on his urban wanderings, especially in the portrait of his childhood:

> I think of an afternoon in Paris to which I owe insights into my life that came in a flash, with the force of an illumination ... I told myself it had to be in Paris, where the walls and quays, the places to pause, the collections and the rubbish, the railings and squares, the arcades and kiosks teach a language so singular.
>
> (Benajmin, 1979a: 318)

To summarise: Benjamin's critique of narrativity was a critical device which also involved a constructive moment. Critique did not merely involve disruption of established meanings but also involved the bringing together of past and present in a new constellation, so allowing the possibility of redemption. Benjamin did not subscribe to a sociological notion of urban experience as a 'way of life', nor did he argue that urban writing was purely a destructive tactic to undermine conventional wisdoms. Rather, he saw the evocation of the urban as allowing the recovery of specific types of experiences which might be resources for present-day action, and in this process place the past and present in a new relationship. I now want to consider what light this perspective throws on his understanding of cities and urbanism.

5 History, aura, and the city

I have argued that Benjamin retained a dialectical perspective on urbanism: by interpreting cities as the site of the new, Benjamin also saw them as antique. I now want to show that the best way of developing this urban vision is through Benjamin's theory of 'aura', which is one of the best-known, and also controversial, elements of his thought (for example, see Mattick, 1993). The concept of aura is developed most fully in his essay 'The work of art in the age of mechanical reproducibility' (1968d) but has been criticised as being too reductionist and technologically deter-minist. However, it can be seen as offering a useful way of developing Benjamin's dual conception of urbanism. For Benjamin, objects possess aura when they have a distance from the viewer and can return his or her gaze. This is possible, Benjamin argues, when the objects are consti-tuted specifically in time and space – that is to say, when they are unique and cannot be reproduced. In this case, unique art objects have an insep-arable relationship with a tradition which sanctifies them. The development of techniques such as printing and photography allows unique objects of art to be reproduced, and they lose their aura – their distinct location in time and space – and with this the role of tradition itself is questioned. The undermining of aura goes hand in hand with the creation of a desire amongst the 'masses' for authenticity. The consequent attempt to recover 'real' aura – in tourism, cultural life, and so forth – becomes an important force in modern societies. It is only when aura is in decline that it can be recognised and desired in its own right (Hansen, 1987).

Although in some of his essays Benjamin seems to welcome the decline of aura (and this is certainly the way that his views have been represented by others, notably Adorno (1977)), his views on this matter were ambivalent (for instance, see his reply to Adorno's critique, Benjamin, 1977a: 140; also, see Hansen, 1987; McCole, 1993). In 'The work of art in the age of mechanical reproduction' he argued that the loss of aura released the work of art from its role in ritual and cult, and allowed art forms to be used to advance specific political causes for the first time. He therefore appeared to welcome the decline of aura, and Adorno in particular took him to task for refusing to recognise the continued critical potential of tradition, 'high' art. The problem with Benjamin's view, as Buck-Morss (1992), echoing Adorno, has argued, is that it cannot distinguish art from propaganda, because the art is now to be used to advance whatever cause is deemed desirable by political activists, whether fascists or communists. Consequently, art comes to be used in the same instrumental, calculating way that in another context Benjamin sees as precisely the damning feature of *Erlebnis*. However, Hansen (1987) has shown that in fact Benjamin did not endorse wholeheartedly the decline of aura, and in many respects attempted to explore how auratic experience could be reconstituted as a form of resistance. She shows, for instance, how Benjamin was interested not just in how cinema brought about nonauratic cultural forms, but also about how it could sustain auratic art.

It is precisely this ambivalence towards the question of aura which is crucial for understanding Benjamin's fascination with cities. Although Benjamin's development of the idea of aura in 'The work of art in the age of mechanical reproduction' takes a Marxist form, it draws on other elements of his thought, and especially romantic themes which influenced him in his early years (McCole, 1993). The young Benjamin, steeped in the romantic ethos of the idealist German youth movements (see McCole, 1993), shared in the romanticisation of the rural and 'nature' which characterised this movement and was generally hostile to the artificiality of modern urban life.[6] Central to this romantic perception of the countryside was the celebration of its 'sublime' quality, its ability to inspire dread and fear. The notion of aura allowed Benjamin to strip the ideological and idealist elements of this romantic conception, which had the effect of fixing sublimity to specific objects transhistorically, through 'naturalising' them, and to allow 'sublimity' itself to be examined historically.

At one level, it is precisely this 'sublime' auratic notion that Benjamin brought to bear on cities. Benjamin's references to his urban 'encounters' are presented in strikingly similar terms. Much of his *A Berlin Chronicle* is concerned with evoking the strangeness and terror of his childhood experiences in a new city. After his visit to Moscow in 1926 he wrote to his friend Scholem that, 'the two months in which ... I had to struggle in and with the city have given me some understanding of things that I could not have achieved in any other way' (in Adorno and Scholem, 1994: 312). He wrote that his essay on Marseilles was dear to him because 'I had to do battle with this city as with no other' (in Adorno and Scholem, 1994: 352). In some cases, this conception of the 'otherness' of cities took on a gendered hue (more generally, see Buci-Glucksmann, 1994). For Paris, he noted that 'I wanted to test the efficacy of a persistent courtship of this city' (in Adorno and Scholem, 1994: 298). In his Passagenwerk he referred to Paris as a 'goddess', 'the goddess of France's capital city, in her boudoir, resting dreamily' (quoted in Witte, 1991: 180). What is striking about these references is both his insistence on the particularity of the cities he wrote about (their nonmechanical reproducibility, in fact) and his construction of them as distant objects to engage in battle, or in courtship, in some sort of risky or uncertain liaison. To express this in terms of his theory of aura, cities retained their distance, 'the ability to look at us in return' (Benjamin, 1968a: 190). Indeed, Paris was 'the city of mirrors' (quoted from Witte, 1991: 181).

It is precisely this quality of cities to retain their specificity which Szondi (1988) emphasises as central to Benjamin's city portraits. Visiting strange cities disrupts one's established routines and habits, allows established conventions to be placed into question, and can restore the childhood experience of wonder, fear, and hope. Such a view places a rather different perspective on Benjamin's analysis of urbanism, which emphasises the extent to which cities were not easily incorporated into his view of modern de-auraticised life. Benjamin was obviously well aware of much that was modern in urbanism, writing about the mass reproducibility of building materials, especially after the invention of iron building techniques and the role of Hausmannisation in modernising cities (see Buck-Morss, 1989: 89ff). But he was also attuned to their old, auratic, qualities. Thus he was attentive to how representations of cities by means of new technical media could not entirely strip the aura from cities themselves. Benjamin suggests, in 'The Paris of the Second Empire in Baudelaire' (1973b), many cultural innovations of the 19th century can be seen as an attempt to grapple with the new urban milieux: the instances he cites include not only the well-known examples of Baudelaire's lyric poetry and photography, but also the detective story, 'panorama literature', the lithograph, and new forms of painting. Baudelaire's poetry was important because he showed that the process of representing adequately the modern city involved searching into the past of the city: 'the modern is a main stress in his poetry . . . but it is precisely the modern which always conjures up pre-history' (Benjamin 1973b: 171).

Benjamin's remarks about urbanism and photography are of particular interest. The ability of photography to allow the reproduction of previously unique views, vistas, and objects fascinated Benjamin: early photographs 'pump the aura out of reality like water from a sinking ship', he remarked (1979c: 250), so permitting urban sights to have their aura stripped away as specific buildings and sights were reproduced (though once again, note the subtlety of Benjamin's phrasing, for, whilst aura is being pumped out of the sinking ship, it is presumably being replaced by fresh auratic water). Thus photographs tended to 'work against the exotic, romantically sonorous names of the cities' (1979c: 250). Nonetheless, Benjamin also suggested the limitations of photography, suggesting that it was only able to record archaic urban sites, such as the railway stations rapidly being supplanted by urban road systems (1979c).

The complexity of the relationship between cities and aura is also revealed by introducing Benjamin's well-known distinction between the reception of works of art in states of concentration and distraction. Auratic art, he claims, is visually examined in states of concentration by devoted followers, with the result that 'a man who concentrates before a work of art is absorbed by it' (1968d: 241). Reception of art in a state of distraction, however, does not involve 'rapt attention [but] noticing the object in incidental fashion' (1968d: 242). Here, vision is less important than the tactile sensing of the object in question. Benjamin makes it clear that architecture offers the best example of an art form which is perceived in distraction, by passers-by (also, see Savage and Warde, 1993). The interpretation of architectural meaning allows the dispersal of aura as distracted passers-by gaze at buildings only in passing.

This formulation is interesting in appearing to contradict Benjamin's other observations about cities. As I have shown, in his essay 'On some motifs in Baudelaire', Benjamin's focus was on how the 'shock experience' led individuals to develop a degree of consciousness inimical to them being able to store memories. However, by suggesting that the built environment itself could be absorbed in a state of distraction, Benjamin implies that images and memories could be absorbed without the intervention of a conscious process, so allowing them to become part of a person's experience. This leads to something of a paradox in terms of Benjamin's theory of aura and urbanism. His own statements indicate that he saw cities as auratic, as sublime. However, he also argued that auratic objects were usually received in a state of concentration, whereas the built environment tends to be experienced in a state of distraction. It is precisely this paradox, however, which explains Benjamin's fascination with the urban landscape. Cities, as built environment, contain the potential for the recovery of memory which is an essential element in redemption, yet they avoid the conservative, cultic, ritualistic elements which usually wrap around the auratic object.

Thus, in *One Way Street*, Benjamin writes of the obelisk in the Place de la Concorde in Paris:

> What was carved in it four thousand years ago today stands at the centre of the greatest of city squares. Had that been foretold to him – what a triumph for the pharaoh! The foremost Western cultural empire will one day bear at its centre the memorial of his rule. How does this apotheosis appear in reality? Not one among the tens of thousands who passes by pauses: not one among the tens of thousands who pause can read the inscription.
>
> (Benjamin, 1979a: 70)

Elsewhere, Benjamin wrote of how 'cultural treasures' can only be contemplated with 'horror', as they have survived historically only because they have serviced and commemorated the ruling classes (1968e: 248). And yet, left in an urban context, these cultural treasures, which are also 'documents of barbarism', lose their meaning to passers-by. Placed in a museum, detached from the urban milieux, the obelisk might be incorporated into the narrative of tradition. Let loose in the city it is sundered from such a context. Items from the urban landscape are auratic but are detached from a sanctifying tradition. They allow the past to be placed in a non-linear relationship to the present. They offer unique resources for the questioning and challenging of modernity.

6 Conclusions

It should now be clear that Benjamin offers a distinctive view of urbanism, stressing the relationship between the built environment, personal and collective memory, and history. Benjamin's insight is that the urban built environment has a number of qualities which allow meanings to be encoded and decoded in ways which are specific to it and cannot usefully be understood by the application of ideas derived from the study of other types of cultural media such as literature or film. In the latter parts of this paper I suggested that the peculiarities of these properties are best revealed by considering how the urban built environment relates to Benjamin's conception of aura. Cities cannot be incorporated easily into his account of mechanical reproducibility, and this fact explains his fascination for their distinct properties and qualities.

Benjamin's conception of aura is interesting I contend, because it is relational, concerned with both the production and the reception of cultural artifacts. In itself this may appear a humdrum statement, as researchers within cultural studies have long emphasised the need to explore the encoding and decoding of texts (notably Hall, 1980). Applied to cities and urbanism, however, Benjamin's arguments have a number of distinctive implications. First, they suggest limitations to work which simply examines the construction of the city as text through the meanings and values encoded in the built environment. This work skilfully analyses the power relations enmeshed in the production of a built fabric, such as that of Paris (Harvey, 1985), or the Sri Lankan city of Kandy (Duncan, 1990), but little serious attention is given to the way that the urban landscape is perceived. It might further be argued that the scholar is bound to analyse the city as text in a state of concentration, an orientation which is unlikely to be shared by most people.

Benjamin's stress is therefore on the urban fabric as it is perceived. This also is an increasing line of inquiry, but Benjamin's position is once again distinctive. He is clearly opposed to humanist readings of the cityscape and built environment, as evident in the work of Bachelard (1958) or Lynch (1960) in which attention centres on how spatial organisation can best be designed to fit the needs of human beings to belong, and to be able to read cities. Benjamin's crucial insight is that processes of perception themselves are historically specific, with the result that attempts to specify transhistorically valid modes of urban perception are doomed to fail. Rather, Benjamin's approach seems much closer to the idea of cognitive mapping discussed briefly by Jameson (1988). Benjamin's analysis also lies much closer to the surrealist influenced urban writers such as de Certeau (1984) and Lefebvre (1991), especially in their concern to examine how spatial practices are related to issues of power and resistance. Benjamin manages to avoid the formalism which permeates Lefebvre's analysis (there is a constant tendency to reification in his distinctions between spatial practices, representations of spaces, and representational spaces; and between dominated and appropriated spaces), and also suggests more concretely how different modes of perception affect the political meanings of the built environment.

Finally, Benjamin's analysis of history is central to his understanding of cities. His concern to undermine teleology and his determination to repudiate epochal theories of history (in which one period 'replaces' another) suggest that current accounts of the rise of 'postmodern' cities need to be treated carefully. Admittedly, it seems possible to utilise Benjamin's account of mechanical reproducibility to suggest that cities themselves are becoming increasingly interchangeable, so suggesting that his romantic, auratic, conception of urbanism has lost its resonance in the age of postmodernised urban landscapes. Buck-Morss (1989: 330) expresses this view clearly:

> the planet's metropolitan population has never been greater. Its cities have never appeared more similar. But in the sense that

Benjamin recorded in the history of the city of Paris, there can be no 'Capital City' of the late twentieth century. The Passagenwerk records the end of the era of urban dream worlds in a way the author never intended.

The rise of mechanically reproduced cities, with their interchangeable fast-food restaurants, road systems, airports, hotels, and shops, has meant that the aura which Benjamin detected in the urban realm has disappeared. Perhaps in an age of constant travel and mobility (see Lash and Urry, 1994) the experience of visiting cities loses its ability to surprise. And it might be possible to use Benjamin's own theory of aura to suggest that much contemporary tourist and 'place-marketing' activity is concerned with trying to reconstruct 'urban authenticity', through carefully restored and marketed tourist spectacles in order to bring 'things closer . . . to the masses' (Benjamin, 1979a: 250; generally, see Urry, 1990). As cities become ever more similar, so people search ever harder for genuine urban distinction, and so such urban specificity becomes artificially constructed by speculative and booster interests (also, see Lefebvre, 1991). It may be the case that these developments have even changed the nature of urban perception itself. Benjamin argued that the attentive tourist saw buildings in states of concentration and therefore in conservative ways (1968d: 233). In an age of distracted tourism and business visiting, the possibility to read cities against the grain may actually be enhanced. Manicured postmodern urban spaces cocoon visitors in safe spaces which reduce the shock experience which Benjamin had detected in the modern city.

There is, however, another side to this story. It remains possible to read the ruins of modern cities in ways which reveal the hopes which they once embodied, as Wright (1991) has shown in his essays on Hackney. The new postmodern consumerist urban culture might be seen as just the latest episode in the 'eversame' which characterises capitalist modernity. Postmodern developments take their place in a pre-existing urban setting which continues to confront the traveller and resident alike with uncertainties and unanticipated encounters. A new sensitivity to 'safe' and 'unsafe' areas, to the geography of risk, threatens to place newer developments in different perspective. The contemporary flâneur might still speculate that the consumer-centred postmodern city is based on unstable foundations.

Acknowledgements

I would like to thank two anonymous referees, Kevin Hetherington, Gordon Fyfe, and, especially, Helen Hills for comments on earlier drafts.

MIKE SAVAGE

Notes

1 I should admit here that I do not speak German and have therefore been restricted to reading Benjamin's work in translation only.
2 Although the following studies are disparate in content and approach they all use some kind of narrative to tell the tales of particular cities: Davis (1990), Soja (1989), and Lynell (1992) on Los Angeles; Sassen (1991), King (1990), Wright (1991), and Zukin (1991) on London and other 'world cities'; and Portes and Stepick (1993) on Miami. Also, see Donald (1992). There is a longer discussion of 'new urban writing' in a precursor to this paper (see Savage, 1993).
3 Benjamin explains this point in terms of a long quotation which points to the chaos and coincidence involved in any event, as described in the language of physics.
4 We can get a sense of what Benjamin was trying to avoid by referring to Anderson's (1983) well-known account of nationalism. Using Benjamin's ideas, Anderson argues that nationalism invokes evoking an imagined community, as only a tiny number of one's cocitizens will be known to you. Anderson argues that narrative novelistic devices play a vital role in constructing such imagined communities, as they allow people with no intrinsic connection to each other to be joined together by their specific roles in a 'plot' constructed by the author and comprehensible to the reader. In the light of Anderson's argument it might be argued that, by repudiating narrative, Benjamin could force open contingency.
5 Recently Cohen (1993) has suggested a common affinity between Benjamin and surrealism in their common use of psychological views of subjectivity. However, the evidence for Benjamin's interest in psychology actually seems rather sparse. Cohen (1993: 38ff) suggests that Benjamin's important notes on the 'mimetic faculty' (in Benjamin, 1979a) draws upon psychoanalytic notions of causality, but Benjamin himself made it quite clear in a later letter that he was 'surprised' to note correlations with Freud (Adorno and Scholem, 1994: 521). Elsewhere he made his lack of familiarity with Freud clear in letters to Adorno (see Cohen, 1993: 25).
6 The exception to this is his early fascination with Paris (Adorno and Scholem, 1994: 26ff).

References

Adorno T, 1977, 'Letter to Walter Benjamin', in *Aesthetics and Politics* (New Left Books, London): 110–133
Adorno T, Scholem G (eds), 1994 *The Correspondence of Walter Benjamin* (University of Chicago Press, Chicago, IL)
Anderson B, 1983 *Imagined Communities* (Verso, London)
Anderson P, 1988, 'Modernity and revolution', in *Marxism and the Interpretation of Cultures* eds C Nelson, L Grossberg (Macmillan Education, Basingstoke, Hants): 194–211
Bachelard G, 1958 *The Poetics of Space* (Beacon Press, Boston, MA)
Bauman Z, 1993, 'Benjamin the intellectual' *New Formations* **20** 47–58
Benjamin A, Osborne P (eds), 1993 *Walter Benjamin's Philosophy: Destruction and Experience* (Routledge, London)
Benjamin W, 1968a *Illuminations* (Schocken, New York)
Benjamin W, 1968b, 'On some motifs in Baudelaire', in *Illuminations* (Schocken, New York): 155–200

50

Benjamin W, 1968c, 'The image of Proust', in *Illuminations* (Schocken, New York): 201–16

Benjamin W, 1968d, 'The work of art in the age of mechanical reproduction', in *Illuminations* W Benjamin (Schocken, New York): 217–52

Benjamin W, 1968e, 'Theses on the philosophy of history', in *Illuminations* (Schocken, New York): 253–64

Benjamin W, 1973a *Charles Baudelaire: A Lyric Poet in the Era of High Capitalism* (New Left Books, London)

Benjamin W, 1973b, 'The Paris of the Second Empire in Baudelaire', in *Charles Baudelaire: A Lyric Poet in the Era of High Capitalism* (New Left Books, London): 9–106

Benjamin W, 1977a, 'Reply', in *Aesthetics and Politics* (New Left Books, London): 134–41

Benjamin W, 1977b *The Origins of German Tragic Drama* (New Left Books, London)

Benjamin W, 1979a *One Way Street and Other Writings* (New Left Books, London)

Benjamin W, 1979b 'Surrealism', in *One Way Street and Other Writings* (New Left Books, London): 225–39

Benjamin W, 1979c, 'A short history of photography', in *One Way Street and Other Writings* (New Left Books, London): 240–57

Benjamin W, 1985, 'Central Park' *New German Critique* **34** 32–58

Benjamin W, 1990 *Moscow Diaries* (Harvard University Press, Cambridge, MA)

Berman W, 1983 *All That is Solid Melts into Air* (Verso, London)

Bradbury R, MacFarlane J (eds), 1976 *Modernism 1890–1930* (Penguin Books, Harmondsworth)

Buci-Glucksmann C, 1994 *Baroque Reason* (Sage, London)

Buck-Morss S, 1989 *The Dialetics of Seeing: Walter Benjamin and the Arcades Project* (MIT Press, Cambridge, MA)

Buck-Morss S, 1992, 'Aesthetics and anaesthetics: Walter Benjamin's Artwork essay reconsidered' *October* **62** (Fall): 3–41

Burgin Z, 1993, 'The city in pieces' *New Formations* **20** 123–44

Clark T J, 1985 *The Painting of Modern Life: Paris in the Art of Manet and his Followers* (Thames and Hudson, London)

Cohen M, 1993 *Profane Illuminations: Walter Benjamin and the Paris of the Surrealist Revolution* (University of California Press, Berkeley, CA)

Davis M, 1990 *City of Quartz* (New Left Books, London)

de Certeau M, 1984 *The Practice of Everyday Life* (University of California Press, Berkeley, CA)

Donald, J, 1992, 'Metropolis: the city as text', in *Social and Cultural Forms of Modernity* eds R Bocock, K Thompson (Polity Press, Cambridge): 220–46

Duncan J S, 1990 *The City as Text* (Clarendon Press, Oxford)

Eagleton T, 1981 *Walter Benjamin* (Verso, London)

Frisby D, 1985 *Fragments of Modernity* (Basil Blackwell, Oxford)

Giddens A, 1981 *A Contemporary Critique of Historical Materialism* (Macmillan, London)

Gregory D, 1994 *Geographical Imaginations* (Basil Blackwell, Oxford)

Hall S, 1980 *Culture, Language and Practice* (Century Hutchinson, London)

Hansen M, 1987, 'Benjamin, cinema and experience: "the blue flower in the land of technology"' *New German Critique* **40** 179–224

Harvey D, 1982 *The Limits to Capital* (Basil Blackwell, Oxford)

Harvey D, 1985 *The Urbanisation of Capital* (Basil Blackwell, Oxford)

Honneth A, 1993, 'A communicative disclosure of the past: on the relation between anthropology and philosophy of history in Walter Benjamin' *New Formations* **20** 83–92

Jameson F, 1988, 'Cognitive mapping', in *Marxism and the Interpretation of Cultures* eds C Nelson, L Grossberg (Macmillan Education, Basingstoke, Hants): 347–57

Jennings M, 1987 *Dialectical Images: Walter Benjamin's Theory of Literary Criticism* (Cornell University Press, Ithaca, NY)

Keith M, Cross M, 1993, 'Racism and the postmodern city', in *Racism, the City and the State* eds M Cross, M Keith (Routledge, London): 1–19

King A D, 1990 *Global Cities* (Routledge, London)

Lash S, Urry J, 1994 *Economies of Signs and Spaces* (Sage, London)

Lefebvre H, 1991 *The Production of Space* (Basil Blackwell, Oxford)

Lury C, 1992 *Cultural Rights: Technology, Legality and Personality* (Routledge, London)

Lynch K, 1960 *The Image of the City* (MIT Press, Cambridge, MA)

Lynell G, 1992 *No Crystal Stair: African-Americans in the City of Angels* (Verso, London)

McCole J, 1993 *Walter Benjamin and the Antinomies of Experience* (Cornell University Press, Ithaca, NY)

McRobbie A, 1992, 'The Passagenwerk and the place of Walter Benjamin in cultural studies: Benjamin, cultural studies, Marxist theories of art' *Cultural Studies* **6** 147–69

Mattick P, 1993, 'Mechanical reproduction in the age of art' *Theory, Culture and Society* **10** 127–48

Mellor R, 1977 *Urban Sociology in an Urbanised Society* (Routledge, London)

Olsen D, 1986 *The City as a Work of Art: London, Paris, Vienna* (Yale University Press, New Haven, CT)

Portes A, Stepick A, 1993 *City on the Edge: The Transformation of Miami* (University of California Press, Berkeley, CA)

Proust M, 1978 *In Remembrance of Things Past* (Penguin Books, Harmondsworth)

Roberts J, 1982 *Walter Benjamin* (Macmillan, London)

Sassen S, 1991 *The Global City: New York, London, Tokyo* (Princeton University Press, Princeton, NJ)

Saunders P, 1981 *Social Theory and the Urban Question* (Century Hutchinson, London)

Savage M, 1993, 'Walter Benjamin and urban writing', WP-3, Department of Sociology and Social Anthropology, University of Keele, Keele

Savage M, Warde A, 1993 *Urban Sociology, Capitalism and Modernity* (Macmillan, London)

Seed J, Wolff J, 1988 *The Culture of Capital: Art, Power and Nineteenth Century Middle Class* (Manchester University Press, Manchester)

Simmel G, 1950, 'The metropolis and mental life', in *The Sociology of Georg Simmel* ed. K Wolff (The Free Press, New York): 409–26

Smith M, 1981 *The City and Social Theory* (Basil Blackwell, Oxford)

Soja E, 1989 *Postmodern Geographies: The Reassertion of Space in Social Theory* (Verso, London)

Szondi P, 1988, 'Walter Benjamin's city portraits', in *On Walter Benjamin: Critical Essays and Reflections* ed. G Smith (MIT Press, Cambridge, MA): 18–32

Tanner D, 1990 *Venice Desired* (Basil Blackwell, Oxford)

Tiedmann R, 1988, 'Dialectics at a standstill: approaches to the *Passagenwerk*', in *On Walter Benjamin: Critical Essays and Reflections* ed. G Smith (MIT Press, Cambridge, MA): 260–91

Tonnies F, 1988 *Community and Society* (Transaction Books, New York)

Turner B, 1994, 'Introduction', in *Baroque Reason* C Buci-Glucksmann (Sage, London): 1–36

Urry J, 1990 *The Tourist Gaze* (Sage, London)

Williams R, 1973 *The Country and the City* (Chatto and Windus, London)

Williams R, 1989 *The Politics of Modernism* (Verso, London)

Witte B, 1991 *Walter Benjamin: An Intellectual Biography* (Wayne State University Press, Detroit, MI)

Wolff J, 1993, 'Memoirs and micrologies: Walter Benjamin, feminism and cultural analysis' *New Formations* **20** 111–22

Wolff K (ed.), 1950 *The Sociology of Georg Simmel* (The Free Press, New York)

Wolin R, 1982 *Walter Benjamin: An Aesthetic of Redemption* (Columbia University Press, New York)

Wright P, 1991 *A Journey Amidst the Ruins* (Radius, London)

Zukin S, 1991 *Landscapes of Power: From Detroit to Disneyworld* (University of California Press, Berkeley, CA)

2

ON GEORG SIMMEL

Proximity, distance and movement

John Allen

Introduction: a sociology of space?

In truth, it hardly seems to be worth mentioning that Georg Simmel knew a thing or two about the social significance of space. After all, it is not that demanding to tease out the spatial significance of his thought in relation to such diverse topics as gender, fashion, style, domination, secrecy, or, as is better known, the remoteness of strangers or the hectic pace of city life. For a sociologist such as Simmel, who understood society as the sum of its social interactions, be they immediate or distanciated, transparent or opaque, the spatial preconditions of sociation, as well as the use to which space was put, represented an integral part of his study of social forms. Indeed, as both David Frisby (1985, 1992, 1997) and John Urry (1994 with Scott Lash, 1995) have consistently pointed out, for Simmel, it was precisely the innumerable forms of social interaction which brought space to life and endowed it with meaning.

Having said that, much of Simmel's actual concern with the *sociology* of space was of a rather abstract nature, concerned to establish the formal preconditions of sociation. In particular, his essay 'The sociology of space' (published in revised form, together with a further essay, 'On the spatial project of social forms', as a chapter in Simmel's *Soziologie* in 1908[1]) graphically explores how various qualities of space condition the possibilities of certain forms of social interaction. The formal significance of boundaries drawn in space, for example, is explored through the insight that such lines connect as much as they divide those on either side of a boundary; an observation that has gained greatly in importance through the work of Edward Said and others writing in a post-colonial vein. Likewise, Simmel's analysis of motion and fixity as categories for comprehending much of what is taken for granted on an everyday basis predates Lewis Mumford's concern for the urban ambivalences thrown up by history. And, of equal importance, is Simmel's analytical treatise of the significance of proximity for social interaction which, as Frisby (1997) notes, represents a pioneering effort in the sociology of space. Nonetheless, for all the analytic refinements and pithy detail, indeed perhaps because of them, Simmel's formal essays on space read as prefigurative textbook sociology. There is little that brings spatiality alive in such accounts. For this, we have to look at Simmel's substantive accounts of modernity, especially in his writings on the rhythms of the metropolis and the impact of a mature money economy on modern culture. Space matters in Simmel's writings on money and modern culture and on the nature of city life, precisely because modernity itself is spatially constituted.

Such a claim amounts to more than saying that space makes a difference; rather it suggests that modern times for Simmel are experienced largely through *changing relations of proximity and distance* and, more broadly, through *cultures of movement and mobility*. Such relations are not intended to be loosely tacked on to an analysis of modernity. On the contrary, they are thought to hold the key to how life in the modern world is experienced and lived. Or, more accurately, they reflect Simmel's understanding of how Berlin city life was experienced and lived at the end of the nineteenth century from which he drew much of his insight and inspiration. In this chapter, therefore, the intention is to develop many of the analytical insights set out in Simmel's rather formal sociology of space (which, as Frisby, 1992, 1997, notes, is perhaps best understood as part of his incomplete study of the abstract preconditions for social interaction, alongside those of number, mass, size, time and so forth) through his more expressive accounts of modernity. As such, the focus will be upon the spatialized theorizing that Simmel adopts when attempting to think through the cultural experience of modern everyday life and its diverse modes of social interaction. This account is divided into two parts. In the first part, the ways in which questions of distance and social distancing are used by Simmel to unravel modern forms of social interaction are considered. The degree of remoteness and proximity in social relationships, the construction of public and private worlds, as well as the being together of 'strangers' are explored in relation to this broad theme. Following that, we turn to the central role that movement and circulation play in Simmel's account of modern culture, especially in relation to claims concerning the fleeting and increasingly abstract nature of forms of sociation.

As well as an exposition of the significance of spatiality in Simmel's thinking and substantive analysis – much of which is drawn from his major work *The Philosophy of Money* (1990 [1900]), together with some of his better known essays such as *The Metropolis and Mental Life* (1950a [1903]), and *The Stranger* (1950b [1908]) – the chapter will also attempt to show how Simmel's thinking on proximity, distance and movement can shed light upon how people make sense of today's complex networks of social interaction, both within and beyond cosmopolitan city life.

Life at a distance

It is probably worth noting at the outset that Simmel's writing style is not to everyone's taste. Once he leaves the ground of formal sociology, in his essays in particular, there is little that is systematic or consistent about his mode of presentation. As far back as 1918, Georg Lukács alluded to Simmel's impressionistic mode of thought; an observation which was subsequently to form the basis of Frisby's (1985) assessment of Simmel's work as unsystematic and fragmented in style. Perhaps the key to understanding Simmel's approach, however, can be gleaned from the sympathetic recognition by one of Simmel's students, Siegfried Kracauer, that in his work

> the phenomena make their appearance in their capacity as complexes of connections ... Since the only significance of the threads spun between the phenomena is to make the hidden connections visible, their paths are quite irregular and arbitrary; they are almost systematically unsystematic. It is utterly insignificant where one ends up when casting them out and fastening them together, so long as one ends up somewhere. This web is not constructed according to a plan, like a firmly established system of thought; instead, it has no other purpose than to be there and to testify through its very existence to the interconnectedness of things. Loose and light, it extends itself far and wide and gives the impression of a world that emits a curious shimmer, like a sunny landscape in which the hard contours of objects have been dissolved and which is now only a single undulation of trembling light veiling individual things.
>
> (Kracauer 1995: 251–2)

With no apparent plan of thought and a concern to illuminate the many sides of seemingly disparate things, Simmel's roving approach, according to Kracauer, nonetheless has much to offer:

> The fruit we reap from this scouting is a growing sensitivity to the intertwinement of the elements of the manifold. We feel it: every phenomenon reflects every other phenomenon, varying a basic melody that also sounds in many other places.
>
> (*Ibid.*)

Simmel's discontinuous train of thought, coupled with his ability to move off into any number of directions, is usefully illustrated by his treatment of social distance in respect of – in no particular order – the studies of the stranger, the circulation of money, and city life. Whilst it is possible to discern a common orientation to issues of social distancing, of proximity and remoteness, in his specific studies the possibility of obtaining any secure meaning to these terms is soon lost as 'the entire plenitude of the world' is poured into them.[2] Living at a distance, then, can, as we shall see, take us into all manner of contemporary and not so contemporary debates.

Negotiating difference, or being with strangers

In his classic essay on *The Stranger*, published in 1908, Simmel tried to convey through this figure a range of ambivalences which have come to haunt us in the practices of negotiating difference. Initially taken up by American sociology in the 1920s, however, the figure of the 'stranger' was unfortunately portrayed all too comfortably as a 'marginal' character: typically a migrant to the US, someone located in between 'old' and 'new' cultures. As such, the figure was quickly reduced to a singular, clear-cut image.[3] Simmel's original intention in writing about the stranger, however, appears to have been rather different. In keeping with the fullness of his approach, Simmel adopted the figure of the stranger to illuminate, or rather to capture, the contradictory experience of what it means to interact socially with someone who is both near in a spatial sense, yet remote in a social sense. As Simmel expressed this dilemma:

> The unity of nearness and remoteness involved in every human relation is organized in the phenomena of the stranger, in a way which may be most briefly formulated by saying that in the relationship to him, distance means that he, who is close by, is far, and strangeness means that he, who is also far, is actually near. For, to be a stranger is naturally a very positive relation; it is a specific form of interaction.
>
> (Simmel 1950b: 402–3)

And to register the point forcefully that we are not talking here about characters separate from, or alien to, the social community, he continues:

> The inhabitants of Sirius are not really strangers to us, at least not in any sociologically relevant sense: they do not exist for us at all; they are beyond far and near. The stranger, like the poor and like sundry 'inner enemies', is an element of the group itself. His position as a full-fledged member involves both being outside it and confronting it.
>
> (*Ibid.*)

The stranger, therefore, is someone who is involved, yet not involved; close to us, yet part of elsewhere. More to the point, Simmel seems to be suggesting that there are traces or degrees of 'strangeness' in all kinds of relationships, not merely those that fit the caricature of the 'outsider' to a group. Traces of it are evident, as Simmel notes for example, in the most intimate relationships of first passion, where estrangement may then follow rejection.[4] It is present in those fleeting encounters between people which take place in the crowded subways or in the throng of busy streets, and equally it is there in city life where all may feel that they belong, yet feel they have to state their difference from others. In that sense, it is perhaps best to view the figure of the stranger as a symbol or an icon through which all manner of social and spatial tensions may be channelled. In the case of the social community, for example, the tension between nearness and distance is something that may be *lived* rather than necessarily resolved. It is something that is experienced as a fact of city life, not something that presents itself as an interminable problem which has to be confronted and dispelled.

Or perhaps, as Rob Shields (1992) has argued, such dualistic dilemmas are the stock-in-trade of a peculiarly modern imagination. On this interpretation Simmel has put his finger, so to speak, on a raw nerve which disrupts our easy relationship with a cultural mapping that reassures us as to what is near and what is far, and who belongs and who does not. The stranger, who is representative of all things from afar, is also present and may thus experience themselves as belonging. In Shields' words:

> As an embodiment of difference, the stranger represents the doubtful existence and dubious truth of what is not spatially present, of what cannot be verified at first hand. Yet the stranger is nonetheless 'here', present, and thus throws the doubtful and flickering quality of absence and non-existence back into the faces of those insiders in the local community, throwing into question the sanctity of presence
>
> (Shields 1992: 189)

Part of Shields' concern here, it should be noted, is to break any steadfast notions that we may hold between proximity and presence. To be close to someone socially does not necessarily require physical proximity and, in a world of disembedded mechanisms and distanciated relations of the kind that Anthony Giddens (1984, 1990) describes, the immediate copresence of subjects is no longer considered to be the necessary basis of community relations. On this view, the boundaries – social as well as physical – which once marked the limits of local relations are now more akin to thresholds across which communication and other forms of distanciated interaction may take place.

To acknowledge the presence of the remote, however, does not undermine Simmel's understanding that the tension between near and far inscribed in forms of social interaction may be lived as *involved* difference. If we come at it from a different angle and consider the contemporary city to be a place where all may be 'strangers' to one another, it takes little imagination to realize that there are no 'host' groups to speak of: everyone belongs, but that does not make everyone the same or alleviate social distance.

A useful illustration of this point can be found in the novel publication, *Roissy Express*, François Maspero's (1994) first hand account of his journey through the Paris suburbs, with a photographer, Anaïk Frantz. What is revealing about this journey, for the reader as much as for the two of them, is that they were as much strangers to the Malian or Portuguese residents they encountered on their 'journey', as the latter were to them and possibly to each other. Through specific forms of interaction, in this case primarily the need to document their journey through photographs, Maspero and Frantz's involvement with the people of their 'own' city exhibited both the cultural misunderstanding and apprehensiveness that comes from the mix of proximity and remoteness they represented and, indeed, projected. On breaking a code of respect by intrusive photographing, for instance, the two 'strangers' were politely but firmly lectured by a Malian grouping on the morality and dignity of social conduct. In particular, for the Malians in a Paris suburb near Aulnay, their difference was neither acknowledged or respected:

> when you take someone's photo, you ask his permission first. The theme is elevated to a principle for life: respect comes first. 'If you had asked me first,' says the first one to speak, a big green-shirted Malian, 'I would have been flattered,' 'And if I asked you now?' replies Anaïk. 'It's too late. Not this time. Some other time maybe.' The big Malian used to be a student. He introduces his brother. And another brother. Another, older man, who has just arrived, inquires in his own language and takes François to one side to repeat the lesson even more severely and politely than before. He would like to think they are not police or journalists, they could even be friends, but friends don't behave like that. 'There are lots of Malians here,' says the former student, who works in Paris. 'In Mali, people have always respected France. But France today isn't what it was – there's no respect for other people. Yet my father fought for France. There are journalists who come to Mali and take photos and then do disgusting reports.' 'Not wishing to be indiscreet, what are you doing round here?' They explain that they're from Paris, that they've just been to see a friend at Rougemont and are walking towards the canal. Just for pleasure. 'And still not wishing to be indiscreet, what do you do for a living?' Anaïk replies that she's a demonstrator in supermarkets. 'You don't

surprise me – just by looking at the way you smile and walk. You're a very attractive woman.' François chooses to state his job as a translator. Otherwise try explaining, in the heat of the moment, the difference between writer and journalist. The professions given are accepted politely, like everything else. With smiles. 'So', insists Anaïk, 'now we've explained ourselves, still no photo?' No. But one day, he promises, perhaps if they meet in Paris. The circle breaks up. They shake hands. The big Malian puts his hand on his heart. 'May God go with you,' he says to Anaïk.

(Maspero 1994: 100)

Looked at from a more abstract angle, such forms of sociation convey or rather express many of the tensions associated with Iris Marion Young's (1990: 234) ideal description of the politics of city life 'conceived as a relationship of strangers who do not understand one another in a subjective and immediate sense, relating across time and distance'.

In Young's view, the multiple group identifications that most people work with today make a travesty of any quest to resolve the tension between involvement and difference in city life. A politics of recognition, which allows difference and belonging to co-exist without the unnecessary strain of seeking mutual identification and commonness amounts, in her terms, to a form of social relations defined as the being together of strangers. Whilst Simmel's figure of the stranger may not have been the philosophical antecedent that she had in mind when arriving at this definition, a sensitivity to living life at a distance is nonetheless surely part of its basic melody.

Negotiating the city, or hiding behind masks

An equally resonant, yet rather different twist on living life at a distance is to be found in Simmel's account of the impact of a city-based money economy upon the urban personality. In such classic essays as *Money in Modern Culture* (1991a [1896]), *The Problem of Style* (1991b [1908]), and above all *The Metropolis and Mental Life*, which was first published in 1903 (1950a), Simmel outlined his interest in the diverse and transitory nature of interactions among people in modern times, many of them conducted at such a hectic pace in the city, that he thought them responsible for producing extreme reactions of detachment, reserve and a kind of blasé indifference to metropolitan life. At the core of his thinking was the assumption that the urban mêlée – its constant and changing stream of impressions, its disorientating shifts in pace and gear, and its many unexpected and unscripted interactions with 'complete' strangers – was simply more than the mass of the urban population could possibly be expected to bear. In short, there was so much going on and around in the city, that it becomes impossible to negotiate all such differences and stimuli without becoming nervous or edgy in disposition.

If, as Simmel maintained, the metropolitan rhythm of events was the cause of this agitated state, the solution or rather the strategy for coping with it lay with people seeking to create a distance between themselves and others, and, more broadly, from the rhythms of the city itself. Some kind of reserve or detachment of feeling was called for, if city life was to be ongoing. Social distancing, as a type of performance in response to the overbearing rhythms of the city, was thought by Simmel to characterize the urban personality in at least three related ways.[5]

One of the most significant acts of distancing described by Simmel is the adoption of a matter-of-fact attitude towards people and things alike. Bound up with the development of a mature money economy, the calculating, rather abstract, colourless nature of economic interactions was seen to position people in ways which generated a series of instrumental practices and rational dispositions within economic life. This form of detachment was considered to be strictly an affair of the head rather than the heart, with the emotions largely withdrawn from the circuit of social exchange. Above all, Simmel believed the medium of money to be responsible for the objectification of social relations to such an extent that a culture of calculation and formlessness confronted the expressive subject in a distanced, alien fashion. On the positive side, however, this heightened form of impersonality was itself seen as a means of creating a space for people to *dis*sociate from city life, to escape its emotive dealings. As such, a personal space, a space for themselves in which things are approached on a matter-of-fact basis, is opened up within an increasingly objectified culture. At its most basic, it is what makes it possible to walk to the bus or tram stop in the morning without feeling obliged to stop and to talk to everyone on the way.

A similar kind of distancing effect is apparent in Simmel's various descriptions of the formation of a blasé attitude in response to the complex rhythms of city life. A sense of indifference which comes about through a satiation of the senses, an ability to be unmoved by all things new, and a flat response to the many differences between things and between peoples, are regarded as just so many ways of keeping the city at a distance. It was considered impossible by Simmel to negotiate the city in all its trappings, movements and differences, to the extent that the cultivation of a social reserve was thought an essential ingredient of urban lifestyles.

The need for inward retreat, in part to preserve a degree of anonymity, somewhat paradoxically finds its expression in what Simmel described as the 'exaggerated subjectivism of the period'. In an interesting twist, Simmel inverts the commonplace meaning of 'style' to insist that it is the concealment of the personal, not its revelation, which leads people to adopt fashionable lifestyles and stylistic modes of expression.

What drives modern men so strongly to style is the unburdening and concealment of the personal, which is the essence of style. Subjectivism and individuality have intensified to breaking point, and in the stylized designs, from those of behaviour to those of home furnishing there is a mitigation and a toning down of this acute personality to a generality and its law. It is as if the ego could no longer carry itself, or at least no longer wished to show itself and thus put on a more general, a more typical, in short, a stylized costume ... Stylized expression, form of life, taste – all these are limitations and ways of creating a distance, in which the exaggerated subjectivism of the times finds a counterweight and concealment.

(Simmel 1991b: 69)

On this view then, conformity to fashion and the adoption of things or affectations in vogue are a means of preserving a sense of self, rather than a means of expressing one's self. Stylizing the self in this context is equivalent, as James Donald (1996) argues, to donning a social mask: little, rather than more of the self, is revealed by conforming to the standard of wider tastes. For Simmel, the mask, Donald suggests, 'represents a phobic reaction to the hyperactivity and overstimulation of the modern metropolis'. In other words, it is a peculiarly urban strategy of distancing: a public display of conformity which is utterly personal yet, at the same time, hides a deeply private life. Elias Canetti (1973) in his work on crowds evoked this potentially liberating space when he spoke of the crowd offering a feeling of relief where its density was greatest. To surrender to the collective identity, on this account, does not so much entail a loss of personal identity, as the freedom to lose oneself in its openness, in its recognizable sociability.

What kind of fractured identity actually lies behind the social mask is not altogether clear in Simmel's work, however, and is itself a matter of debate. Donald, for instance, infers that Simmel assumed the presence of a deeper, more complex personality, but perhaps Richard Sennett (1969) comes closer in his recognition that a connection between the public and private worlds of the citizen does not necessarily imply that one is reducible to the other. There may well be no complex inner world of subjectivity for Simmel, but there is an objectified, impersonal metropolis which people negotiate on a daily basis. And it is this external negotiation which, in Nikolas Rose's (1996) terms, is folded into our subjective selves – and which forms the fragile basis of what it is *to be* in the city.[6]

It is also tempting in this context to speculate, nearly a century after Simmel's essay on the metropolis was published, on the processes of distancing which may presently shape modes of interaction across information or networked cities. In the so-called 'real-time' cities, where the communications technologies are said to open up new forms of encounter and copresence, talk of hiding behind electronic masks is likely to be less revealing than accounts which consider the meaning of blasé in the face of new sources of overstimulation and immediacy. If today, we really are witnessing in Manuel Castell's (1996) terms a network society in which cities are part of the 'space of flows', then relations between cities – as much as relations within cities – have to be negotiated in situations where new forms of mobility and flux are evident. If the seeming simultaneity of events leads to what Barbara Adam (1995) has referred to as the 'global present', then the manner in which they are experienced and objectified may well generate attitudes of indifference to the speed and pace at which complex forms of social interaction are played out. It is obviously difficult to say with any certainty, but as more and more people are drawn into the worlds of circulation through mediated and unmediated contacts, then perhaps Scott Lash and John Urry are prophetic in their observation that 'new forms of social distance [will] have to be learnt within the confined contexts of mobility' (1994: 255).

In other words, so much may be felt to be known about what is going on in the world – almost as it happens – that people, no matter how partial their actual knowledge of affairs, may become blasé to such happenings simply as a means of coping. To bring Simmel up-to-date in this respect, perhaps there is now as much a need to live the 'global' intensity of relationships and their effects at a distance, as there is the complex rhythms within cities.

Cultures of movement and mobility

So far, the significance that Simmel attached to the workings of a mature money economy has been noted only in passing as symptomatic of modern forms of interaction. In fact, it is possible to draw out from his work the claim that the immediacy and pace of metropolitan life, as well as the empty, calculating nature of many modern forms of interaction are directly traceable to the characteristics of a mature money economy. In his essay *Money and Modern Culture*, for instance, it is easy to see how money can be considered as the medium through which cultural relationships are formed, whereby the indeterminacy of money is mirrored in a formless culture without qualities. Drained of colour, the abstract character of a money economy was thought by Simmel to bring with it a culture of calculation and a levelling of all things to matters of quantity, not quality.

With the benefit of hindsight, it is now commonplace to draw attention to the potential cultural determinism implicit in such an outlook and in particular to retreat from the idea of a universal cultural experience. Notwithstanding such criticisms, there is more however to Simmel's observations than a series of congruences between the circulation of money and that of the everyday world. As Nigel Dodd (1994) in particular has argued, Simmel's preoccupation with money was not with money as a technical feature of a modern economy, but rather with the *idea* of money: what cultural associations money generated and how people positioned themselves in relation to its movement and circulation.

Making sense of movement, or experiencing life at a pace

Reference has already been made to Simmel's belief that the oscillations of a mature money economy generated a hard-headed, matter-of-fact attitude towards people and things. However, as a means of coming to terms with the overbearing rhythms of a city's economic life, such a practice only makes sense as a particular interpretation of the movement and mobility of money in everyday life.[7] Above all, it suggests that what people *imagine* themselves to be involved in when they engage in exchange transactions actually gives meaning to objectified notions of rhythm and movement. There is, as Nigel Thrift (1996) has argued for example, no essential property of speed which is somehow conferred upon its users, only different senses in which that movement is made meaningful.

In a similar manner, it could be argued that one of the concerns that Simmel was attempting to illuminate in his major work, *The Philosophy of Money*, published at the beginning of this century, was precisely how it is that cultural notions of movement become objectified through the medium of money. If the medium of money really does increase the complexity of culture, especially that of city cultures, then its *effects* on how people experience and give meaning to their surroundings and relations with others should be apparent. And indeed, where the first part of *The Philosophy of Money* provides an insight into the meanings that have been attributed to the role of money and the exchange relationship, the second part sets down what Simmel considers to be the cultural consequences of a money economy. As befits Simmel's writing style, the two parts of the text are not tidily exclusive and themes raised in one part have a habit of re-appearing at random in the other, but the final chapter of the book does draw attention to a number of ways in which our impressions of space and time are said to have been shaped by the workings of a mature money economy.[8]

One concerns the broad issue of distancing that we have already considered, although here a different set of connections are illuminated which draw attention to money's ability to overcome distances and bring places closer together. This, of course, is close to David Harvey's (1985, 1989) concerns with the process of capital's annihilation of space through time and the consequent speed-up in circulation which, in turn, is said to be manifest in the (postmodern) cultural overload of signs, images, and other sensory stimuli. For Simmel, however, the overcrowded proximity and friction of modern cultural life was also thought to have the effect of re-working our senses of what is near and what is far. In line with his (later) analysis of the proximate stranger, the abstract, colourless nature of money was thought to exacerbate the social distance between relationships close at hand and to draw near those that are distant. The blasé reaction in this context would thus amount to an embrace of the far-off in the search for ever-new stimulations, whilst at the same time blocking those relationships nearby.

Perhaps the most striking illustration offered by Simmel, however, of the ways in which our sensitivity to movement has been affected by the circulation of money lies in his account of the accelerated pace and rhythm of modern life. As the pace at which money circulates around the urban economy alters, so Simmel believed did the pace of life – and with it our experience of space and time. By this observation, however, Simmel was only tangentially concerned with a potential speed-up in human affairs. As the symbolic representation of the cultural character of particular societies at particular times, money as a malleable, formless substance detached from any one-sided interests 'measures itself against the number and diversity of inflowing and alternating impressions and stimuli',[9] and it is in this sense that the pace of life may be said to have increased. There were simply more things to experience in the cultural economy, much of them with greater intensity, which people had to come to terms with through their flows and movements. As Simmel put it:

> money contributes to determining the form and order of the contents of life. It deals with the *pace* of their development, which is different for various historical epochs, for different areas of the world at any one time and for individuals of the same group. Our inner world extends, as it were, over two dimensions, the size of which determines the pace of life. The greater the differences between the contents of our imagination at any one time – even with an equal number of conceptions – the more intensive are the experiences of life, and the greater is the span of life through which we have passed. What we experience as the pace of life is the product of the sum total and depth of its changes. The significance of money in determining the pace of life in a given period is first of all illustrated by the fact that a *change* in monetary circumstances brings about a *change* in the pace of life.
>
> (Simmel 1990: 498, emphasis in original)

It should be evident from this extract, therefore, that in referring to an accelerated pace of life, Simmel did not have in mind some set of financial technologies which confer their properties of mobility on to their users.

On the contrary, the pace of culture which is said to confront the individual in objectified form is experienced and interpreted in different ways, depending upon how people are positioned in relation to its flows and movements. Hence, the meaning of the 'global present' for, say, those financial dealers involved in today's quick-fire space of international finance is likely to be very different from those for whom fast risk and effortless gain lack any relevant reference points. For the latter, say, the pensioner with a few stocks and shares, the world may even have appeared to slow down, but it is their interpretation of its accelerated pace and intensity which matters most.[10]

In fact, what Simmel was able to convey through the example of money was *both* the objectivity of social relationships – that is, their independence from those who constituted them – and their diverse subjective interpretation. In choosing the example of money, however, he also objectified the rhythm and pace at which things circulate in everyday life, as well as portraying how people experienced and gave meaning to notions of money and its movement and mobility. In truth, he may not have set out to develop these aspects of space–time, but he did nonetheless provide an insight, and a legacy, into how we may begin to spatialize cultural forms in a constitutive rather than a formal manner. A legacy which can be adapted to widely different contexts as can be seen, for example, in Michael Watts' (1994) account of the symbolic meaning of oil money in 1970s Nigeria. In a scenario vastly different from nineteenth century Berlin, Watts shows:

> how the infusion of oil monies in an industrializing capitalist state in Africa provides a vantage point from which one can show that money contributes to, and reflects, how social integration and disintegration are at work simultaneously. In this light, some of the symbolic, cultural and socio-political expressions of money in Nigeria seem to endeavour to hold money operations within certain limits. In other respects, oil money – as social power, as state corruption and degeneracy, as blind ambition and illusion – has eroded sociability, turning everything it touches into shit: oil money as deodorized faeces that has been made to shine. Oil money provides a means to pry open the black box of society, while the structure of society provides the entry point into understanding the complex ways in which money simultaneously mediates social relations and provides a fundamental means of experiencing them.
>
> (Watts 1994: 442–3)

Drawing upon Simmel to demonstrate how the circulation of oil monies in Nigeria led to the development of an objectified culture which largely eroded sociability, Watts was able to show how the fetishistic quality of oil money, its occult form, not only positioned various social groupings, but was also read differently by the Muslim and other groupings involved. When those in the Muslim community for instance saw themselves as caught up in the accelerated pace of modern life brought about by the circulation of oil monies, the rejection was of a *particular* rhythm and movement of money – as symbolized by the Nigerian petro-naira reducing all things to a base moral level. For Watts, as for Simmel, money is a medium through which social relations are experienced, as well as an objective presence shaping cultural economies.

Conclusion

As mentioned at the beginning of this chapter, Simmel's writing is a particular taste. His attention to the fine detail of social interaction, in all its nuanced forms, can be a sheer delight until you realize that much of the historical context has been left out. Moreover, just at the point when you think you have grasped his line of thought and pieced together the reasoning, he has the ability to slide the frame of meaning out from beneath you and to leave you wondering where it will all lead. Kracauer was undoubtedly right when he said that it goes against the grain for Simmel to be systematically analytic or precise conceptually.

He was also right, however, to stress that what Simmel had to offer was an understanding of the simplest phenomenon, the most commonplace interaction, in all its fullness and variety. His breadth of interests in the ostensibly self-evident was, by any measure, quite astounding. Earlier, reference was made to the topics of gender, fashion, domination and the like, but to that you could add an interest in the role of the senses (modes of seeing, hearing, and smelling others), the phenomenology of the meal, the experience of adventure, the meaning of adornment, the style of exhibitions, the social role of faith – and this is merely to scratch the surface of his 'cultural' interests and publications. Simmel's writings are certainly not easily classified in disciplinary terms, although various commentators have slotted his works into aesthetics, philosophy, history, metaphysics, social psychology and sociology, as well as the more recent label of cultural studies. But to be quite candid, following Kracauer, one could just as easily label him 'a philosopher of the soul, of individualism, or of society' (Kracauer 1995: 225), and still fail to convey the magnitude and scope of his work. In whatever academic slot his interests and concerns are placed, perhaps the one recurring theme, however, is his focus on the nature of social interaction (or rather sociation), especially in its commonplace, apparently superficial forms.

And indeed, a major benefit of such a focus, which has often passed unremarked, is the attention that is duly paid to how people live their lives spatially – close to, nearby, remote from, detached, on the move, at a particular pace – and how they make sense of the experience. Whether it is the Berlin of the 1890s, the Paris suburbs of the 1990s, Lagos in the 1970s, or wherever, the ability to think the everyday culture spatially has its obvious attractions. Two of which, on the basis of this chapter, are the need to think through the implications of proximity and distance and of movement and mobility for many of today's modes of interaction. At the present moment, for example, when the world's cities are set to grow dramatically in size, their role as meeting places for a diverse range of ethnicities, cultures and peoples highlights the significance of how difference is negotiated in the city, and with that the importance of how proximity is experienced and the distance between ourselves and others understood and interpreted. The potentially awkward juxtaposition of feelings, the close intensity of relationships in certain contexts, may well provoke distancing effects of the kind that would have tested even Simmel's roving imagination.

Equally, if we are to entertain the view (in certain cities at least) that there has been an acceleration in the pace of modern life; that there is more to negotiate, more to absorb, because of a greater degree of immediacy to social relationships, then certain consequences follow. Even allowing for an element of exaggeration in this scenario, it remains incumbent to think through the effects of the different rhythms and intensities involved. This is not – to stress the point – an issue of fast speed, of social relationships moving up a gear, but rather a concern to unravel how people negotiate the fact that the gap between event and experience is felt to have diminished. It is not necessary to tie this culture of movement to the dynamics of money however, as Simmel did, to appreciate the fact that how people negotiate this immediacy and diversity will shape their everyday conduct and social interaction. For the moment though, all that we can do is to speculate on what commonplace antics may be involved, blasé or otherwise.

Acknowledgements

Jo Foord, Steve Pile, Michael Pryke and Nigel Thrift read and commented on an earlier version of this chapter. The current version owes much to their insights.

Notes

1 The former essay is reproduced in Frisby and Featherstone's (1997) recent collection of Simmel's writings on culture. In the introduction to this text, as well as in Frisby (1992), the latter essay reviews Simmel's more formal concerns with the significance of boundaries, frontiers, location and the like. Interestingly, (in Frisby, 1992), he also draws attention to Emile Durkheim's critique of Simmel's 'spatial geometry', who, it turns out, preferred the earlier spatial formulations

of the political geographer, Friedrick Ratzel to those of Simmel. See also, Durkheim's 'The Realm of Sociology as a Science' in Frisby (1994).

2 Kracauer 1995: 257.

3 The writings of the Chicago School in the 1920s, in particular Robert Park, were influential in presenting the figure of the 'stranger' in this one-dimensional way. See Levine (1977) for a review and critique. At much the same time, the concept of 'social distance' was rendered 'manageable' and converted into a set of variables which could be measured in the positivistic style of the day. See Levine, Carter and Miller Gorman (1976).

4 Simmel 1950b: 406: A trace of strangeness in this sense easily enters even the most intimate relationships. In the stage of first passion, erotic relations strangely reject any thought of generalization: the lovers think that there has never been a love like theirs; that nothing can be compared either to the person loved or to the feelings for that person. An estrangement – whether as cause or as consequence it is difficult to decide – usually comes at the moment when this feeling of uniqueness vanishes from the relationship.

5 It is not entirely clear that social distancing is a peculiarly urban phenomenon, as each of the three ways referred to also appear in *The Philosophy of Money* and clearly relate to modern culture in general. At the end of *The Metropolis and Mental Life* however, Simmel informs his readers that the content of the essay does not derive from any citable literature but rather from the arguments and ideas developed in *The Philosophy of Money*. At minimum, then, it is possible to say that money and city life are both cause and consequence of social distancing, although that is not to say that it is a city-bound social performance.

6 Rose 1996: 142: The concept of the fold or pleat suggests a way in which we might think of human being without postulating any essential interiority, and thus without binding ourselves to a particular version of the law of this interiority whose history we are seeking to disturb and diagnose. The fold indicates a relation without an essential interior, one in which what is 'inside' is merely an infolding of an exterior.

The concept of the fold is borrowed from the work of Gilles Deleuze (1988, 1993).

7 This argument is developed further in Allen and Pryke (1999).

8 Entitled, 'The Style of Life', the final chapter of *The Philosophy of Money* is credited by Habermas (1996) as a key influence in bringing about a shift in the conception of 'culture' from an expressive, subjective dimension to that of a material, objective process.

9 Simmel 1990: 505.

10 For an elaboration, see Pryke and Allen, (2000) 'Monetized time-space: Derivatives – money's "new imaginary"?'.

References

Allen, J. and Pryke, M. (1999) 'Money Cultures after Georg Simmel: Mobility, Movement and Identity' in *Environment and Planning D: Society and Space*, 17, 1: 51–68.

Adam, B. (1995) *Timewatch: The Analysis of Social Time* (Cambridge, Polity).

Castells, M. (1996) *The Rise of the Network Society* (Oxford, Blackwell).

Canetti, E. (1973) *Crowds and Power* (Harmondsworth, Penguin Books).

Deleuze, G. (1988) *Foucault* (London, Athlone Press).

Deleuze, G. (1993) *The Fold: Leibniz and The Baroque* (Minneapolis, University of Minnesota Press).

Dodd, N. (1994) *The Sociology of Money: Economics, Reason and Contemporary Society* (Cambridge, Polity).

Donald, J. (1996) 'The citizen and the man about town' in Hall, S. and du Gay, P. (eds) *Questions of Cultural Identity* (London, Sage).

Frisby, D. (1985) *Fragments of Modernity: Theories of Modernity in the Works of Simmel, Kracauer, and Benjamin* (Cambridge, Polity).

Frisby, D. (1992) *Simmel and Since: Essays on Georg Simmel's Social Theory* (Routledge, London).

Frisby, D. (1994) *Georg Simmel: Critical Assessments* (London and New York, Routledge).

Frisby D. (1997) 'Introduction to the texts' in Frisby, D. and Featherstone, M. (eds) *Simmel on Culture: Selected Writings* (London and New Delhi, Sage).

Frisby, D. and Featherstone, M. (eds) (1997) *Simmel on Culture: Selected Writings* (London and New Delhi, Sage).

Giddens, A. (1984) *The Constitution of Society: Outline of a Theory of Structuration* (Cambridge, Polity Press).

Giddens, A. (1990) *The Consequences of Modernity* (Cambridge, Polity).

Habermas, J. (1996) 'Georg Simmel on Philosophy and Culture: Postscript to a Collection of Essays' in *Critical Inquiry*, 22: 403–14.

Harvey, D. (1985) *Consciousness and the Urban Experience* (Oxford, Basil Blackwell).

Harvey, D. (1989) *The Condition of Postmodernity* (Oxford, Basil Blackwell).

Kracauer, S. (1995) *The Mass Ornament: Weimar Essays* (Cambridge, Mass., Harvard University Press).

Lash, S. and Urry, J. (1994) *Economies of Signs and Space* (London, Sage).

Levine, D. (1977) 'Simmel at a Distance: On the History and Systematics of the Sociology of the Stranger' in *Sociological Focus*, 10, 1: 15–29.

Levine, D., Carter, E. B., Miller Gorman, E. (1976) 'Simmel's Influence on American Sociology' in *American Journal of Sociology* 81: 813–45 and 1112–32.

Maspero, F. (1994) *Roissy Express: A Journey Through the Paris Suburbs* (London and New York, Verso).

Pryke, M. and Allen, J. (2000) 'Monetized time-spaces: Derivatives – money's "new imaginary"?' in *Economy and Society* 29: 2.

Rose, N. (1996) 'Identity, Genealogy, History' in Hall, S. and du Gay, P. (eds) *Questions of Cultural Identity* (London, Sage).

Sennet, R. (1969) *Classic Essays on the Culture of Cities* (New York, Appleton–Century–Crofts).

Shields, R. (1992) 'A Truant Proximity: Presence and Absence in the Space of Modernity' in *Environment and Planning D: Society and Space*, 10, 2: 181–98.

Simmel, G. (1950a) 'The Metropolis and Mental Life' in Wolff, K. H. (ed.) *The Sociology of Georg Simmel* (New York, The Free Press).

Simmel, G. (1950b) 'The Stranger' in Wolff, K.H. (ed.) *The Sociology of Georg Simmel* (New York, The Free Press).

Simmel, G. (1991a) 'Money in Modern Culture' in *Theory, Culture and Society* 8, 3: 17–31.

Simmel, G. (1991b) 'The Problem of Style' in *Theory, Culture and Society* 8, 3: 63–71.

Simmel, G. (1990) *The Philosophy of Money* (ed.) Frisby, D. (London and New York, Routledge).

Thrift, N. (1996) 'New Urban Eras and Old Technological Fears. Reconfiguring The Goodwill of Electronic Things' in *Urban Studies* 33: (8) 1463–1493.

Urry, J. (1995) *Consuming Places* (London and New York, Routledge).

Watts, M. (1994) 'Oil as Money: The Devil's Excrement and the Spectacle of Black Gold' in Corbridge, S., Martin, R. and Thrift, N. (eds) *Money, Power and Space* (Oxford, Basil Blackwell).

Young, I. M. (1990) *Justice and the Politics of Difference* (Princeton, New Jersey, Princeton University Press).

3

MIKHAIL BAKHTIN
Dialogics of space

Julian Holloway and James Kneale

Introduction

The aim of this chapter is twofold. Firstly we seek to describe and delimit the ways in which the Russian thinker Mikhail Bakhtin 'thinks' space: to draw out and exemplify ways in which he understood and wrote about space and spatial relations. Through delineating Bakhtin's 'geographical imagination' the second aim of this chapter can be achieved. Specifically this involves taking steps towards a thoroughly *dialogical theory of space*. In embarking tentatively toward this goal we have found that the path is difficult to traverse. In particular, journeying towards a dialogics of space means encountering difficulties arising from Bakhtin's differing notion of *context*. The recognition of this varying conception of context is significant as it structures the argument and the organisation of this chapter. In other words, this chapter travels from the material and phenomenological to a wider social notion of context in Bakhtin's work, and in doing so we move towards a dialogical theory of space.

Yet crucially the travels presented here do not seek to arrive at a pre-ordained destination. To arrive in such a place and to understand its contours fully, in terms of the limitations of a chapter of this length, would be nothing short of miraculous. Second, to arrive at such a place would mean abandoning the fundamental tenets of Bakhtinian thought. As we shall see Bakhtin's philosophy is one of open-endedness and becoming. To reach a point where the opportunity for further travel, or more precisely for continuing dialogue, is denied is a position that does not exist in Bakhtin's thought.

A useful point of departure is the remarkable biography of Mikhail Mikhailovich Bakhtin (1895–1975), since a brief outline of his life illustrates some of the overriding notions of dialogism. For example the Russian towns of Vilnius and Odessa where he spent his pre-university days are noted by Holquist (1990: 1) to be 'unusually heterogeneous in their mix of cultures and languages', thus reflecting and inspiring Bakhtin's interest in many-languagedness, or in his terminology, *heteroglossia*. After leaving St. Petersburg University in 1918 in the aftermath of the revolution he settled in the towns of Nevel and Vitebsk until 1924. This period is often denoted as the first significant period in Bakhtin's oeuvre, characterised by his engagement with neo-Kantianism and thus more (traditional) philosophical works, many of which have been published posthumously (for example *Toward a Philosophy of the Act*, 1993).

Here Bakhtin became a member of a group of intellectuals with whom he shared many conversations, debates and dialogues. Included in what has come to be known as the 'Bakhtin circle' were Voloshinov and Medvedev. These two figures are of utmost importance in Bakhtin's biography and work, not only in terms of their exchanges, but because of the two works attributed to them: *Marxism and the Philosophy of Language* (MPL, 1973, originally 1929) and *The Formal Method in Literary Scholarship* (FMLS, 1985, originally 1928) respectively. It has been alleged that these texts are not the work of the authors named on the original manuscript, but of Bakhtin himself. Commonly known as the 'authorship dispute', this controversy persists and is unlikely to be resolved – Bakhtin never affirmed or denied his authorship of the 'disputed texts'. In our opinion the debate over original authorship, and thus who owns the words in these texts, is exemplary of Bakhtin's dialogism: these texts can be seen as the products of dialogical encounters and interactions between Bakhtin, Medvedev, Voloshinov and others in the 'Bakhtin circle'.

In 1929 Bakhtin was arrested and exiled in Kazakhstan. This signalled a shift in the orientation of his work to cultural history and the evolution of the novel, yet with the metaphysical questions of his early years still very much in mind. Many of Bakhtin's more well known works, especially his treatise on Rabelais, were written during and just after this exile period, some of which were lost, destroyed or even, with cigarette papers in short supply, smoked by the author himself! After the war Bakhtin taught at Saransk University, until he moved to Moscow in the 1960s, where his prominence as a thinker dramatically soared with the publication of the second edition of *Problems of Dostoevsky's Poetics* [PDP] (1984a, originally written in 1929), and his 'discovery' and promotion by three scholars at the Gorky Institute. In his final years Bakhtin's writing returned to the philosophical focus of his earlier work. This focus, as well as the task of rewriting and editing older manuscripts, marks the third period of his oeuvre.

From this time on Bakhtin's thought has been drawn upon, utilised and extended in a variety of different fields and disciplines.[1] Yet it is both the applicability and appropriation of Bakhtin's concepts and ideas in such a multitude of arenas that makes, in part, the aim of describing his work a difficult task. Thus, in order to represent his thought, and elucidate its geography, we must enter a dialogue not only with a thinker of enormous breadth and variety, but with a host of interpreters from across the social sciences and humanities. We have been made constantly aware that *a* Bakhtin does not exist, and thus any attempts to draw his work into one overriding category of description tells of a centripetal force that he sought to challenge through revealing, and often championing, those centrifugal forces of diversity and heterogeneity. Once again the aims of this chapter run the constant risk of going against the arguments that Bakhtin himself developed.

This hazard, wherein we '"monologize" the singer of "polyphony"', has not been heeded by many in the social sciences and humanities (Clark and Holquist 1984: 4). Thus, all too often we have Bakhtin defined as *only* a theorist of literature, a folklorist or social critic. As such 'the last few years have witnessed . . . a kind of posthumous wrestle over the political soul of Bakhtin' (Stam 1988: 117). Therefore, entering into an analytical and theoretical dialogue not only with the work of Bakhtin himself, but also his appropriators and interpreters, we run another risk of reifying one type of Bakhtinian thought. With this pitfall in mind we fully admit to have taken two (Western) versions of his thought as central to our argument. The first half of the chapter is informed by the overview provided by Holquist (1990). His is a liberal reading of Bakhtin, seen through the ethical and epistemological themes of Self and Other, which for Pechey (1989), denies the socio-political themes and ramifications of his work. To incorporate the latter, the second half of this paper moves from the phenomenological to the social, with the notion of social speech genres and carnival taking precedence. Here a more 'left' Bakhtinianism is utilised, particularly that of Hirschkop (1989). Moreover, the mutual articulation, or again more precisely the dialogue, between these two Bakhtins furnishes the possibility of a dialogical theory of space, or at least the initial steps on that journey. Here we begin these travels where Bakhtin began his: the notion of Self and Other.

Self and Other in Bakhtin

The philosophy of Self and Other in Bakhtin's work holds central significance for his thinking. One of the most succinct and revealing statements on this topic comes from *The Problems of Dostoevsky's Poetics*:

> I am conscious of myself and become myself only while revealing myself for another, through another, and with the help of another. The most important acts constituting self-consciousness are determined by a relationship toward another consciousness (toward a *thou*) . . . The very being of man (both external and internal) is the *deepest communion. To be* means *to communicate* . . . To be means to be for another, and through the other for oneself. A person has no internal sovereign territory, he is wholly and always on the

boundary: looking inside himself, he looks *into the eyes of another* or *with the eyes of another* ... I cannot manage without another, I cannot become myself without another.

(Bakhtin 1984a: 287, emphasis in original)

The emphasis here upon visuality and sight reveals the first way in which Bakhtin thinks space. For Bakhtin, drawing upon neo-Kantianism and post-Newtonian revelations in physics, particularly Einstein's relativity theory, the categories of time and space are fundamental to our perception of the world. I organise the world through time and space categories from my unique place in existence. This organisation of the world through the categories of space and time are unique to me in that no-one else can inhabit the (physical) place that I do: no two bodies can occupy the same space. This is known as the *law of placement*. However, this unique placement I have in existence is shared, since everyone else also has a unique place in existence. In other words, we are presented with the paradoxical and almost contradictory idea of *differences in simultaneity*, that is best summed up in Bakhtin's phrase 'the unique and unified event of being'. As Holquist (1985: 227) puts it, the 'resulting paradox is that we all share uniqueness'.

To further explain the law of placement we must utilise Bakhtin's concrete example of two people facing each other. It is here that the emphasis of seeing and vision in the above passage allows us to begin to articulate the relation between Self and Other. If I face you there are certain things that I can see that you are unable to see and vice-versa: the wall behind your back, the clouds in the sky, your own forehead. We both possess a 'surplus of seeing'. Thus, I place you as a whole in a certain position in space, as you do to me. However, as I cannot see myself as a whole (I cannot see my own forehead), I am unable to position myself without the assistance of your sight. This example organises Bakhtin's notion of the Self/Other relation. Stated simply I need the Other in order to create a sense of Self. The Self therefore is nothing in itself. Self means nothing without the alterity or outsideness that is provided by the Other: 'I cannot become myself without another'. Being in Bakhtin's thought is in effect co-Being. In turn this refutes the possibility of a monadic and privileged centre to the Self, denying the possibility of a static, immutable, least of all transcendental essence: there is 'no internal sovereign territory' to the Self. Yet this does not mean that the Self merges with the Other, in some sort of Hegelian dialectical synthesis: the law of placement precludes this. Being is unique *and* unified, different *and* simultaneous. There is a fundamental non-coincidence between Self and Other, and thus the two never merge.

The potential of positing some form of (humanist) centre or interior to the Bakhtinian Self arises from the unique perceptual place it has in space and time, in which no Other can exist. Again, however, I cannot see everything from this position. Because of this we are always responsive and answerable to this outsideness. As Bakhtin puts it 'there is no alibi in existence'. Alterity is fundamental to the 'not-I-in-me'. The implication of this is that for self-authorship through outsideness to proceed, in order to discover the 'not-I-in-me', in some way the Self must complete the Other. In other words, it must fix or better still objectify the Other in time and space. For Bakhtin the Self attempts an *architectonics* (the ordering into wholes) of the Other. As mentioned above we see them and temporally and spatially position them as wholes in relation to other people and different objects.

The recognition of this difference through the performance of such an architectonics is precisely the significance of alterity and outsideness. Yet while the Self completes the Other, the Self will never be brought into stasis and fixity. The Self will always exceed that which it necessarily derives from alterity, precisely because its place in existence is unique. In addition this place is an *event*. The ontology of the Bakhtinian Self is one which is characteristically always open and in a constant state of Becoming. Put differently, the Self can know no limits; it is not after all a locus of primary meaning, it has 'no alibi'.[2] In overview we must conceive of the Bakhtinian Self 'as a multiple phenomenon of essentially three elements (it is – at least – a triad, not a duality): a centre, a not-centre, and the relation between them' (Holquist 1990: 29). Bakhtin's therefore is very much a *relational* approach to ontology and philosophy – an approach which we now substantiate further.

The (dialogical) utterance

In the above quote Bakhtin states that '*to be* means *to communicate*'. In other words, once we stop responding to the world, if we cease being addressed by the environment and the others around us, we simply cease to be.[3] At this point we must ask how does this (co-)Being manifest itself? In what form does this communication occur that is so central to our ontology? Bakhtin answers this question by endowing the *sign* with central and overriding importance: '*consciousness itself can arise and become a viable fact only in the material embodiment of signs*' (Voloshinov 1973: 11, emphasis in original).

Consciousness, thought ('inner speech'), experience and understanding, all of which pertain to the (infinite) addressivity and responsibility to the world, only exist through the semiotic material of the sign. In order to express outwardly an experience or an understanding in this ongoing event of perpetual addressivity we must objectify it in the sign. Thus, the 'potentialities of expression' are the potentialities of the sign, and the 'possible routes and directions' that this expression may take are *always social* in their forms (Voloshinov 1973: 91). For Bakhtin it is crucial to take communication or language in its *concrete socio-historical context*. The emphasis of linguistics should be the situated and concrete speech performance: the historically generative process of everyday discourse rather than a hypo-statical set of self-identical norms.[4] In other words, language, in what Stewart (1986: 43) calls Bakhtin's *'anti-linguistics'*, should be taken in its dynamism and mutability: the 'living impulse' of language.

From this Bakhtin's social semiotics takes the *utterance* as its basic unit of analysis. The boundaries of the utterance are delimited by the *'change of speaking subjects*, that is, a change of speakers' (Bakhtin 1986: 71). This 'relinquishing of the floor' gives the utterance, in its variable size (from the 'single word rejoinder' to the 'scientific treatise') a beginning and an end. Yet because of its very situatedness the utterance can never be analysed or understood in isolation, as it is never in of itself. The utterance is *always* situated in a relation, it is always shaped by the relationship it has with *other* utterances: its boundaries while being recognisable are never imper-vious. Therefore, the work of signification or meaning *always* occurs as part of a *dialogue* between (at least) two utterances.

We would like to illustrate this in two ways, that subsequently develop what has gone before and introduce another aspect of Bakhtin's thought. Firstly, the dialogical utterance can be exemplified through the commu-nicative act between Self and Other as two situated interlocutors. The articulated utterance of the Self from its inception is always placed in a relation to that of the Other via the referencing, understanding and aware-ness of the Other's past, present and potential future utterances. The utterance is 'double voiced' in the sense that both the Self's and Other's voices interpenetrate the utterance: the utterance is thus 'internally dialo-gized'. The subjects' own utterance meets the (alien) word of the other, as the latter is always anticipated and/or incorporated into the former (Danow 1991). 'Any utterance – the finished, written utterance not excepted – makes response to something and is calculated to be responded to in turn. It is but one link in a continuous chain of speech performances' (Voloshinov 1973: 72).

The second useful way of illustrating the dialogical utterance is through introducing Bakhtin's concept of *novelness*. Novelness refers to the poten-tial for dialogue latent in all art but which is most often found in particular examples of the novel. For Bakhtin the work of Dostoevsky and Rabelais (and here we concentrate on the former) possesses 'novelness' in abun-dance because it is open to dialogue (not closed like the monological novel where the author has the final word), and as such these novels can be seen as textualisations of Self–Other relations. Thus, Dostoevsky's novels contain relations between various consciousnesses (author and hero, one character and another) which remain 'unmerged . . . with equal rights and each with its own world' (Bakhtin 1984a: 6). Moreover, the communication between these different consciousnesses takes the form of dialogized utterances. Thus, Bakhtin traces those utterances which answer others, which take up

76

and transform other points of view, those which are 'double-voiced' or which contain a 'sideward glance' at the position of the Other. Taking Dostoevsky's *Poor Folk* as an example, Bakhtin states:

> Discourse here is double-voiced . . . Not only the tone and style but also the internal semantic structure of these self-utterances are defined by an anticipation of another person's words . . . In *Poor Folk* Dostoevsky begins to work out the 'degraded' variety of style – discourse that cringes with a timid and ashamed sideward glance at the other's possible response, yet contains a muffled challenge.
>
> (Bakhtin 1984a: 205)

Through the 'orchestration' of different and multiple co-existing voices, Dostoevsky produces *polyphony* and achieves novelness. The polyphonic novel, then, is characterised by the articulation of many voices that remain unmerged. Yet through dialogical utterances these voices glance sideways at each other, thus recognising the need for the other's voice in the production of meaning.

The speech genre

What arises from a discussion of the dialogical utterance is the need for a way of understanding *how* the other's voice (or more precisely their past, present and future utterances) is recognised and registered into the utterance. The answer to this has been hinted at above, but let us take a step back to fully achieve this. The utterance as the basic unit of speech communication is always situated in the context of *social* time and space: 'Each rejoinder, regardless of how brief and abrupt, has a specific quality of completion that expresses a particular *position* of the speaker, to which one may respond or may assume, with respect to it, a responsive *position*' (Bakhtin 1986: 72, our emphasis).

Position here refers to the placing of the speaker in an ideological terrain. In other words, the speaker deploys utterances which embody a particular world-view or social interest, what we can call a *positionality*. The diversity and manifold variety of these different points of view or ideologies, in competition and conflict, is termed heteroglossia (many-languagedness). However, now we face the question as to how this social interest and positionality is registered in the utterance. Bakhtin answers this through the concept of the *speech genre*. Thus, through the deployment of certain *ways* of talking, the enunciator's position (in the contested ideological terrain of heteroglossia) is revealed. Speech genres are (relatively) stable and conventional forms of 'content, linguistic style and compositional structure' (Gardiner 1992: 81). In the speech performance the social interest, the position of the speaker, is registered by the enunciation of these recognisable forms of speech. Thus, the many voices, the polyphony, of the social world, are bound to the many languages, the many speech genres, of heteroglossia. Bakhtin takes a further step in his description of speech genres by differentiating between *primary* and *secondary* speech genres. This is 'understood not as a functional difference', but one of complexity (Bakhtin 1986: 61–2). Primary speech genres are performed in the everyday sphere of 'unmediated speech communion', whereas secondary genres are more complex and organised forms, such as 'novels, dramas, all kinds of scientific research, major genres of commentary' (Bakhtin 1986: 61–2).

We are now able to understand how the utterance becomes 'double-voiced'. The speech performance is a process of *evaluation* of the Other's speaking position that becomes known through the use of different speech genres. The enunciator thus recognises the generic form the Other's utterance takes and incorporates this understanding into his or her own utterance. Put differently, the speaker's voice contains or is interpenetrated by the other interlocutors' (past, present or potential) voices through the evaluation of their way of speaking. Identification of speech genres as social languages internally dialogises the utterance. However, on close inspection of this process of evaluation and the double-voiced utterance, in the work of the Bakhtin circle, certain difficulties arise. Specifically these revolve around the notion of *context*.

As Hirschkop (1989) argues there are two conceptions of context in Bakhtin's work. The first is the concrete verbal situation of two speakers in dialogical interaction. This is the phenomenological context of self and other, organised according to the law of placement, that was described earlier. Yet context also appears in Bakhtin's work in the 'wider' sense of heteroglossia. Here we have a social context, replete with competing ideologies and interests, or more precisely 'the other languages against which the utterance "must define itself"' (Hirschkop 1989: 15). There is then a kind of *gap* between the phenomenological and the social meaning of context in Bakhtin's work: 'We are thus confronted with an awkward analytical choice: do we define context as the immediate material situation . . . or do we define it as heteroglossia, a more spacious conception, but one which restricts the context to the stuff of language?' (Hirschkop 1989: 16).

Moreover, Bakhtin often appeals to the *uniqueness* of the material dialogical context and the concrete utterance enunciated therein. This concrete situation is depicted as unrepeatable and distinct. Yet the notion of heteroglottic context suggests some form of repeatability. For evaluation and 'double voicing' to be possible utterances must take generic forms and thus the utterance 'tastes of the context and contexts in which it has lived its socially intense life' (Bakhtin 1981: 293). The varying speech genres of heteroglossia thus form something resembling an extra-verbal structure that determine the value of the utterance. Therefore, if we are to retain any sense of the unique verbal context, the non-reiterative utterance, we end up facing, in our opinion, the well-rehearsed problems of structure and agency in language: structure becomes the heteroglottic context and agency the material/concrete context of Self and Other in dialogue. Let us partially resolve (or probably more accurately shamelessly side-step for lack of space) this issue by suggesting that context here should be seen as a 'developed-developing' event (Shotter 1993). By this we mean that the heteroglottic context constrains the utterance by accentuating it with a socially located view on the world, but never fully determines the material/concrete utterance, which is in turn endowed with the possibility of re-defining and re-developing that very same constraining heteroglottic context. Bakhtin did retain the possibility to 're-accentuate genres' and so this we believe is still within the parameters of Bakhtin's, admittedly varying, conception of 'context' (Bakhtin 1986: 78, 79). Thus, the heteroglottic context becomes the social or 'third' element 'in between' the Self and Other placed in the material/concrete context.

Indeed, this conception informs the following moves. For it is here that we suggest that the social terrain of heteroglossia can be argued to be a *socio-spatial* landscape. In other words, if speech genres carve up the social then they can also be seen to carve up space.

Carnival's 'second world': space and speech genre

It is time to discuss this wider social notion of context. Our discussions so far have discussed the ways in which Bakhtin's thought possesses a spatial dimension in terms of Self–Other relationships as relational positions. In the last section we noted that these utterances take place within, and may transform, a wider socio-linguistic context (the speech genre). We now turn to the spatial aspects of these speech genres, which are most clearly explored in Bakhtin's writings on Carnival. Bakhtin returned to Carnival again and again; apart from *Rabelais and his World* (1984b), significant parts of the second edition of *Problems of Dostoevsky's Poetics* (1984a) and 'Forms of Time and of the Chronotope in the Novel' (in *The Dialogical Imagination*, 1981) also consider Carnival and its relations to literature. Here we will concentrate on those aspects of Carnival which stress the relationship between space and speech genre.

Although Carnival is presented as a set of images, retrieved from the writings of Rabelais and others, Bakhtin was concerned with the social and linguistic *practices* of early modern popular culture. Bakhtin wrote that Carnival creates and draws upon 'a second world and a second life outside officialdom' (1984b: 6), the inevitable rejoinder to monological utterances, which attempt to deny dialogue by having the 'last word'[5]: 'No dogma, no authoritarianism, no narrow-minded seriousness can co-exist with Rabelaisian images; these images are opposed to all that is finished and polished, to all pomposity, to every ready-made solution in the sphere of thought and world outlook.' (*Ibid.*: 3).

We can see how dialogue addresses the *asymmetry* of power relations in the way that Carnival challenged the utterances of 'official culture':

> . . . Carnival celebrated temporary liberation from the prevailing truth and from the established order: it marked the suspension of all hierarchical rank, privileges, norms, and prohibitions. Carnival was the true feast of time, the feast of becoming, change and renewal. It was hostile to all that was immortalised and completed.
>
> (*Ibid.*: 10)

Further delineating the nature of this 'second world', Bakhtin described the disparate forms and practices of Carnival as aspects of 'grotesque realism', because they emphasise *renewal through degradation*. They invert the hierarchies of official culture in a way which expresses a cosmic philosophy, a cycle of death and rebirth which is utopian because it is always oriented to the future.[6] As a consequence, they establish a *unity* between the people, setting the stage for freer social relations.

The material body is vital to this second world because all Carnival practices 'turn their subject into flesh' (*Ibid.*: 21), dragging high culture down to 'the sphere of earth and body in their indissoluble unity' (*Ibid.*: 19–20). This process of renewal emphasises its nature as a body of becoming. Crucially, this grotesque body is 'open to the outside world' (*Ibid.*: 26) through its orifices and protuberances, especially those of the 'material lower bodily stratum': genitalia, buttocks, anus, belly, breasts.[7] These are points of contact with the social world, which mark it as a body open to dialogical relations, just as we have already noted that the Self is open to the words of the Other: 'It is not a closed, complete unit: it is unfinished, outgrows itself, transgresses its own limits' (*Ibid.*: 26); '[it] is blended with the world, with animals, with objects' (*Ibid.*: 27). The classical body celebrated by the Renaissance, in contrast, is smooth, closed, finished: it attempts to monologically deny the role of Others in its own constitution.[8] Because the grotesque body is open, it is also *the body of the people* in more than one sense: '[The body] is presented not in a private, egotistic form, severed from the other spheres of life, but as something universal, representing all the people. . . a people who are continually growing and renewed' (*Ibid.*: 19).

Carnival's second world is built upon dialogical social relations in these ways; but is more than just a metaphorical space. 'The language of the marketplace', Bakhtin's phrase for the speech practices of the markets, streets, and public spaces of the people, is literally rooted in space. This language, translated into English by Hélène Iswolsky as 'Billingsgate', was both an important speech genre located in (and producing) a specific social space and a dialogical answer to the monologue of the elite. These speech practices develop an important unity between Carnival's participants. In a well-known passage Bakhtin wrote 'Carnival is not a spectacle seen by the people; they live in it, and everyone participates because its very idea embraces all of the people. While Carnival lasts, there is no other life outside it.' (*Ibid.*: 7).

In this sense, Carnival created a special world of language and inter-action, 'permitting no distance between those who came into contact with each other' (*Ibid.*: 10). The marketplace, home of Billingsgate, 'was a world in itself, *a world which was one*' (*Ibid.*: 153, emphasis added) because 'the exalted and the lowly, the sacred and profane are levelled and are all drawn into the same dance' (*Ibid.*: 160). It is this sense of openness and unity which creates Carnival's progressive force.

There is no room here to discuss the political efficacy of Carnival, which has been extensively discussed since the publication of *Rabelais and his World* in English in 1984. We hope to address this question elsewhere, but we hope that our spatial reading of Carnival avoids some of the problems identified by others (see, for example, Bristol, 1985; Burke, 1994; Darnton, 1984; Davis, 1987; Le Roy Ladurie, 1980). We would emphasise that Carnival is not an abstract 'force' but a set of practices which do not determine its consequences; that these practices are located in specific contexts; and that if we move away from seeing Carnival as an *inversion* of order (Davis, 1987; Sibley, 1995) we can avoid an episodic view of cultural politics, where disorder and transgression are restricted to rare, large-scale outbursts of popular feeling.

Bakhtin made it plain that Carnival was not simply to be found in revelry or riots, but also in everyday speech, conceptions of the body, and so on. As the dialogical Other of official culture, Carnival must always be present; it contaminates the supposedly monological utterances of the powerful. Carnival may be a weakened force, but its currents still run through popular culture. In this sense, we should be looking for elements of everyday life which can become 'Carnivalised', just as novelness refers to Carnivalised literature: open to the play of dialogue, resisting the 'last word'.

Once we have reconceived Billingsgate as the performance of spatialised social relations (including linguistic ones) we can see that space and speech genre can be mutually constitutive. The speech performances of Billingsgate draw upon the dialogical social relations of the marketplace. As in the novel, this speech genre has the potential to rewrite language and social space; it represents a centrifugal opposition to the centripetal, ordering attempts of monologues. As a result, we should not be looking for temporary or liminal inversions of hierarchies, but the ways that Carnival constantly attempts to undermine these monologues in all spaces.[9]

The chronotope

We finish with Bakhtin's most obviously spatial concept: the chronotope. This is a trope of literature which governs the representation of time and space in the novel.

> We will give the name *chronotope* (literally, 'time space') to the intrinsic connectedness of temporal and spatial relationships that are artistically expressed in literature ... In the literary artistic chronotope, spatial and temporal indicators are fused into one carefully thought-out, concrete whole. Time, as it were, thickens, takes on flesh, becomes artistically visible; likewise, space becomes charged and responsive to the movements of time, plot, and history.
>
> (Bakhtin 1981: 84)

Chronotopes take generic form, so that each genre displays a different conception of the relations between time and space, but Bakhtin was keen to stress the history of these conventions. Bakhtin's examples, from Greek romances of the second to sixth centuries AD to the novels of Flaubert, Stendahl, and Balzac, show a range of chronotopic arrangements of time and space, tied principally to the closing of the open Self (leading to a concern with ways of representing private, interior spaces) and changing conceptions of personal time. This 'chronotopic analysis' therefore offers great scope to geographers interested in the constitution of novelistic space.[10]

There are two other ways in which the chronotope is of use to geographers. Mireya Folch-Serra (1990) perceptively points out that the chronotope offers a tool for analysing the constitution of spaces beyond literature. Folch-Serra combines Bakhtin's ideas on language and the novel to suggest, in effect, a dialogical method for the study of landscape, region, and place. Space is constructed by the constant dialogical interaction of a multiplicity of voices; at any point in space and time it is possible to see a chronotope which is more or less fixed depending upon the strength of competing centripetal (monological) and centrifugal (dialogical) forces.

> The Bakhtinian conceptual landscape goes beyond the visual criteria that made the geographer an interpreter of natural conditions. It strives, rather, at ongoing historical developments that alternately 'anchor' and destabilize the 'natural harmony' of a given region through constant interaction between meanings. These meanings are spawned, of course, by conversation. A dialogical landscape indicates the historical moment and situation

(time and space) of a dialogue whose outcome is never a neutral exchange. Landscape becomes not only 'graphically visible' in space but also 'narratively visible' in time, in a field of discourses all attempting to account for human experience.

(Folch-Serra 1990: 258)

Developing this idea is an ambitious exercise, and one which needs careful attention to Bakhtin's ideas of dialogue and the chronotope. Its value, though, is as a working method which does not privilege discourses or fix representations, but instead depends upon a recognition of their relative weight in dialogue.

Finally, the chronotope essay also offers the possibility of tracing the spaced and timed constitution of the self. As has already been noted, Bakhtin's 'historical poetics' illustrate changes in Western senses of time and space.[11] This broad cultural history could be used to explore the chronotopes of the Self, which is 'timed' as well as 'spaced' through its position within both the material and heteroglottic contexts.

Conclusion

In this chapter we have attempted to draw out the spatial aspects of Bakhtin's work, from the relations of Self and Other to the larger scale of the chronotope. This represents only the beginning of a geographical dialogue with this work, and we want to end by sketching out some of the more interesting paths others – including geographers – have taken. If we have one general comment here, it is that dialogue in the widest sense needs to be made central to Bakhtin's work.[12]

The first avenue of enquiry concerns the hybridity of identities and places. Postcolonial writings on diasporas have stressed the multiple constitution of cultural identity through the figure of the migrant or exile, who falls between two worlds. This is a thoroughly dialogical notion, though we should remember that the multiple identities of the white traveller are very different from those of the exile (Cresswell 1997). The theme of movement and displacement is an important one, and it is significant that Paul Gilroy's study of Black Atlantic 'double consciousness' develops through the identification of the chronotope of the ship 'as a chance to explore the articulations between the discontinuous histories of England's ports, its interfaces with the wider world' (Gilroy 1993: 17). The ship allows us to trace a number of issues: time–space representations of the Atlantic; the relationship between spaces and identities, constituted by discrete movements across the ocean; the hybrid communities of the ships themselves; and the asymmetrical dialogues between Europe, Africa, and the Americas which the ships facilitated. In fact, if we think of the ships as mobile utterances it is possible to apply Bakhtin's ideas to the way these 'conversations' formed these places. Thinking dialogically stresses the complex processes which make up social spaces, which bind local and global together in different forms in different places. Another example here is Joseph Sciorra's (1996) study of Puerto Rican *casita de madera* in New York, which reads them as chronotopes of memory and national identity, grounding identity and community in space. The *casitas* are also hybrids, mixtures of pre- and post-colonial forms made by bricoleurs as part of the 'caribbeaniza- tion of Nueva York' (*Ibid.*: 66).

The second area of study concerns ideas of space and transgression. This has already received some attention from geographers and others writing about spaces of carnival (Cresswell 1996; Jackson 1988; Lewis and Pile 1996; Shields 1991; Stallybrass and White 1986), as well as discus- sions of the political meanings of historical Carnival. This work has enormous potential to enrich our understandings of cultural politics, but we feel that geographers need to be sensitive to the wider principles of dialogism, rather than interpreting *Rabelais and his World* as a study of inverted hierarchies and 'safety valves'.

Third, an important area of study is being opened up by feminist engage- ment with and criticism of Bakhtin's ideas, and particularly the gender of Carnival's grotesque body. From initial accusations of misogyny in *Rabelais* (see Booth 1986 and Russo 1986) feminists have begun to work through the ambivalence of Bakhtinian concepts like the grotesque body. Many of the best examples of this (for example, Ginsburg 1993) also draw upon psychoanalysis, and this is another potentially exciting area for geograph- ical research.[13]

Finally, if we can accept dialogism as a method, we can begin to think about strategies for writing and doing geography. The use of humour in writing has been briefly but thoughtfully considered by David Matless (1995b), whose starting point is Foucault's observation that 'Genealogy is history in the form of a concerted carnival' (1986: 94). Although Bakhtin isn't mentioned, Matless's elaboration of the politics of humorous criticism chimes in with the former's observation that 'every act of world history was accompanied by a laughing chorus' (1984b: 474).[14] There is certainly scope for a carnivalised geography beyond the more narrow concept of poly- phonic writing (Crang 1992), and in fact Matless' own work offers some interesting examples (1995a: 114–18). Similarly, Marc Brosseau's (1995) geographical treatment of Bakhtin aims to initiate and develop a dialogical relationship between geography and literature to examine geographies of the novel. Considered dialogically, geography and literature can be mutu- ally articulated 'without having to melt both identities in the process' (Brosseau 1995: 92). This then is more of a methodological utilisation of Bakhtin's relational approach, wherein two modes of representation can be realised together without reduction or the loss of difference. Brosseau also hints at a dialogical theory of space: for example, through revealing how

the novel expresses the ephemeral and contingent process of the (re)pro-
duction of city-spaces via dialogical encounters.

These brief reviews hopefully indicate that there are many possible
directions that a dialogical study of space could take. We have written this
in dialogue with many other writers beyond Bakhtin; we hope we have
contributed to this ongoing discussion. And since it is impossible to have
the last word in dialogue, we expect this utterance to provoke others.

Notes

1 For a useful bibliography see Holquist, 1990: 195–200.
2 Any limits that the Self can experience that may bring it into stasis, such
   as death, it cannot know: I do not experience my own death, only Others
   do.
3 This is for Bakhtin an *ethical* point which is considered in depth in *Toward a
   Philosophy of the Act* (1993) – see Gardiner (1996), Morson and Emerson (1993).
4 Compare with Saussure's *parole*, which is rendered for the most part 'accessory'
   and 'random': 'a purely individual act' juxtaposed 'to the system of language
   as a phenomenon that is purely social and mandatory to the individuum' (Bakhtin
   1986: 81, see also FMLS).
5 'Official culture' is therefore a hybrid rather than a monolithic mass, 'contaminated' by its dealings with its Other.
6 It is worth noting that Bakhtin's use of the term 'utopian' is the very antithesis
   of those monological closed systems of rational thought associated with literary
   utopias (after More), and the utopian blueprints of modernist planning. Bakhtin's
   conception of the novel is anti-utopian because it refuses to accept a final word
   and truth (Vice 1997: 78); Carnival is utopian because it dares to imagine a
   future beyond these monological certainties.
7 The gendering of the grotesque body is a complex issue which we cannot explore
   in full here; see the references in the final section for discussions of this theme.
8 'All attributes of the unfinished world are carefully removed [from the body],
   as well as all the signs of its inner life' (1984b: 320).
9 One fruitful avenue to explore in this regard would be a comparison of De
   Certeau's 'tactics' (1984) with the playful but deadly serious performance of
   Carnival.
10 Many commentators on the chronotope, like Holquist (1990), tend to stress its
   temporal aspects. This probably reflects the importance of time in the novel;
   Bakhtin's essay is a radical development of the Russian Formalists' concern with
   fabula (story) and *sjuzhet* (plot). In some chronotopes space does seem to be subordinated to time – the 'adventure time' of the Greek romance is the clearest
   example – but even here Bakhtin's writings represent the fullest engagement of
   literary theory with the textualisation of space in the novel.
11 See the section on time in the classical biography and autobiography (1981:
   130–46) or the time-space of the chivalric romance (151–8) for examples.
12 For example, David Harvey's use of Bakhtin as part of a project towards a
   dialectical/relational view of time and space ironically appropriates the latter
   as a philosopher of Self and Other, akin to the liberal reading made by Holquist
   (1990). In particular the 'perspectival' situatedness of Self and Other gains ascendancy in this reading of Bakhtin, although the way in which this 'point of view'
   is socially interpolated does receive mention:

the perspectival view then merges into a more general relational view of space and time by virtue of the continuous shifts of social practices that put value upon both the 'I' and the 'others' by creating particular space–time nexus between them.

(Harvey 1997: 271)

Similar to our endeavours, Harvey here attempts to move from the material/phenomenological context to a more socially 'spacious' conception of context.

13 While the Bakhtin circle was explicitly opposed to Freudianism (Voloshinov 1976), it has been suggested that the encounter between Bakhtin and Lacan could be much more productive.

14 However, we should also take note of Matless's warning that humour can serve many different ends; in Rabelais' carnival, women are often the butts of masculine laughter.

References

Bakhtin, M. M. (1981) *The Dialogical Imagination*, edited and translated C. Emerson and M. Holquist, Austin: University of Texas Press.

Bakhtin, M. M. (1984a) *Problems of Dostoevsky's Poetics*, edited and translated C. Emerson, Minneapolis: University of Minnesota Press.

Bakhtin, M. M. (1984b) *Rabelais and his World*, translated H. Iswolsky, Bloomington: Indiana University Press.

Bakhtin, M. M. (1986) *Speech Genres and Other Late Essays*, edited C. Emerson and M. Holquist, translated V. W. McGee, Austin: University of Texas Press.

Bakhtin, M. M. (1993) *Toward a Philosophy of the Act*, edited V. Liapunov and M. Holquist, translated V. Liapunov, Austin: University of Texas Press.

Booth, W. C. (1986) 'Freedom of interpretation: Bakhtin and the challenge of feminist criticism', in G. S. Morson (ed.) *Bakhtin: Essays and Dialogues on his Work*, Chicago and London: University of Chicago Press.

Bristol, M. (1985) *Carnival and Theater: Plebeian Culture and the Structure of Authority in Renaissance England*, New York: Methuen.

Brosseau, M. (1995) 'The city in textual form: Manhattan Transfer's New York', *Ecumene*, 2, 1: 89–114.

Burke, P. (1994) *Popular Culture in Early Modern Europe*, Aldershot: Scolar Press.

Clark, K. and Holquist, M. (1984) *Mikhail Bakhtin*, London: Harvard University Press.

Crang, P. (1992) 'Politics and polyphony: Reconfigurations of geographical authority', *Environment and Planning D: Society and Space* 10: 527–49.

Cresswell, T. (1996) *In Place/Out of Place: Geography, Ideology and Transgression*, Minneapolis: University of Minnesota Press.

Cresswell, T. (1997) 'Imagining the nomad: Mobility and the postmodern primitive', in G. Benko and U. Strohmayer (eds) *Space and Social Theory: Interpreting Modernity and Postmodernity*, Oxford: Blackwell.

Danow, D. K. (1991) *The Thought of Mikhail Bakhtin: From Word to Culture*, London: Macmillan.

Darnton, R. (1984) *The Great Cat Massacre, and Other Episodes in French Cultural History*, New York: Vintage.

Davis, N. Z. (1987) *Society and Culture in Early Modern France*, Cambridge: Polity.

De Certeau, M. (1984) *The Practice of Everyday Life*, translated S. Rendall, Berkeley: University of California Press.

Folch-Serra, M. (1990) 'Place, voice, space: Mikhail Bakhtin's dialogical landscape', *Environment and Planning D: Society and Space* 8: 255–74.

Foucault M. (1986), 'Nietzsche, genealogy, history', in P. Rabinow, *The Foucault Reader*, Harmondsworth: Penguin.

Gardiner, M. (1992) *The Dialogics of Critique: M. M. Bakhtin and the Theory of Ideology*, London: Routledge.

Gardiner, M. (1996) 'Alterity and ethics: A dialogical perspective', *Theory, Culture And Society* 13, 2: 121–143.

Gilroy, P. (1993), *The Black Atlantic: Modernity and Double Consciousness*, London and New York: Verso.

Ginsburg, R. (1993) 'The pregnant text. Bakhtin's ur-chronotope: the womb', in D. Shepherd, *Bakhtin: Carnival and Other Subjects*, Amsterdam and Atlanta, Georgia: Rodopi.

Harvey, D. (1997) *Justice, Nature, and the Geography of Difference*, Oxford: Blackwell

Hirschkop, K. (1989) 'Introduction: Bakhtin and cultural theory', in K. Hirschkop and D. Shepherd (eds), *Bakhtin and Cultural Theory*, Manchester: Manchester University Press.

Holquist, M. (1985) 'The carnival of discourse: Bakhtin and simultaneity', *Canadian Review of Comparative Literature*, 12, 2; 220–34.

Holquist, M. (1990) *Dialogism: Bakhtin and His World*, London: Routledge.

Jackson, P. (1988). 'Street life: the politics of carnival', *Environment and Planning D: Society and Space* 6: 213–27.

Le Roy Ladurie, E. (1980) *Carnival: A People's Uprising at Romans, 1579–1580*, translated M. Feeney, London: Scolar Press.

Lewis C. and Pile S. (1996) 'Woman, body, space: Rio carnival and the politics of performance', *Gender, Place and Culture* 3, 1: 23–41.

Matless, D. (1995a) 'The art of right living: landscape and citizenship, 1918–39', in S. Pile and N. Thrift (eds), *Mapping the Subject: Geographies of Cultural Transformation*, London: Routledge.

Matless, D. (1995b) 'Effects of history', *Transactions of the Institute of British Geographers* 20, 4: 405–9.

Medvedev, P. N. (1985) *The Formal Method in Literary Scholarship: A Critical Introduction to Sociological Poetics*, translated A. J. Wehrle, Cambridge, MA and London: Harvard University Press.

Morson, G. S. and Emerson, C. (1993) 'Imputations and amputations: reply to Wall and Thomson', *Diacritics* 24, 4: 93–9.

Pechey, G. (1989) 'On the borders of Bakhtin: dialogisation, Ddecolonisation', in K. Hirschkop and D. Shepherd (eds) *Bakhtin and Cultural Theory*, Manchester: Manchester University Press.

Russo, M. (1986) 'Female grotesques: carnival and theory', in T. de Lauretis (ed.) *Feminist Studies/Critical Studies*, Bloomington: Indian University Press.

Sciorra, J. (1996) 'Return to the future: Puerto Rican vernacular architecture in New York city', in A. King (ed.) *Re-Presenting the City: Ethnicity, Capital and Culture in the 21st Century Metropolis*, Basingstoke: Macmillan.

Shields, R. (1991) *Places on the Margin: Alternative Geographies of Modernity*, London and New York: Routledge.

Shotter, J. (1993) *Conversational Realities: Constructing Life Through Language*, London: Sage.

Sibley, D. (1995) *Geographies of Exclusion: Society and Difference in the West*, London and New York: Routledge.

Stallybrass, P. and White, A. (1986) *The Politics and Poetics of Transgression*, London: Methuen.

Stam, R. (1988) 'Mikhail Bakhtin and left cultural critique', in E. A. Kaplan (ed.), *Postmodernism and its Discontents: Theories, Practices*, London: Verso.

Stewart, S. (1986) 'Shouts on the street: Bakhtin's anti-linguistics', in G. S. Morson (ed.), *Bakhtin: Essays and Dialogues on his Work*, Chicago: University of Chicago Press.

Vice, S. (1997) *Introducing Bakhtin*, Manchester and New York: Manchester University Press.

Voloshinov, V. N. (1973) *Marxism and the Philosophy of Language*, translated L. Matejka and I. R. Titunik, London: Harvard University Press.

Voloshinov, V. N. (1976) *Freudianism: A Marxist Critique*, translated I. R. Titunik, New York: Academic Press.

4

WITTGENSTEIN
AND THE FABRIC OF
EVERYDAY LIFE

Michael R. Curry

In a discipline that has sometimes seemed inexhaustibly voracious in its appetite for new philosophical delicacies, there has long – or so it has seemed to me – been a mystery. And that mystery has been the virtual absence of interest in the work of Wittgenstein. With few exceptions[1], geographers have simply had nothing to say about his work. This is, though, quite in contrast to the situation in other areas. In philosophy his work is widely discussed; the last ten years alone have seen the publication of over one hundred books – and about eight hundred articles – devoted to it. Moreover, others whose work has been often cited by geographers have themselves seen Wittgenstein as a central figure. In sociology, Anthony Giddens (1979) appealed to Wittgenstein's work as a cornerstone of his own; and Bourdieu used a quotation from Wittgenstein's *Vermischte Bemerkungen* (1977) as an epigraph to his *An Invitation to Reflexive Sociology* (1992). Indeed, and as Thrift (1996) has pointed out, where social theorists have claimed in chapter one that the work of Wittgenstein is the foundation of their own, geographers have tended to begin their appropriations with chapter two.

The invisible man

One can of course come up with a number of explanations for this silence on the part of geographers. Perhaps the social theorists in question were merely currying favor with philosophers. Or perhaps it is a matter of the nature of Wittgenstein's work itself. It is, after all, notoriously difficult to summarize. If like most philosophers he is not partial to footnotes, in the case of his work more than that of others one needs – at the outset – to have a strong sense of the philosophical terrain within which he is operating. And in the end it is difficult to characterize his position. Is he a realist? An idealist? Interpreters have a myriad of views.

Yet whatever those views, his patent concern with language has made his work suspect among those concerned with 'material conditions' and the like. Indeed, some would argue that he can *only* be viewed as an idealist, as someone operating at the level of the superstructure. Here his concern with what he termed 'language games' seems, too, to suggest that his work is profoundly relativistic. And the validity of this interpretation has, in fact, been suggested by the ways in which his work has been used, by relativists like Peter Winch (1990 {Original, 1958}; 1964; 1959) and Richard Rorty (1979; 1982; 1983). At the same time – perhaps paradoxically – some have seen his work not as relativistic, but rather as dangerously conservative. Here, claims such as 'What has to be accepted, the given, is – so one could say – forms of life' (PI, II: 226)[2] have led some commentators to see his work as a sort of Oakeshottean traditionalism (Nyìri 1982; Wheeler 1988). Finally, of course, his work – and particularly his later work – might be seen as having very little to do with geography. What, after all, do statements like 'Thought can as it were *fly*, it doesn't have to walk' (Z § 273) have to do with geography? Perhaps, in the end, those geographers who have skipped to chapter two of Bourdieu and Giddens have been right; Wittgenstein's work is simply too abstract, too far removed from the everyday practice of geography to make a difference.

It seems to me, quite to the contrary, that Wittgenstein might be seen as *the* geographical philosopher. Indeed, and notwithstanding forays by others – I have in mind here Foucault's silly 'Of Other Spaces' (1986) – in this century Wittgenstein has been the philosopher whose work has most deeply and dramatically addressed problems that have exercised geographers. And he has addressed these problems – of the role of space in philosophy, social theory, and common sense; of the importance of places; and of the nature of the natural – in a truly radical way, in a way that gets to the root of the matter. But here we can best see his work not as that of the traditionally Olympian and architectonic philosopher, standing outside the world – and humanity – and legislating a new and better system for encompassing the whole. Rather, we need to see it as, in an important sense, the product of an empirical researcher who at every turn found evidence that philosophical problems arise out of the everyday activities of common people. Indeed, for Wittgenstein the history of Western philosophy can be seen as the result of this Olympian urge, to go beyond one's own social context, the context within which actions and utterances make sense, to stand outside, to see the world from a point of view that is not a point of view, and to see more clearly than do the rabble. By contrast, Wittgenstein promoted a view in which the rabble – men and women, children, adults, and the aged, the bright and the feeble-minded – need to be heard.

An excursus

To say that Wittgenstein's work has been little understood by geographers is not, whatever I may have just said, to single out geographers, for the work *is* difficult, and in fact, that interpretations of his work have over the last eighty years undergone a sea change or two shows how difficult it is. It will be useful to think of those changes in terms of longstanding trends in the history of philosophy, and of social theory as well. As far back as Plato, one very important strand of philosophy has been based on the belief that the clarification of discourse is an important task. On this view many, perhaps all, of the problems that we think of as 'philosophical' derive from confusions in thinking. And the dialectical and dialogical become important tools for the clearing away of those confusions, those myths and prejudices that prevent us from 'seeing' the truth.[3] At the same time, many philosophers, from the Aristotle of the *Metaphysics*, through the Hegel of the *Phenomenology*, and on to today, have believed that more is needed, that the philosopher needs to construct a system. Here philosophy is seen as a science, but a very special sort of science, whose subject matter is not the 'real' world but rather the world behind it, of thought and ideas.

From the outset, this way of understanding philosophy infected the understanding of Wittgenstein's work. On the one hand, Bertrand Russell, in the introduction to Wittgenstein's *Tractatus Logico-Philosophicus* (1961), saw the work as an attempt to describe what an ideal language would look like. On this view Wittgenstein was operating in the tradition of people like the Aristotle of the *Metaphysics* (1941), the Descartes of the letters to Mersenne (1970), the Port-Royal Grammar and logic (Arnauld and Lancelot 1975; Lancelot, Arnauld, and Nicole 1816), and the philosophical language of Wilkins (1668).

Notoriously, Wittgenstein – who at the time did not have a PhD or an academic appointment, while Russell was at the top of his career – considered withdrawing the *Tractatus* from publication, just because he believed Russell to have misrepresented it so badly. In fact, he believed that what he had done in the *Tractatus* was to *clarify* the nature of factual assertions, and the reasons that they made sense. At the same time, he believed the assertions in his own work to be, strictly speaking, beyond the realm of sense:

> My propositions serve as elucidations in the following way: anyone who understands me eventually recognizes them as nonsensical, when he has used them – as steps – to climb up beyond them. (He must, so to speak, throw away the ladder after he has climbed up it.)
>
> (TLP § 6.54)

Just as there was a debate, and within the traditional discursive structure, about the meaning and purpose of Wittgenstein's *Tractatus*, so too was there such an argument about his later work, work that began in the 1930s with his *Blue and Brown Books* (1958), and that culminated in his *Philosophical Investigations* (1968). Adding to the debate, though, was the formalization of the division between the clarificatory function of philosophy and the system-building, or architectonic. Drawing in part on Wittgenstein's own work, transmitted in the form of oral accounts and informal transcriptions of class notes, Anglo–American philosophy came increasingly, during the 1940s and especially the 1950s and early 1960s, to be associated with the view that the true purpose of philosophy is strictly one of clarifying the use of language, and that all metaphysics consists merely of linguistic miscues. Personified in the work of J. L. Austin (1975; 1970), who said that the first task of a philosopher faced with a problem was to resort to the dictionary, Anglo–American philosophy largely sundered its ties with the architectonic project.

That project, though, remained alive in two places. On the one hand, it remained in Anglo–American circles in the tradition – now centered around the philosophy of science – that arose from early logical atomists, then logical positivists, and finally logical empiricists (Frege 1952; Ayer 1952; Russell 1956). From Russell on, many of them saw themselves as intellectual heirs of Wittgenstein's *Tractatus*; as Gustav Bergmann put it (Bergmann 1971), the *Tractatus* was the 'glory' of Wittgenstein and the *Philosophical Investigations* the 'misery.' On the other hand, it remained alive in continental philosophy, which was seen among Anglo–American philosophers, by and large, as incomprehensible myth-making.

The analysis of Wittgenstein's work remained through the 1950s bound by the continued hegemony in Anglo–America of this split. On one side advocates of philosophy as a clarifying project preferred the *Investigations*; on the other remained the architectonic logical-empiricists, whose allegiance was to the *Tractatus*. Both sides, though, shared an inability to see the elements of Wittgenstein's work that were not firmly within the Anglo–American mainstream. But a breakdown began in the late 1950s, with the publication in the same year of Peter Winch's *The Idea of a Social Science* (1990 {Original, 1958}) and Norwood Russell Hanson's *Patterns of Discovery* (1958), the first an application of Wittgenstein's later work to the social sciences and the latter an application of those ideas to the natural sciences. The two were soon followed by what came to be the longest-lived of the genre, Kuhn's *The Structure of Scientific Revolutions* (1970 {Original, 1962}). In effect, each of those works took Wittgenstein's later project into the heart of scientific orthodoxy. Each attacked the possibility of science as a disinterested view from nowhere, equally enthralled with and in thrall to an equally disinterested philosophy. Still, scholarship on Wittgenstein's *Philosophical Investigations* remained, by and large, locked into the view that his was a critical project (Pitcher 1964).

However, in 1973 the tide turned. In a remarkable work, Janik and Toulmin (1973) redrew Wittgenstein, as an alienated, exiled Viennese, and as one whose philosophical roots were much closer to Schopenhauer than to Frege. And followers of this interpretation, now increasingly the orthodox one, have come to see in Wittgenstein's *Investigations* strong echoes of the Continental hermeneutic – and architectonic – project (Chew 1982; Gadamer 1976). This stream has, in turn, led to the use of Wittgenstein's work as the underpinning of a number of projects, perhaps most notably in the sociology of science, where his work is widely cited. For some there, this work, this Continent-inspired architectonic project, could lead only in one direction, to the view that all is conversation, that conversation is permanent, that all standards are equal (Rorty 1979; 1982), or that scientific knowledge is not better than any other (Bloor 1981; 1983).

But if the literature on Wittgenstein, and especially on his later work, has increasingly seen it not simply as clarifying, but rather as an architectonic project, it seems to me that this may not be much of an improvement. Indeed, locked into the view that philosophy is one or the other, it fails to see the way in which his work is, in fact neither, but rather a very different project indeed. In what follows I shall lay out the lineaments of this view, through a consideration of a series of central questions about the nature of space, of rules, and of forms of life.

On space

In the *Tractatus* Wittgenstein laid out a conception of the relationship between propositions and the world, a conception that is nothing if not spatial. There:

> The fact that the elements of a picture are related to one another in a determinate way represents that things are related to one another in the same way.

> Let us call this connexion of its elements the structure of the picture, and let us call the possibility of this structure the pictorial form of the picture.
>
> (TLP § 2.15)

> Pictorial form is the possibility that things are related to one another in the same way as the elements of the picture.

> *That* is how a picture is attached to reality; it reaches right out to it.
>
> (TLP §§ 2.151–2.1511)

According to Norman Malcolm, Wittgenstein

> was in a trench on the East front, reading a magazine in which
> there was a schematic picture depicting the possible sequence of
> events in an automobile accident. The picture there served as a
> proposition; that is, as a description of a possible state of affairs.
> It had this function, owing to a correspondence between the parts
> of the picture and things in reality.
>
> (Malcolm 1966: 7–8)

If representation, here, involves a kind of mapping of propositions onto
the world, both the propositions and the world are seen as occupying a
kind of space: 'The facts in logical space are the world . . . Each thing is,
as it were, in a space of possible states of affairs. This space I can imagine
empty, but I cannot imagine the thing without the space (TLP § 1.13,
§ 2.013).

Now, if 'The totality of true propositions is the whole of natural science
(or the whole corpus of the natural sciences' (TLP § 4.11), and if 'Logic
pervades the world: the limits of the world are also its limits' (TLP § 5.61),
it might seem that Wittgenstein is promoting a view of space as infinite
and pre-existing, a kind of Newtonian space. It might, that is, appear as
though for Wittgenstein we are locked in a universe of atoms, a universe
whose constituents are in turn locked in the embrace of the propositions
that mirror them.

And, indeed, this view, of Wittgenstein as ready, like Hume before
him, to say of a work 'Does it contain any abstract reasoning concerning
quantity or number? No. Does it contain any experimental reasoning
concerning matter of fact and existence? No. Commit it then to the flames:
for it can contain nothing but sophistry' (Hume 1975 {1777}: 165) has
been supported, some would argue, by his assertion that '*The limits of my
language* mean the limits of my world' (TLP § 5.6). Olsson, for example,
put it this way:

> As my language changes so does my view of the world, because
> Heidegger (1968: 277)[4] was correct in his claim that the being of
> man is found in his language. Conversely, as my view of reality
> changes so does my mode of expression. What counts, therefore,
> is both my conception of the facts and the facts themselves,
> for facts cannot exist outside of conception and my conception
> reflects the particular language I am using. Since language is the
> medium in which the mind operates, the issue is not the collec-
> tion of facts but the communication of how these facts are ordered
> in the mind.
>
> (Olsson 1980: 6b)

But Wittgenstein's approach to the issue of space in the *Tractatus* is more complex than that, and in a way that presages – as does so much of the *Tractatus*, on a more contemporary reading – his later work, and what I want to argue is a rich and fertile conception of place. For in fact, the assertion here is not that the limits of language *are* the limits of my world, but rather that those limits *mean* the limits of my world, and for Wittgenstein language to which the term 'meaningful' can be rightly applied is language that is factual. Indeed, for the Wittgenstein of the *Tractatus* there is much about our lives that cannot be put in the language of facts – and of science. For '*How* things are in the world is a matter of complete indifference for what is higher. God does not reveal himself *in* the world' (TLP § 6.432). In fact, 'The whole modern conception of the world is founded on the illusion that the so-called laws of nature are the explanations of natural phenomena' (TLP § 6.371).

> Thus people today stop at the laws of nature, treating them as something inviolable, just as God and Fate were treated in past ages.
>
> And in fact both are right and both wrong: though the view of the ancients is clearer in so far as they have a clear and acknowledged terminus, while the modern system tries to make it look as if everything were explained.
>
> (TLP § 6.372)

And so

> My propositions serve as elucidations in the following way: anyone who understands me eventually recognizes them as nonsensical, when he has used them – as steps – to climb up beyond them. (He must, so to speak, throw away the ladder after he has climbed up it.)
>
> He must transcend these propositions, and then he will see the world aright.
>
> (TLP § 6.54)

Here, then, we see at the end of the *Tractatus* that Wittgenstein had in mind a very different way of thinking about space. This is not the infinite space of Newton, but rather a space that is finite and delimited. And it is this view, from the end of the *Tractatus*, that begins to be elaborated in Wittgenstein's *Philosophical Investigations*. Early in the work he describes some very basic languages, one for example used by a builder and the builder's assistant, consisting only of a few words, 'block,' 'slab,' and so on.

Do not be troubled by the fact that languages (2) and (8) [i.e., the builder's language] consist only of orders. If you want to say that this shews them to be incomplete, ask yourself whether our language is complete; – whether it was so before the symbolism of chemistry and the notation of the infinitesimal calculus were incorporated in it; for these are, so to speak, suburbs of our language. (And how many houses or streets does it take before a town begins to be a town?) Our language can be seen as an ancient city: a maze of little streets and squares, of old and new houses, and of houses with additions from various periods; and this surrounded by a multitude of new boroughs with straight regular streets and uniform houses.

(PI § 18)

Now we are beyond the image of space, to one of place. And it is an image wherein language may legitimately take on a variety of functions, well beyond the one of making factual assertions. According to Malcolm, the decisive moment in this change of mind was the following:

Wittgenstein and P. Sraffa, a lecturer in economics at Cambridge, argued together a great deal over the ideas of the *Tractatus*. One day (they were riding, I think, on a train) when Wittgenstein was insisting that a proposition and that which it describes must have the same 'logical form', the same 'logical multiplicity', Sraffa made a gesture, familiar to Neapolitans as meaning something like disgust or contempt, of brushing the underneath of his chin with an outward sweep of the finger-tips of one hand. And he asked: 'What is the logical form of *that*?' Sraffa's example produced in Wittgenstein the feeling that there was an absurdity in the insistence that a proposition and what it describes must have the same 'form'. This broke the hold on him of the conception that a proposition must literally be a 'picture' of the reality it describes.

(Malcolm 1966: 69)

Whatever the reason, Wittgenstein over the next years developed a very different way of thinking about philosophy and philosophical problems. And that view had, at its heart, the rejection of what Malcolm called the 'proposition' – though it would be better to call it the 'image' – that propositions are pictures of reality. With it went the purified idea of logical space, as language, logic, and even mathematics were rethought, exposed to the ethnographic eye, and seen at their heart to be possible only when embodied in the actions of real people in real places. The older view, the view of the *Tractatus*, came to be seen not so much to be a view from nowhere as a view from a very distinct place, the academy. And here, as it turns out, Wittgenstein does agree with Heidegger's assertions in *Hölderlin and the Essence of Poetry*. For there, notwithstanding Olsson's interpretation, Heidegger avers that

> We – mankind – are a conversation. The being of men is founded in language. But this only becomes actual in *conversation* [emphasis in original]. Nevertheless the latter is not merely a manner in which language is put into effect, rather it is only as conversation that language is essential. What we usually mean by language, namely a stock of words and syntactical rules, is only a threshold of language.
>
> (Heidegger 1965: 277)

We imagine, Heidegger suggests, that we can think about something that is 'just language', a set of words and rules that is neither written nor spoken, but rather a pure system. But this is just an image.

On following a rule

Like Heidegger, Wittgenstein notes that when we think about language we typically imagine it as a system, and a simple one at that. If we may be inclined to think of that view of language as a modern one, Wittgenstein in fact suggests that we see it as far back as Augustine, who in the *Confessions* related:

> When they (my elders) named some object and accordingly moved towards something, I saw this and I grasped that the thing was called by the sound they uttered when they meant to point it out ... Thus, as I heard words repeatedly used in their proper places in various sentences, I gradually learnt to understand what objects they signified; and after I had trained my mouth to form these signs, I used them to express my own desires.
>
> (Quoted in PI § 1)

Wittgenstein notes,

> These words, it seems to me, give us a particular picture of the essence of human language. It is this: the individual words in language name objects – sentences are combinations of such names. – In this picture of language we find the roots of the following idea: Every word has a meaning. This meaning is corre- lated with the word. It is the object for which the word stands.
>
> (PI § 1)

But in fact, this view of language both assumes and leaves out a great deal. It assumes a model of language. And it leaves out a great deal of what counts as language. Moreover, it renders language impossible. To begin – and this of course takes us back to the incident on the train – Wittgenstein notes that contrary to the image propounded by empiricists like Hume (1975 {1777}) and Ayer (1952), language is complex indeed,

> But how many kinds of sentence are there? Say assertion, question, and command? – There are countless kinds: countless different kinds of use of what we call 'symbols', 'words', 'sentences'
> . . .

> It is interesting to compare the multiplicity of the tools in language and of the ways they are used, the multiplicity of kinds of word and sentence, with what logicians have said about the structure of language. (Including the author of the *Tractatus Logico-Philosophicus*.).
>
> (PI § 23)

Indeed, he suggests, we need not only to see the 'the multiplicity of kinds of word and sentence', we need to see that we can treat those various kinds separately. We might, he suggests, see them as very much like games, what he termed 'language games'.

> Review the multiplicity of language games in the following examples, and in others:
>
> Giving orders, and obeying them
> Describing the appearance of an object, or giving its
>   measurements
> Constructing an object from a description (a drawing)
> Reporting an event
> Speculating about an event
> Forming and testing a hypothesis
> Presenting the results of an experiment in tables and
>   diagrams
> Making up a story; and reading it
> Play-acting
> Singing catches
> Guessing riddles
> Making a joke; telling it
> Solving a problem in practical arithmetic
> Translating from one language into another
> Asking, thanking, cursing, greeting, praying.
>
> (PI § 23)

Now, if we think of concepts in the traditional way, one that we have inherited via Aristotle from Plato – and one that is very much built into common-sense ways of thinking about science – we imagine that it is possible to define a given concept in terms of a set of defining characteristics, or an essence. Is this the case with language games – or with language more generally? Wittgenstein denies that it is.

> For someone might object against me: 'You take the easy way out! You talk about all sorts of language-games, but have nowhere said what the essence of a language-game, and hence of language, is: what is common to all these activities, and what makes them into language or parts of language . . .

> And this is true. – Instead of producing something common to all that we call language, I am saying that these phenomena have no one thing in common which makes us use the same word for all, – but that they are related to one another in many different ways.
>
> (PI § 65)

Here he takes the ethnographic stand. 'Consider for example' he says,

> the proceedings that we call 'games'. I mean board-games, card-games, ball-games, Olympic games, and so on. What is common to them all? – Don't say: 'There *must* be something common, or they would not be called 'games'' – but *look and see* whether there is anything common to all – For if you look at them you will not see something that is common to *all* but similarities, relationships and a whole series of them at that. To repeat: don't think, but look!
>
> (PI § 66)

Here again we see the importance of not being misled by the sort of spatial imagery that dominated the *Tractatus*:

> For how is the concept of a game bounded? What still counts as a game and what no longer does? Can you give the boundary? No . . . 'But then the use of the word is unregulated, "the game" we play with it is unregulated.' – It is not everywhere circumscribed by rules; but no more are there any rules for how high one throws the ball in tennis, or how hard; yet tennis is a game for all that and has rules too.
>   How should we explain to someone what a game is? I imagine that we should describe games to him, and we might add: 'This

and similar things are called "games"'. And do we know any more about it ourselves? Is it only other people whom we cannot tell exactly what a game is? – But this is not ignorance. We do not know the boundaries because none have been drawn. To repeat, we can draw a boundary – for a special purpose. Does it take that to make the concept usable? Not at all! (Except for that special purpose.) No more than it took the definition: 1 pace = 75 cm. to make the measure of length 'one pace' usable. And if you want to say 'But still, before that it wasn't an exact measure', then I reply: very well, it was an inexact one. – Though you still owe me a definition of exactness.

(PI §§ 68–9)

In the end, 'we see a complicated network of similarities overlapping and criss-crossing: sometimes overall similarities, sometimes similarities of detail' (PI § 66). So Wittgenstein argues that far from being a matter of mapping an abstract system onto the world, language in fact consists of sets of practices – some spoken, some not. There is the language game of ordering a pizza, giving an academic lecture, arguing with one's spouse, and so on. Indeed, one could say that to acquire culture, to become civilized, is just a matter of learning the appropriate language games, of learning what to say, where, and when. And this brings up what is surely a central issue for Wittgenstein, the notion of a 'rule.'

It is a commonplace, one drummed into us all from grammar school on, that language operates in accordance with rules. And it is just as much a commonplace that the way in which rules work – even if they have exceptions – is relatively straightforward. The number of the subject and predicate of a sentence need to agree, or the gender of a noun and an adjective; one learns the rule and then applies it. Yet when we put the matter in this way, a problem immediately arises. For where *is* the rule? In the modern age we are likely to say, 'In my mind, of course.' And indeed, this has been very much the way in which rules, and the idea of culture, have been thought out in the twentieth century: they are in one's head. Or in the collective head of the group to which one belongs. But if this is the case, and if rules define what others are doing, how can we ever know what that is? With respect to others we fall into what Stanley Cavell has called a Manichean view, where you have your rules, I have mine, and never the twain shall meet (Cavell 1969). It is a view in which the other is truly, irrevocably the other. Indeed, it is a view in which *I* am unknowable to myself; as William Lyons (1986) has shown, the view of the mind as something that an individual can know, a view whose origins extend back through Descartes (1983) to Augustine (1963), has fallen distinctly out of favor in this century. Further, this view of rules appears not to be able to give a plausible account

100

of the ways in which rules actually work. Consider the following example, from Saul Kripke's controversial *Wittgenstein on Rules and Private Language*:

> Let me suppose, for example, that '68 + 57' is a computation that I have never performed before. Since I have performed – even silently to myself, let alone in my publicly observable behavior – only finitely many computations in the past, such an example surely exists . . .
>
> I perform the computation, obtaining, of course, the answer '125'. I am confident, perhaps after checking my work, that '125' is the correct answer.
>
> Now suppose I encounter a bizarre sceptic . . . Perhaps, he suggests, as I used the term 'plus' in the past, the answer I intended for '68 + 57' should have been '5'! . . . After all, he says, if I am now so confident that, as I used the symbol '+', my intention was that '68 + 57' should turn out to denote 125, this cannot be because I explicitly gave myself instructions that 125 is the result of performing the addition in this particular instance. By hypothesis, I did no such thing . . . In the past I gave myself only a finite number of examples instantiating this function . . . So perhaps in the past I used 'plus' and '+' to denote a function
>
> $$x y = x + y, \quad \text{if } x, y < 57$$
> $$= 5 \quad \text{otherwise.}$$

Who is to say that this is not the function I previously meant by '+'?
(Kripke 1982: 8–9)

And if this seems a bizarre example, consider another: I ask you to 'continue the following series of numbers': 11, 9, 7, . . . You continue, 5, 3, and then stop. Well, I say? I'm done you reply. Well, what of one? In classical Greece it was not a number, but 'unity.' And zero? A recent invention. Negative numbers? More recent still. Still, we imagine, Wittgenstein suggests, that the rule has built into it its own application. But the number of numbers, like the number of sentences, is infinite; in fact, as Kripke's example shows, any pattern that I have created might be seen to be in accord with an unlimited number of rules. Yet, Wittgenstein notes, if you say

> But how can a rule shew me what I have to do at this point? Whatever I do is, on some interpretation, in accord with the rule. . . . This was our paradox: no course of action could be determined by a rule, because every course of action can be made out to accord with the rule. The answer was: if everything can be

made out to accord with the rule, then it can also be made out to conflict with it. And so there would be neither accord nor conflict here.

(PI § 198, § 201)

But, he continues,

> It can be seen that there is a misunderstanding here from the mere fact that in the course of our argument we give one interpretation after another; as if each one contented us at least for a moment, until we thought of yet another standing behind it. What this shews is that there is a way of grasping a rule which is *not* an *interpretation*, but which is exhibited in what we call 'obeying the rule' and 'going against it' in actual cases.

(PI § 201)

Indeed, 'To obey a rule, to make a report, to give an order, to play a game of chess, are *customs* (uses, institutions) (PI § 199).' So in the end, we need to see that to obey a rule – in mathematics or language, chess or football or the workplace – is not – or not merely – to act in accord with some image. Rather, it is to do something within a broader social context. Rules are defined and maintained only, as David Bloor (1997) has forcefully argued, within institutions.

And so, if we return to our belief about the mathematical series, that 'All the steps are really already taken,' we see that that description 'only made sense if it was to be understood symbolically. – I should have said: *This is how it strikes me* ... My symbolical expression was really a mythological description of the use of a rule' (PI §§ 219–21).

It may appear that these 'mythological descriptions' are doing the work, rather like a computer program is said to guide the workings of the computer. But 'Remember that we sometimes demand definitions for the sake not of their content, but of their form. Our requirement is an architectural one; the definition a kind of ornamental coping that supports nothing' (PI § 217).

In the end, the explicit formulation of a rule is not 'a visible section of rails invisibly laid to infinity' (PI § 218), not an appeal to a Tractarian image of space.

Forms of life

The notion of context has suggested to many commentators a further concept, that of 'forms of life'. Wittgenstein used it only a few times – five in the *Philosophical Investigations* and here and there elsewhere. Yet to many this concept, for better or worse, constituted a kind of foundation to his later work, a new and perhaps better way of thinking about context.

Wittgenstein uses the concept early in the *Investigations*. Noting that 'It is easy to imagine a language consisting only of orders and reports in battle' (PI § 19), but that 'the speaking of language is part of an activity' (PI § 23), he asserts that 'to imagine a language means to imagine a form of life' (PI § 19). And

> It is what human beings say that is true and false; and they agree in the language they use. That is not agreement in opinions but in forms of life.
>
> If language is to be a means of communication there must be agreement not only in definitions but also (queer as this may sound) in judgments.
>
> (PI §§ 241–42)

Indeed, 'what has to be accepted, the given, is – so one could say – forms of life' (PI: 226).This is more or less all that he has to say about forms of life, but here as elsewhere in his work a cottage industry has grown up, devoted to its interpretation. In his characteristically tough-minded way, Ernst Gellner put the problem this way: Wittgenstein, he said, has 'switched to a cult of *Gemeinschaft*, in the very curious disguise of a theory of language and philosophy' (Gellner 1988: 18–19). Gellner's attack, like that of Stephen Turner (1994) on Kripke's Wittgenstein, focuses on the appeal to something that must be shared. Kripke, for example, says that

> The set of responses in which we agree, and the way they inter-weave with our activities, is our form of life ... Wittgenstein stresses the importance of agreement, and a shared form of life, for his solution to his sceptical problem.
>
> (Kripke 1982: 96)

But are forms of life indeed shared? Well, on the face of it they are; after all, Wittgenstein has argued that language and rules must be public. And, indeed, many analysts have drawn just that conclusion. Malcolm, for example, suggested that 'I believe that [Wittgenstein] looked on religion as a 'form of life' (to use an expression from the *Investigations*) in which he did not participate' (Malcolm 1966: 72). And Peter Winch, too, offered such an understanding of forms of life,

> [C]riteria of logic ... arise out of, and are only intelligible in the context of, ways of living or modes of social life ... For instance, science is one such mode and religion is another; each has criteria of intelligibility peculiar to itself.
>
> (1990 {Original, 1958}: 100; see also 1964)

Finally, and most wildly, Popperian Peter Munz appealed to Wittgenstein's claim that the aim of philosophy is 'to shew the fly the way out of the fly-bottle' (PI § 309) in arguing that 'the bottle in which the fly found itself was hermetically sealed and not transparent.' With the idea of a form of life, then, Wittgenstein provided 'a philosophical foundation for the totalitarian claims of the sociology of knowledge,' because

> it is established that each speech community is a law unto itself because it prescribes the rules which determine the meaning of the sentences permitted in it. This conclusion is by itself quite stultifying for it permits the espousal or perpetration of any nonsense and mischief provided one can perform it within a speech community or find a speech community which has adopted rules or which is already sporting rules which will allow such acts or such thought behaviour. All outside criticism and any scrutiny in terms of external standards is automatically eliminated.
>
> (Munz 1987: 75)

Positive or negative, these interpretations of the concept share an appeal to Cavell's 'Manichean' understanding, one in which a form of life is metaphorically a region, an enclosed arena within which something is shared among a group of people.

Now there is a difficulty with this idea of a shared form of life, and a difficulty that has long been recognized by students of culture. (One need not stop with Mitchell's 'There's no such thing as culture: Towards a reconceptualization of the idea of culture in geography' (1995), but can trace the concept back, certainly, to Malinowski some sixty years before (1931).) The problem, simply put, is that to appeal to something 'shared' seems to be, right at the outset, to appeal to a concept just as ineffable as 'rule' is, at least on the usual mentalistic understanding. In fact, though, it seems to me here that Gellner and Kripke have misunderstood Wittgenstein, and that Turner's position is, in the end, much closer to Wittgenstein's than he believes.

There have, actually, been alternative interpretations, which on the face of it appear more consistent with other elements of Wittgenstein's work. For example, according to J. F. M. Hunter, a form of life is

> 'something typical of a living being': typical in the sense of being broadly in the same class as the growth of a living organism. . . . I shall therefore sometimes call this the 'organic account' . . . [since it involves activities that flow] from a living human being as naturally as he walks, dances, or digests food
>
> (Hunter 1971: 278–9)

In fact, 'however a person does something, it is his simply functioning that way which is a form of life' (*Ibid.*: 293). This view does seem to draw support from Wittgenstein's discussion of the nature of rules.

> 'How am I able to obey a rule?' – if this is not a question about causes, then it is about the justification for my following the rule in the way I do.
>
> If I have exhausted the justifications I have reached bedrock, and my spade is turned. Then I am inclined to say: 'This is simply what I do.'
>
> (PI § 217)

But in fact, this very statement suggests a different reading of the concept. Consider the sorts of concerns that Wittgenstein's later work evinces. First, he is concerned about the propensity that people have to extend the application of concepts beyond their legitimate scope – and then to be puzzled by the results. This often happens when we are misled by grammatical similarities among statements. So, for example, from the fact that I can say 'I have a toothache' and 'I have your book,' we imagine that we ought to be able to say 'I have your toothache' – and are puzzled about a person's relationship to his or her body when that statement makes no sense. Similarly, we imagine that we can go from 'People seek happiness' to 'Plants seek light,' with no problems. A second area of concern was the propensity to create reified abstractions. Certainly central here was the way in which people commonly go from the assertion that words have meanings to the assertion that there must be something *called* a meaning, that exists somewhere 'out there'. We imagine that because we can talk about 'equilibrium' or 'capital', that they must be things that somehow exist in the world. Or from 'I think' we conclude that there must be an 'I' that thinks.

In the end, these two propensities lead us at once to find the explicable inexplicable and the inexplicable explicable. The nature of the infinite comes to be a simple issue, resolvable using set theory. While the nature of the mind, and how it can be connected to the body, baffles us all. In the latter case we are tempted to embrace metaphysical answers, to create theories – and to imagine that if we just create the right set of basic elements, like culture or forms of life, everything will fall into place, and the mystery will be removed.

In part, the problem here is that those who have seen forms of life as basic elements have failed to see what is at issue when Wittgenstein asserts that 'If I have exhausted the justifications I have reached bedrock, and my spade is turned. Then I am inclined to say: "This is simply what I do"' (PI § 217).

For them, Wittgenstein's argument runs something like this: We usually imagine that there must be solid justifications for what we say. When asked why objects fall to earth we refer to gravity; when asked why a compass works, to magnetism. As he put it in the *Tractatus*, 'The whole modern conception of the world is founded on the illusion that the so-called laws of nature are the explanations of natural phenomena' (TLP § 6.371).

And in fact, when one asks a scientist about the nature of gravity, say, one is referred to further phenomena, variously to apparatuses and laws and institutions and practices. But, Wittgenstein is saying, at some point, we are all in the position of the parent faced with a two-year old who insists on asking 'Why?' If most people 'today stop at the laws of nature, treating them as something inviolable, just as God and Fate were treated in past ages' (TLP § 6.372), the scientist too is forced, in the end, to say, 'That's just how it is.' Or again, 'Then I am inclined to say: "This is simply what I do".' And when we say this, when we appeal to what we take to be the bedrock in our lives, we are appealing to 'what must be accepted', to 'forms of life'. So to say that a form of life, for Wittgenstein, must be accepted is just to say that something *becomes* a form of life by virtue of having that role, that function. Now it may seem that this is a transparent and unproblematic process: You ask me a series of probing questions about my actions, and at some point I say, 'This is just the way we do it around here,' or 'This is just the way we do it in our family,' or 'It's a women's thing.' The suggestion is that both the asking and the answering are undertaken with the motive of finding the truth. Yet as social scientists we all know that when we go into the field people often dissemble; they often attempt to put a good face on things. We know that the everyday images, descriptions, and stories that surround our customary activities are often window dressing, or as some would have it, ideology, or bad faith, or wishful thinking, or self promotion, that they themselves are elements of particular practices in particular contexts.

And we also know that within these contexts, it is not simply that people 'share' the same attitudes and beliefs, that within some given context we find a homogeneous set of actions and beliefs. Quite to the contrary, we need only consider almost any situation in which there are inequalities of authority. I am driving and am stopped by a well-armed police officer. We are certainly acting within a well-defined context; I know what to say and he knows what to say. But that does not mean that we share the same beliefs about the situation, or would say the same things about it. In the posthumously collected *Zettel*, Wittgenstein pointed to this fact, when he asserted that

> What determines our judgment, our concepts and reactions, is not what one man is doing now, an individual action, but the whole hurly-burly of human actions, the background against which we see any action.

> Seeing life as a weave, this pattern (pretence, say) is not always complete and is varied in a multiplicity of ways ... And one pattern in the weave is interwoven with many others.
>
> (Z § 567–69)

What could be farther from Munz's 'hermetically sealed fly-bottle' than this 'hurly-burly', this 'weave', what Andrew Pickering (1993) has more recently termed 'the mangle'? But we cannot begin to see this until we see that whatever their differences, Munz is agreeing with Winch and Malcolm, and with Hunter, and even with Kripke; they agree that a form of life is something from which one constructs a world, something very much like a culture.

But in using this concept Wittgenstein is being *critical* of the idea that a form of life is 'something typical of a living being', and particularly where that seems to imply that a person, for example, could be said to be the sum of his or her parts. Rather, the focus here is on the ways in which what Foucault (1972) would later term 'discursive formations' come to exist. Wittgenstein uses the concept of forms of life – and uses it rarely – to note that although we live in a world of difference, where no event is ever exactly repeated, 'we, in our conceptual world, keep on seeing the same, recurring with variations. That is how our concepts take it. For concepts are not for use on a single occasion' (Z §§ 567–9).

If life is 'a weave', that weave is at once evanescent and enduring. And the concept of 'forms of life' is meant to undercut the temptation to ignore that fact, to create a home of new linguistic 'boroughs with straight regular streets and uniform houses.' But at the same time, it is meant to show that we ought not to be taken in by, to romanticize the 'maze of little streets and squares, of old and new houses, and of houses with additions from various periods.'

Notice, though, that the very statement, 'That's how we do things,' and Wittgenstein's framing – 'If I have exhausted the justifications I have reached bedrock, and my spade is turned. Then I am inclined to say: "This is simply what I do" (PI § 217)' – presumes a question, a disruption, the quest for a reason. We articulate those justifications when in the face of those disruptions we lose, as Yi-Fu Tuan (1980) has put it, the ability to be rooted.

For in fact, the key concept here is surely not 'sharing', but rather 'fitting', or 'belonging'. Most people do, in their everyday lives, go about their business with little reflection. Whether shopping or driving the children to school or pounding nails or giving a lecture, much of human life is routine, customary. We may live our lives among people whom we don't know, and with whom we may feel that we have little in common, but we by and large manage to fit in with them; not to do so is in the end to be marginalized, to be judged a misfit or worse. Yet as Wittgenstein has shown, the very fact that we use language introduces into our lives a kind of metaphysical urge, a constant temptation to escape the bounds of our situation, criticizing it or generalizing about it, comparing it to others or theorizing about it. We live our lives in a world in which, as Naomi Scheman (1996) has so memorably put it, our words are in a state of diaspora, constantly exiled from their natural places.

In fact, Wittgenstein viewed much of the Platonist discourse in terms of which we describe the world as misleading, as positing without evidence a world of ideas or concepts that are free-floating guarantors of the structure of the world. For him the very possibility of understanding the actions of others – which we patently do – required that we abandon this way of looking at the world, and see the human world as one of habits and practices, one of customs. But the application of the methods of philosophy, the use of reason to recognize and overcome the tendency of words to escape their appropriate contexts, at the same time, he believed, leads us to see the world in a different way. We live not in the bifurcated world, partly human, partly sacred, of the middle ages; neither are we the isolated individuals in absolute space of the modern age. Rather, we are actors within the weave, the hurly-burly of life.

Consider an example: I am presiding over a seminar at a university in England. I have asked that people read material beforehand, and some have. There is the usual give and take; some people are quiet and some voluble. Now, many of us have been in a similar situation, and we know that there are certain ways that people act, and certain ways in which most don't. One view would be that we somehow share a set of values or expectations or dispositions. But on Wittgenstein's view, we need to see the situation as a complex one. As a guest I am surprised, or at least displeased, if I am not treated with a certain degree of respect. I wouldn't quite say that beforehand I 'expected' that, but if it is absent I am likely to say, 'Well I certainly didn't expect to be treated that way.' Further, some of what goes on makes sense not because I am a guest at this university, but because I am a member of an academic community, or a visitor to England, or a male of a certain age, or an American. And so on. Indeed, we can say the same about every member of the seminar. The critical point is that while in one sense we can be said to be doing one thing – engaging in a seminar – we are in fact doing a whole range of other things as well. And when Wittgenstein refers to the reaching of the end of justifications, he is speaking of the justification for one of those things. What I say about actions associated with my being a professor, or an American, or a male are sure to be different one from another.

Moreover, how 'far' one must go to reach the end of those justifications will vary; behind some actions there is a long story, behind others not much at all. Most Americans would answer the question, 'Why do you salute the flag' with 'Because I am an American,' and would be done with it. So if we need to see life as a weave of interrelated activities, we also need to see the terrain of life as various in its textures; some is 'thick' and some is 'thin'. Similarly, some activities are longstanding, and some not. The practice of saluting the American flag is relatively old – and likely to be seen as simple and straightforward, outside of the South; in contrast, some activities, like watching 'Melrose Place,' may be just as basic – 'I watch it because I like it, that's all' – but are likely to be a bit more transient. Finally, and notwithstanding these differences, in texture and longevity, none of these actions is intrinsically more basic or central or fundamental than the others. There is no 'real' bedrock of capital or consumer preferences or emotional drives, beyond that which is granted that status. Equally, no intellectual activity is more basic; philosophy or literary studies, the quadrivium and the trivium, are social enterprises, whose relationship with other social enterprises is contingent.

Does this mean, though, that 'everything is relative?' Must we conclude that because what counts as reason or logic or truth arises out of human actions in particular contexts, that everything is up for grabs? From a practical perspective, Wittgenstein would say everything is certainly *not* up for grabs. Indeed, if we see our lives as making sense because of the foundations on which they rest, then they are only as secure as those foundations. If we view the theory of genetics as the underpinning of biology, then the entire edifice is only as secure as that foundation. On the other hand, the metaphor of a 'weave' functions to point attention to the interconnectedness of people's actions, where a change here can reverberate through the system, and where there may be a great many impediments to that change.

Looking at the matter from another perspective, though, Wittgenstein would point out that there is a basic problem with the formulation of the question. For in formulating the question of relativism, in saying that 'All truths are relative to a social context,' we are imagining that we can speak of 'all truths' in the same way that we speak of 'all blue-eyed babies', as though we could take a census, and come up with the economists' 'perfect information'. But recall that 'If language is to be a means of communication there must be agreement not only in definitions but also (queer as this may sound) in judgments' (PI §§ 241–2). The claim of the truth of relativism must extend beyond concepts to judgments, actions, even technologies and institutions. In the end, the claim is empty; it is an assertion that looks as though it makes sense, but it is like a car with no engine.

The place of Wittgenstein

This leads us to a final question, on the place of Wittgenstein within geography, and perhaps within the social sciences more broadly. I suggested at the outset that his work had been remarkably uninfluential within geography, but that is only partly true. Certainly on the evidence of citations and publications we find little within geography that deals explicitly with his work. Yet there are other forms of influence, and on those measures his is certainly far stronger. For there can be no doubt that his work has been influential in a broad range of works that themselves have been extremely influential within geography. In philosophy, it was central to the construction of an alternative to the empiricist philosophy of science that was hegemonic through the 1950s. Thomas Kuhn (1970 {Original, 1962}), and especially Norwood Russell Hanson (1958), drew on his work in developing alternative accounts of the nature of science. We find echoes his work in David Bloor (1983; 1997; 1976) and other advocates of the 'strong program' in the sociology of science; more recently, his work is prominent in Latour and Woolgar (1979), Shapin and Schaffer (1985), and Pickering (1992; 1993).

In the social sciences, works by Peter Winch (1990 {Original, 1958}) and A. R. Louch (1966) filled the same function. At the same time, in anthropology Geertz (1973; 1980; 1983) and Marcus (1992) have been influenced by his work. And I have already mentioned, in sociology, Anthony Giddens (1979). It seems to me, though, that his work has something to say more directly to geographers. For right at the heart of it is a deep appreciation of the nature of places and their role in everyday lives. And, too, there is a powerfully argued view, in which those places, far from being carved out of a pre-existing spatial container, are created and maintained through the everyday actions of everyday life. More than any other recent thinker, Wittgenstein managed to cut through the welter of spatial metaphors in which we live – level, scale, container, hierarchy – and see the extent to which all arise out of a human life that is carried out in places.

Notes

1   The exceptions are Gunnar Olsson's *Birds in Egg/Eggs in Bird* (1980), on Wittgenstein's early *Tractatus*, a couple of little-noticed papers by myself (Curry 1989; Curry 1991), a discussion paper by Joe May (1980), and most recently, and visibly, a recent work by Nigel Thrift (1996).
2   In keeping with conventional practice, references to Wittgenstein are abbreviated as follows: TLP – *Tractatus Logico-Philosophicus* (references are to section numbers); PI = *Philosophical Investigations* (references in Part I are to section numbers, in Part II to page numbers); RFM – *Remarks on the Foundations of Mathematics* (references are to section numbers); and Z – *Zettel* (references are to section numbers).
3   It is perhaps odd that this dialogical approach, where conceptual clarity emerges from face-to-face argument, leads to knowledge that is characterized in terms of visual metaphors, like 'seeing'; here we might see Plato's *Republic* and his story of the cave as the fountainhead of much confusion.
4   The reference here is to Heidegger (1965).

References

Aristotle. 1941: Metaphysica. In *The Basic Works of Aristotle*, edited by Richard McKeon. Translated by R. P. Hardie and R. K. Gaye. New York: Random House: 689–934.

Arnauld, Antoine and Claude Lancelot. 1975: *The Port-Royal Grammar: General and Rational Grammar*. Translated by Jacques Rieux and Bernard E. Rollin. The Hague: Mouton.

Augustine, Saint. 1963: *On the Trinity*. Translated by Stephen McKenna. Vol. 45, *The Fathers of the Church*. Washington, D.C: Catholic University of America Press.

Austin, J.L. 1970: *Philosophical Papers*. Second edition. Translated by J. O. Urmson and G. J. Warnock. London: Oxford University Press.

Austin, J.L. 1975: *How to Do Things with Words*. Second edition. Translated by J. O. Urmson and Marina Sbisa. Cambridge: Harvard University Press.

Ayer, A. J. 1952: *Language, Truth, and Logic*. Second edition. New York: Dover.

Bergmann, Gustav. 1971: The glory and the misery of Ludwig Wittgenstein. In *Essays on Wittgenstein*, edited by E. D. Klemke. Urbana: University of Illinois Press: 25–43.

Bloor, David. 1976: *Knowledge and Social Imagery*. London: Routledge and Kegan Paul.

Bloor, David. 1981: The strengths of the strong programme. *Philosophy of the Social Sciences* 11: 199–213.

Bloor, David. 1983: *Wittgenstein: A Social Theory of Knowledge*. New York: Columbia University Press.

Bloor, David. 1997: *Wittgenstein, Rules, and Institutions*. London: Routledge.

Bourdieu, Pierre and Loïc J. D. Wacquant. 1992: *An Invitation to Reflexive Sociology*. Chicago: University of Chicago Press.

Cavell, Stanley. 1969: The availability of Wittgenstein's later philosophy. In *Must We Mean What We Say?* Cambridge: Cambridge University Press: 44–72.

Chew, Sing C. 1982: From Dilthey to Habermas: Reflections on Verstehen, hermeneutics, and language. *Current Perspectives in Social Theory* 3: 57–72.

Curry, Michael R. 1989: Forms of life and geographical method. *Geographical Review* 79: 280–96.

Curry, Michael R. 1991: The architectonic impulse and the reconceptualization of the concrete in contemporary geography. In *Writing worlds: Discourse, Text, and Metaphor in the Representation of Landscape*, edited by James Duncan and Trevor J. Barnes. New York: Routledge: 97–117.

Descartes, René. 1970: *Descartes: Philosophical Letters*. Translated by Anthony Kenny. Oxford: Clarendon Press.

Descartes, René. 1983: Discourse on the method of rightly conducting the reason. In *The Philosophical Works of Descartes*. Translated by Elizabeth S. Haldane and G. R. T. Ross. Cambridge: Cambridge University Press: 79–130.

Foucault, Michel. 1972: *The Archaeology of Knowledge*. Translated by A. M. Sheridan Smith. New York: Pantheon.

Foucault, Michel. 1986: Of other spaces. *Diacritics* 16: 22–7.

Frege, Gottlob. 1952: On sense and reference. In *Translations from the Philosophical Writings of Gottlob Frege*. Translated by Peter Geach and Max Black. Oxford: Basil Blackwell: 56–78.

Gadamer, Hans-Georg. 1976: The phenomenological movement. In *Philosophical Hermeneutics*. Translated by David E. Linge. Berkeley: University of California Press.

Geertz, Clifford. 1973: Thick description: Toward an interpretive theory of culture. In *The Interpretation of Cultures*. New York: Basic Books: 3–30.

Geertz, Clifford. 1980: Blurred Genres: The Refiguration of Social Thought. *American Scholar* 49: 165–79.

Geertz, Clifford. 1983: *Local Knowledge: Further Essays in Interpretive Anthropology*. New York: Basic Books.

Gellner, Ernst. 1988: The Stakes in Anthropology. *American Scholar* 57: 17–30.

Giddens, Anthony. 1979: *Central Problems in Social Theory*. Berkeley: University of California Press.

Hanson, Norwood Russell. 1958: *Patterns of Discovery*. Cambridge: Cambridge University Press.

Heidegger, Martin. 1965: Hölderlin and the essence of poetry. In *Existence and Being*. Translated by Douglas Scott. Chicago: Henry Regnery: 270–91.

Hume, David. 1975 {1777}: An enquiry concerning human understanding. In *Enquiries Concerning Human Understanding and Concerning the Principles of Morals*. Third edition. Oxford: Oxford University Press.

Hunter, J. F. M. 1971: 'Forms of life' in Wittgenstein's Philosophical Investigations. In *Essays on Wittgenstein*, edited by E. D. Klemke. Urbana: University of Illinois Press: 273–97.

Janik, Allan and Stephen Toulmin. 1973: *Wittgenstein's Vienna*. New York: Simon and Schuster.

Kripke, Saul. 1982: *Wittgenstein on Rules and Private Language*. Cambridge: Harvard University Press.

Kuhn, Thomas S. 1970 {Original, 1962}: *The Structure of Scientific Revolutions*. Second, enlarged edition. Chicago: University of Chicago Press.

Lancelot, Claude, Antoine Arnauld, and Pierre Nicole. 1816: *A New Method of Learning with Facility the Latin Tongue [The Port Royal Latin grammar]*. Translated by T. Nugent. London: Printed for F. Wingrave and J. Collingwood.

Latour, Bruno and Steve Woolgar. 1979: *Laboratory Life: The Social Construction of Scientific Facts*. Beverly Hills: Sage Publications.

Louch, A. R. 1966: *Explanation and Human Action*. Berkeley: University of California Press.

Lyons, William. 1986: *The Death of Introspection*. Cambridge: The MIT Press.

Malcolm, Norman. 1966: *Ludwig Wittgenstein: A Memoir*. London: Oxford University Press.

Malinowski, Bronislaw. 1931: Culture. In *Encyclopedia of the Social Sciences*. New York, Vol. 4: 621–46.

Marcus, George E. 1992: *Rereading Cultural Anthropology*. Durham: Duke University Press.

May, J. A. 1980: Wittgenstein and Social Studies. Toronto: Department of Geography, University of Toronto.

Mitchell, Don. 1995: There's no such thing as culture: Towards a reconceptualization of the idea of culture in geography. *Transactions, Institute of British Geographers* 20: 102–16.

Munz, Peter. 1987: Bloor's Wittgenstein or the fly in the bottle. *Philosophy of the Social Sciences* 17: 67–96.

Nyìri, J. C. 1982: Wittgenstein's later work in relation to conservatism. In *Wittgenstein and his Times*, edited by Brian F. McGuinness. Chicago: University of Chicago Press: 44–68.

Olsson, Gunnar. 1980: *Birds in Egg/Eggs in Bird*. London: Pion.

Pickering, Andrew. 1992: *Science as Practice and Culture*. Chicago: University of Chicago Press.

Pickering, Andrew. 1993: The mangle of practice: Agency and emergence in the sociology of science. *American Journal of Sociology* 99: 559–89.

Pitcher, George, ed. 1964: *The Philosophy of Wittgenstein*. Englewood Cliffs, NJ: Prentice-Hall.

Rorty, Richard. 1979: *Philosophy and the Mirror of Nature*. Princeton: Princeton University Press.

Rorty, Richard. 1982: *The Consequences of Pragmatism (Essays: 1972–80)*. Minneapolis: University of Minnesota Press.

Rorty, Richard. 1983: Postmodernist bourgeois liberalism. *The Journal of Philosophy* 80: 583–88.

Russell, Bertrand. 1956: Logical atomism. In *Logic and Knowledge: Essays 1901–1950*, edited by Robert C. Marsh. London: George Allen and Unwin: 175–82.

Scheman, Naomi. 1996: Forms of life: Mapping the rough ground. In *The Cambridge Companion to Wittgenstein*, edited by Hans D. Sluga and David G. Stern. Cambridge; New York: Cambridge University Press: 383–410.

Shapin, Steven and Simon Schaffer. 1985: *Leviathan and the Air Pump: Hobbes, Boyle, and the Experimental Life*. Princeton: Princeton University Press.

Thrift, Nigel. 1996: *Spatial Formations*. London: Sage.

Tuan, Yi-Fu. 1980: Rootedness versus sense of place. *Landscape* 24: 3–8.

Turner, Stephen P. 1994: *The Social Theory of Practices*. Chicago: University of Chicago Press.

Wheeler, Samuel C. III. 1988: Wittgenstein as conservative deconstructor. *New Literary History* 19: 239–58.

Wilkins, John. 1668: *An Essay Towards a Real Character, and a Philosophical Language*. Menston, Yorkshire: Scolar Press.

Winch, Peter. 1959: Nature and convention. *Proceedings of the Aristotelian Society*: 231–52.

Winch, Peter. 1964: Understanding a primitive society. *American Philosophical Quarterly* 1: 307–24.

Winch, Peter. 1990 {Original, 1958}: *The Idea of a Social Science and its Relation to Philosophy*. Second edition. New York: Humanities Press.

Wittgenstein, Ludwig. 1958: *The Blue and Brown Books: Preliminary Studies for the 'Philosophical Investigations'*. New York: Harper and Bros.

Wittgenstein, Ludwig. 1961: *Tractatus Logico-Philosophicus*. Translated by D. F. Pears and Brian F. McGuinness. London: Routledge and Kegan Paul.

Wittgenstein, Ludwig. 1967: *Zettel*. Translated by G. E. M. Anscombe and Georg Henrik Von Wright. Oxford: Basil Blackwell.

Wittgenstein, Ludwig. 1968: *Philosophical Investigations*. Third edition. Translated by G. E. M Anscombe. New York: Macmillan.

Wittgenstein, Ludwig. 1977: *Vermischte Bemerkungen*. Frankfurt: Suhrkamp Verlag.

Wittgenstein, Ludwig. 1983: *Remarks on the Foundations of Mathematics*. Revised edition. Translated by Georg Henrik Von Wright, Rush Rhees, and G. E. M. Anscombe. Cambridge, MA: MIT Press.

Part 2

REFORMULATED SPACES

Decolonisation, the wake of '68

5

UN-GLUNKING GEOGRAPHY

Spatial science after Dr Seuss and Gilles Deleuze

Marcus A. Doel

a man will be *imprisoned* in a room with a door that's unlocked
and opens inwards; as long as it does not occur to him to
pull rather than push it

Ludwig Wittgenstein, *Culture and Value*

Meanwhile – Cats, Glunks, werewolves, and other poststructuralists

Could she Un-thunk the Glunk alone? . . .
 It's very doubtful whether.
 So I turned on MY Un-thinker.
 We Un-thunk the Glunk together.
 (Dr Seuss, 1969: no pagination)

Gilles Deleuze was a philosopher, a creator of concepts. Some have called him a 'philosopher of difference' and a remorselessly 'horizontal thinker'. Dr Seuss was an author, a creator of children's books. He wrote especially for 'beginners'. The Cat in the Hat is one of Dr Seuss' best known characters. He is also wonderfully Deleuzean. In *The Cat in the Hat* and *The Cat in the Hat Comes Back* (Dr Seuss, 1957 and 1958, respectively), we see him practising nomad thought, schizoanalysis, rhizomatics, becomings of every persuasion, and chaosmosis. Like a gust of fresh air, the Cat in the Hat's antics sweep through the sedentary and Oedipalized scenes of domestic banality. With a wave of his paws everything that once appeared to be settled and fixed into places become once again mobile elements in a delirious movement of immanent and expressionistic creation. Whatever is given as ready-made – boat, fish, spade, cake, dress, snow, pinkness, earth, etcetera – is deterritorialized from its habitual actuality and sent cascading through ever-shifting contexts of reproduction and

117

rearticulation along a hundred thousand lines of flight. But the Cat in the Hat never works alone. He plays along with Sally and me, the fish and the dish, Thing One and Thing Two, and the stack of little cats (tagged A to Z) that hang out in his hat. In short, the Cat in the Hat is a small region of continuous variation in the wider chaosmos that we like to call a World. (Yet Dr Seuss, like Mr Magoo, prefers to naturalize the scenes of sedentary and domestic banality.) Nothing can resist the disarranging force of the Cat in the Hat – not even the world that he and his cats paint pink. But then we come across the Glunk, a perfect enactment of immutable oneness, identity, and presence that just *is*. And as every*one* knows, the metaphysics of presence and the ontology of being cannot be un-glunked or un-thunked. What is a philosopher of difference to do?

Here is how it happened, at least according to Dr Seuss. Once upon a time, not so long ago, the Cat in the Hat's little sister grew weary of thinking up friendly little things with smiles and fuzzy fur. So she turned up her Thinker-Upper as fast as it would go, and summoned forth a Glunk. . . . Needless to say, the actions of the Glunk spelt disaster for the domestic bliss of the household. Worse still, and as everybody knows, a Glunk *cannot* be un-thunk. Once summed, it will remain eternally unmoved. Yet on the brink of the abyss, the Glunk *did* un-glunk. Miraculously, the Cats in the Hats managed to un-thunk it. What achieved the un-glunking was neither the Cat in the Hat nor his sister alone, but their contingent alliance and joint action as a line of flight: 'there is an AND between the two, which is neither one nor the other, nor the one which becomes the other, but which constitutes the multiplicity' (Deleuze and Parnet, 1987: 34–5). In due course I will draw out the implications of this multiplicity, but for the moment suffice it to note the articular prizes open the seemingly intractable presence of inalienability and immutability. For 'it's along this line of flight that things come to pass, becomings evolve, revolutions take place. . . . an AND, AND, AND which each time marks a new threshold, a new direction of the broken line, a new course for the border' (Deleuze, 1995: 45). In short, when it comes to un-glunking and un-thunking, the active 'figure is never *one*' (Derrida, 1989: 5). The active figure is the interval, which by definition has a habit of splaying things out. This is why Deleuze and Guattari (1988: 478) are so adamant that 'the interval takes all, the interval is substance.' For example, the Leibnizian fold may now sweep away the longstanding, Aristotelian prejudice against joints, against everything that articulates and comes between. And with this new-found emphasis on the affective power of joint-action, we are already in the domain of geography and spatial science: the affective power of space and spacing.

Now, if Dr Seuss' Glunk exemplifies the realist and materialist inalienability of existence pure and simple (it just '*is*', rooted to the spot, and no amount of idealist un-thunking will budge it), then the interval joining the Cats in the Hats together (the '*and*') enacts its unhinging and deconstruction. Between them, the force of essentialism is swept away into a contingent variation of immanent consistency: an assemblage holds together. Such is the stammering of ontological constructivism and expressionism: no more givens, just shape-shifting ways of being. Hereinafter, identity is just a habit or habitus: it is an effect of embedment and conjunction. Something transpears 'as a result of contingency rather than necessity, as a result of an ambience or milieu rather than an origin, of a becoming rather than a history, of a geography rather than a historiography, of a grace rather than a nature' (Deleuze and Guattari, 1994: 96–97). Moreover, with this nod towards ceaseless plasticity, becoming, and metamorphosy, one can be sure that sorcery, alchemy, and lycanthropy are at the door. Deleuze and Guattari (1988: 275) casually note that 'Of course there are werewolves and vampires, we say this with all our hearts.'

Geography and the Glunk With No Name

No more certainties, no more continuities. We hear that energy, as well as matter, is a discontinuous structure of points: punctum, quantum. Question: could the only certainty be the *point*?

(Tschumi, 1994: 219)

I don't like points.

(Deleuze, 1995: 161)

Meanwhile, geography, the art of spatial science, turned up its own Thinker-Uppers as fast as they would go, and found that it had thunked up a legion of Glunks. There are Marxist Glunks, humanist Glunks, positivist Glunks, feminist Glunks, postcolonial Glunks, postmodern Glunks, radical Glunks. ... These Glunks have many names, which include Relativism, Nihilism, Perspectivism, Reflexivity, Doubt Paralysis, Undecidability, and Idealism. Each is sticky and tricky in its own way, and none shows any hint of un-glunking. Upon them, most of geography has become (un)stuck. However, in what follows I want to snuggle up with just one Glunk – a devilish fiend: The Glunk With No Name. At least, I am not aware of anyone giving it a name, although Krell (1997: 66) aptly refers to a certain 'Punctilious spirit'. No doubt the absence of a proper name is a sure sign that it has evaded domestication and Oedipalization. Unlike, say, Relativism or Perspectivism, it will not have to respond obediently to the masters' call of its name. For the sake of convenience I will lend the Glunk With No Name a tag, as the graffiti artist might say. (The word 'tag' itself is of unknown origin. Like the origin and background of the Glunk With No Name, it is lost in the folds of spacetime. A tag is both a mark of identification and a relay or trace, a trail or attachment, putting out or stringing along. Tag (s)plays out an endless and proliferating pursuance.)

The seemingly un-glunkable, un-thunkable, and un-domesticable Glunk that I want to engage with can be tagged: *Pointillism*. Everywhere one looks, geography and geographers are hung up on points: sites, places, nodes, integers, integrands, wholes, digits, identities, differences, the self, the same, the other, positions, op-positions, bifids, trifids, and so on and so forth. Lines are run between points. Surfaces are extended from lines. Volumes are unfolded from surfaces. And then there is the networking, not to mention the hybridization, othering, thirding. . . . Etcetera. In sum, spatial scientists have suspended themselves between all manner of points, and that is their undoing. On the one hand, it ruins and annuls pointillism, which, as a metaphysics of presence, is always already (in)stalled – even at its origin and from the off. On the other hand, the undoing of pointillism unfastens, opens up, and splays out that which pointillism has sought to repress: *the differential relations of expressionism*. What poststructuralist geography bears witness to is, first and foremost, the return of the repressed.

Simplifying to the extreme, poststructuralist geography amounts to the un-glunking of pointillism in geography; to the release of all of those articular intervals that open up the forced stabilization and self-identity of what appear to be points. Accordingly, Gilles Deleuze and Félix Guattari, like Jacques Derrida and Jean-François Lyotard, are exemplary un-glunkers of pointillism, of the metaphysics of presence, and of the essentialist ontology of being. A hundred thousand disadjusted and disjoined '*ands*' take flight from the opening up of the '*iss*'. In the passage from pointillism to expressionism, from the logic of identity to the rhythm of difference-producing repetition, space and spacing are (s)played out. Poststructural geography as interminable dislocation, distortion, and contortion. It effects becomings that are otherwise than being (Doel, 1996, 1999).

Deleuze's geophilosophy

Speaking always as geographers

(Deleuze, 1983: 83)

effects of conjuncture (and that is the world)

(Derrida, 1994: 18)

Gilles Deleuze was a philosopher, and wrote nothing but philosophy. He wrote on the back of a host of figures in the history of philosophy, such as Bergson, Hume, Kant, Leibniz, Nietzsche, and Spinoza. He wrote on the back of artists and writers, such as Francis Bacon, Lewis Carroll, Kafka, Proust, and the Marquis de Sade. He wrote alongside other creative practices, such as the cinema, and he wrote in his own name, no less than he did with his friends, most notably Félix Guattari. Amongst spatial scientists he is perhaps best known for the two volumes of *Capitalism and Schizophrenia* that he wrote with Guattari, entitled *Anti-Oedipus* and *A Thousand Plateaus* (Deleuze and Guattari, 1984, 1988).

Prior to dipping into Deleuze, let us be clear on two points: Deleuze insists that there is nothing 'difficult' about (his) philosophy – it is 'pop philosophy' – ; and that crazy talk is not enough. (For certain contexts and readerships, it may have sufficed to simply write several thousand 'ands. . . .' For in a certain sense, that truly *is* the play of the world.) Now, in creating his philosophy, a bewildering range of bits and pieces from all over the place came within Deleuze's orbit. No taxonomy could possibly bring them under control, and I am often reminded of Foucault's 'shattering laughter' when, having read the seemingly demented classification of a 'certain Chinese encyclopaedia' as relayed by Borges, he comments on 'the stark impossibility of thinking *that*' (Foucault, 1970: xv). For a flavour of this variety, and a glimpse of some geographical motifs, let us simply list the first few entries in the Index to *A Thousand Plateaus*, omitting the various authors whose dispersion is no less wide ranging:

> Aesthetics: and smooth and striated space. *See also* Art; Epistemology.
> Affect: and becoming-animal; and body; definition of; and haeccity; and war machine.
> Afrikaans: as major language.
> Agriculture: West as.
> Alembert's equation.
> America: as flow; as rhizome.
> Analogy: and representational thinking; and resemblance. *See also* Representation.
> 'And': and linguistic variation; vs. 'to be'.
> Aphorism: as plateau.
> Arborescent schema: and becoming; critique of; of evolution; as hierarchy; of language; and line and point; and rhizome; and segmentarity; and territorial assemblage; of thought; and tracing; and writing. *See also* Rhizome; State apparatus; Stratification.
> Archimedes: and nomad science.
> Architecture: and consistency; and State science. *See also* Geometry; Science.
>
> (Deleuze and Guattari, 1988: 589)

From the other direction, let us similarly list the first few entries in the Contents:

1   Introduction: Rhizome
    Root, radicle, and rhizome – Issues concerning books – The

one and the Multiple – Tree and rhizome – The geographical directions, Orient, Occident, America – The misdeeds of the tree – What is a plateau?

2   1914: One or Several Wolves?
Neurosis and psychosis – For a theory of multiplicities – Packs – The unconscious and the molecular

3   10,000 B.C.: The Geology of Morals (Who Does the Earth Think It Is?)
Strata – Double articulation (segmentarity) – What constitutes the unity of a stratum – Milieus – The diversity within a stratum: forms and substances, epistrata and parastrata – Content and expression – The diversity among strata – The molar and the molecular – Abstract machine and assemblage: their comparative states – Metastrata

(Deleuze and Guattari, 1988: v)

If this were not enough to leave the would-be-reader flummoxed and in stitches, Deleuze and Guattari encourage one to approach the book as if it were a *map*: there are entrances and exits everywhere; fold it however you want; follow whatever trajectory takes your fancy, etcetera. Treat the book as if it were a map – or as if it were a record, a tool kit or a machine. Treat it every which way, if you can. (This is the good way to approach a work.) It's still philosophy. A book, a work, an event: they all vary in and of themselves. Each new context calls for another lending of consistency to its chaosmotic virtuality. Hence the reference to multiplicities, rhizomes, and schizos. So, nothing simply 'is' as it would appear to 'be'. Hence the setting off of the variable 'and' in place of the constant toing and froing of the sedentary 'is' and 'is not': identity–difference; self–other; being–nothingness; etcetera. Every 'one', every 'each', every 'a' is packed with innumerable others that are bursting to get out for a breath of fresh air, a taste of the outside, and a stroll in the open. Hereinafter, geographers who count on the stability of points should beware. They may not be what they are since they are always already becoming-other, becoming-undecidable, and becoming-imperceptible. Fade to grey: metamorphosy without origin or end.

To cut a long story short, poststructuralist geography emerges from the deconstruction of pointillistic articulations of space, time, and place; with the joyful realization that oneness simply lacks consistency. In accordance with this image of thought, Jean Baudrillard, Hélène Cixous, Jacques Derrida, Michel Foucault, Félix Guattari, Luce Irigaray, Jean-François Lyotard, Henri Michaux, Gunnar Olsson, and numerous others have all endeavoured to break with what I have tagged 'pointillistic' representations of space and time (Doel, 1999). Now, I should warn the reader in advance that my Deleuze is not – and could never be – *the* Deleuze. And if you do not like this MAD-Deleuze, there are infinitely more Deleuze's to be differentiated, actualized, and lent consistency from the pack-animal that bears and counter-signs his name (cf. Malabou, 1996). In its eternally returning, difference-producing repetition – its stuttering and stammering; its ceaseless becoming-other-than-what-it-will-have-been –, Deleuze's philosophy gives rise to 'events for everyone'. Yet do not expect 'each' to stay the same. Like the characters in the film *Jacob's Ladder*, everything vibrates. Neither one nor many, Deleuze is *a* manifold, *a* multiplicity, *a* rhizome. Deleuze is *a* storm, in the sense that *it* rains. Lashing strokes and blurry traces, each composition of which amounts to a unique combination or singularity. Hereinafter, Deleuze rains over geography (Doel, 1996). Why is this so? It is because the task of philosophy is the creation of *concepts*, and for Deleuze creation is inseparable from the lending of consistency to a certain milieu of heterogeneous bits and pieces, of differential relations. One cannot think without 'spacing', nor can one space without 'thinking'. Such is the gist of geophilosophical materialism and immanent expressionism.

Consistency is what is left when constancy is dissimilated, deconstructed, and discharged. Differential repetition opens things up: the given or actual is lent an experience of the virtual (which is far from ideal or imaginary insofar as the virtual has a real consistency all of its own: Doel and Clarke, 1999). It opens 'each' to an experience of the 'all'. Everything is cracked, fissured, and fractal – not in terms of self-similarity and affine redundancy (the image of thought carried by structuralism), but in terms of self-dissimilarity and heterological disadjustment (the image of thought carried by poststructuralism). What are typically treated as constants now function as portions of consistency that have had their movements of dissimilation, deterritorialization, and differentiation forcibly blocked, looped, or quilted. Consequently, to open things up to variation Deleuze enacts a repetition not of the same, but of the differential at play within the same (the 'ands' in each 'is': the articular joint action animating every Glunk). In this way the common-or-garden structural differenciation between one and every other gives way to the incalculable differentiation of the one-all in and of itself. Little wonder, then, that Ansell Pearson (1997) should call Deleuze 'the difference engineer', or that Boundas and Olkowski (1994: 3) should call him a '*Stutterer, thinker of the outside.*' In a way that resonates with the practice of Derridean deconstruction, Deleuze follows a 'destabilization on the move in, if one could speak thus, "the things themselves",' while 'dissimilating the givens' as he goes (Derrida, 1988: 147; Lyotard, 1990: 76, respectively). The figure 'One' – of identity, presence, being, essence, and the same – no longer *holds* together. Only the interval, the joint, and the 'and' are capable of holding *together*, while maintaining and affirming the disjointure of that which they articulate and express (cf. Derrida, 1994). Hereinafter, constancy lacks consistency. The discourse of the 'is' should be placed under erasure, or at the very least rendered vibratory. Take 'place' as a convenient example.

Solids, liquids, and vapours – amidst the space of flows

> Matter thus offers an infinitely porous, spongy, or cavernous texture without emptiness, caverns endlessly contained in other caverns: no matter how small, each body contains a world pierced with irregular passages, surrounded and penetrated by an increasingly vaporous fluid, the totality of the universe resembling a "pond of matter in which there exist different flows and waves" [Leibniz].
>
> (Deleuze, 1993b: 5)

It is commonplace to suggest that more or less everything that once appeared solid and securely formed is in remorseless disintegration and dissolution. Fixity gives way to fluidity. In a very little while, the Earth may have become one amorphous hydrosphere: flow against flow – sometimes liquid, at other times gaseous, perhaps with vestiges of insoluble and inert matter left floating under suspension. Some weep at such an image of thought; others gloat. Still others protest that many important solids will continue to resist dissolution (e.g. the State) or that below a certain critical threshold, vapory and liquidity once again solidify, such that the flows return to ground (e.g. 'world cities' as basing points for the flows of the contemporary global economy, or capital itself, insofar as it crystallizes out as infrastructure etcetera). Nevertheless, geographers now routinely speak of 'spaces of flows' and testify to their growing power to affect: the flows of money, desire, capital, pollution, information, resources, ideas, images, people, etcetera.

In a space of flows the stretching or distanciation of interaction across spacetime is at once extensive and intensive: the fluid networks distend, establishing more and more connections, while the folding of flow upon flow heightens the complexity of both the system as a whole and the nodes through which the system is interlaced with itself, to the point where specific places find themselves increasingly disembedded, unhinged, and scattered to the wind. Place and placelessness are no longer opposed, as the humanistic geographers believed. Hereinafter, a place is both NowHere and NoWhere. Its taking place is undecidable and splayed out. Placement, like spacing, happens on a Möbius strip – a double articulation of incompossibilities: the smooth and the striated; territorialization and deterritorialization; stabilization and destabilization; constancy and consistency; etcetera.

Within such complex networks, the understanding of *localities* as unglunkable places that endure over time and are the subject or measure of change has given way to a notion of *glocalization* that endures only so long as the structure retains its current configuration. In this way, glocalities are no longer the subject or measure of change, but are themselves in continuous variation. A place is not a constant undergoing change, but a differential equation: flow upon flow; variation upon variation; differential upon differential. The local and the global no longer represent a difference in kind, nor in scale; but only in terms of the degree of folding and constriction. Glocalization is an effect of origami. And yet both the 'old' and the 'new' ways of dealing with space – rigid and fluid spaces, gridded and networked spaces, absolute and relative-cum-relational spaces, Euclidean and non-Euclidean spaces, abstract and lived spaces – invariably rest upon an inconsistent, unbecoming, and ill-mannered image of thought: *pointillism*.

Pointless geography

once again, the interval takes all, the interval is substance.
(Deleuze and Guattari, 1988: 478)

A vacillation of space . . . space hesitates about its identity.
(Lyotard, 1990: 106)

It would be better to approach space as a verb rather than as a noun. *To space* – that's all. Spacing is an action, an event, and a way of being. There is neither space 'behind' something, functioning as a backcloth, ground or continuous and unlimited expanse (absolute space), nor space 'between' something, as either a passive filling or an active medium of (ex)change (relative, relational, diacritical, and dialectical spaces). There is just spacing (differentials). The 'points' – as things, events, terms, positions, relata, etcetera – that are supposedly played out 'upon' and 'alongside' space are illusory. Space is immanent. It has only itself. And yet the event of geography – of spacing – is manifold, variegated, and ramified to the nth degree. Spacing is what happens and takes place: it is the differential element within everything that happens; the repetitious relay or protracted stringiness by which the fold of actuality opens in and of itself onto the unfold of virtuality. Space is what reopens and dissimilates the givens.

When thinking of poststructuralist spatiality there has been a tendency to assume that it breaks up a given space into so-many fragments, such that the parts no longer add up. Hence the qualification of poststructuralist spatiality in terms of non-totalization (there is no longer a unifying whole or final form), incommensurability (there is no longer a common measure or homologon), and incompossibility (the various detached pieces occupy different universes). Or else poststructuralist spatiality is said to multiply a given space, *à la* cubism, such that each dimension, plane, or point of view becomes radically overdetermined, relativized, and thereby unhinged. Hence the further qualification of poststructuralist spatiality in terms of heterotopia (a manifold space without common measure), cacophony (multiple components bereft of regularized orchestration), and dissemination (endless differentiation, deferral, and referral of meaning, value, reference, intentionality, and sense).

Simplifying to the extreme, poststructuralist space is broken up into discontinuous elements that are ramified without a unifying frame of reference: vapory, dustiness, and chaosmosis. Space is not a jigsaw: there is nothing to add up, integrate, subl(im)ate, or summate. Space knows only of differentials. And so the question is whether or not these differentials are polarized with positive and negative charges, so that they may come to be moved and affected. Hence the dispute between poststructuralists and dialecticians. Whilst the latter bank on integration (the recovery of the One from the Many), the former ceaselessly decline integration – in the etymological sense of *de-clinare*: bending away from. By declining integration, the multiple takes on a consistency all of its own – Multiplicity – , a consistency which no longer depends on the toing and froing of the One and the Many; totalization and fragmentation; self and other; universality and particularity. And if it were not for the fact that the perpetual 'swerve away from integration' eschews self-similarity and affine redundancy, one could call poststructuralist space *fractal* – infinite disadjustment, disjointure, and destabilization. Yet in all of this it is vital to realize that what is broken up and opened up, and what is multiplied and ramified, is not pointillistic but articular. Irigaray (1991: 59) characterizes the situation beautifully: 'Metamorphoses where no whole [*ensemble*] ever consists, where the systematicity of the One never insists.'

So, the basis of poststructuralist spatialization can be stated very simply: the minimal element is not the enclosed, charged, and polarized *point*, but the open *fold*; not a given One, but a differential relation; not an 'is' but an 'and'. Accordingly, 'The model for the sciences of matter is the "origami" ... or the art of folding' (Deleuze, 1993b: 6). Space knows nothing of points, integers, and identities: it knows only of manifolds. The fold is, precisely, what can be folded in many ways. This is why the figure is never *one*. There are only ever 'packets of singularities, packets that come undone in their turn,' says Lyotard (1990: 79). Or as Deleuze and Guattari (1988: 350) put it: 'we have no system, only lines and movements.' Schizoanalysis.

Such is the Deleuzean way of creating concepts, engineering difference, and folding space (Deleuze, 1990, 1993b, 1994; Deleuze and Guattari, 1994). There is no longer a stratification of the world *à la* Plato's Timaeus – who skimmed off being from becoming, the eternal from the fleeting, inalienable forms from degraded copies. Instead, one gets swept up by the laying out of a plane of immanence that can be folded, unfolded, and refolded in many ways. The multiplicity or manifold is planar not because it is homogeneous, uniform, and detached from volumity, but precisely because it immediately expresses all heterogeneity and variability by way of its own differential composition. The plane of immanence and consistency is a non-Euclidean and cracked surface that permits of no supplementary dimensions to those that are folded in on itself: there are neither dimensions 'below ground' nor ones 'above ground' to serve as foundations, essences, possibilities, or ideal forms. Immanence sweeps away transcendent paradigms, hylomorphism, and *Khora*. If there appear to be such supplementary dimensions hovering above, below, or alongside the plane of immanence, then they are not given in advance of the plane. To the contrary, they are constructed as a special effect of a certain folding of the plane, which corresponds to a seizure of power within the multiplicity, such that a particular fold in the order of things no longer expresses a pass-word (line of flight) but an order-word (quilting point of identification, recognition, and normalization). Phallogocentrism may rise to dominate and overcode the plane, but it never detaches itself. Nor can it prevent its falling back onto the plane from whence it came. In other words, whatever passes itself off as supplementary, as participating without belonging, and of existing without becoming caught up in the play of differential-repetition, is duplicitous. At this juncture we are once again very close to Derridean deconstruction, which employs the duplicitous and undecidable force of supplementarity to open up a line of flight in the op-positional givens (Doel, 1994, 1995).

Accordingly, 'it is not enough to say, "Long live the multiple," difficult as it is to raise that cry,' say Deleuze and Guattari (1988: 6). 'The multiple *must be made*, not by always adding a higher dimension, but rather in the simplest of ways, by dint of sobriety, with the number of dimensions one already has available, always $n - 1$ (the only way the one belongs to the multiple: always subtracted).' In short, consistency is expressed through the discharge of constancy. Henceforth, think, write, and act to the $n - 1$th dimension. This is how the one returns to the manifold.

Scrumpled geography – manifold spacing

One does indeed find folds everywhere

(Deleuze, 1995: 156)

You will not arrive at a homogeneous system that is not still worked on by immanent, continuous, and regulated variation.

(Deleuze, 1993a: 210)

Space is folded in many ways. It is manifold and multiplicitous. Space has no points of constancy, only folds that lend consistency. What appear as points or constants are really folds upon folds. 'Folds are in this sense everywhere, without the fold being a universal. It's a "differentiator," a "differential"' (Deleuze, 1995: 156). Yet if one were to insist on retaining the notion of a point, then it would be more consistent to think of it not in nounal terms of position without magnitude, but in verbal terms of direction and orientation. Like the vanishing point in perspectival painting, such a point points into that which it vanishes. And since a point, no less than a space, is folded in many ways, this directional aspect takes on an infinite complexity and intensity. Point-fold. Point-schiz. Point-tag. And when a point borders on infinity, it becomes a singularity – not in the sense of 'oneness', but in the mathematical sense of a break-point, such as when a function takes on an infinite value or when matter becomes infinitely dense (cf. Clarke, Doel, and McDonough, 1996).

> Singularities are the precise points at which all of the variations in (of) the field are copresent, from a certain angle of approach, *in potential.* That copresence in potential is '*in*tension' as opposed to extension ..., closer to a 'virtuality' ... *absolutely real.* It is absolute in that it is nowhere in the space-time coordinates of extension, and yet it is perspectival, because the variation of the field is ever on the approach, from a certain angle, to a singularity of its own copresence. It is real, yet incorporeal.
>
> (Massumi, 1996: 397)

Singular points are virtual, while their corresponding actualizations appear as point-folds, bifurcation-points, and point-schizzes. And while it may appear as though linear unfoldings string together these singularities, as if a continuum of infinitesimal and extensive variation were bordered by two termini of maximal and intensive variation, this line would itself be composed of an infinite number of singularities. There is neither a continuum of ordinary points, nor a discontinuum of extraordinary points. Instead, there is a dissimilatory fractal of singular point-folds, reminiscent of Cantor dust. Space is an infinitely folded chaosmos – the space of *a* concept, no less than of *a* life (cf. Guattari, 1992). The trick is not to join the dots, banking on identity and integration, but to construct an endurable line of flight across the manifold. Thus, 'multiplicity is the real element' and 'immanence is constructivism' (Deleuze, 1995: 146). The consistency of folding always remains to be established: it is never given, pre-formed or ready-made. Fortunately, the art of folding bends to every occasion. It is pliancy pure and simple, although it complicates everything. Bereft of transcendent rules or final solutions, a folded thing always opens up to (an experience of) infinity.

Each act of folding creates a distinct singularity whilst also expressing an incalculable multiplicity. This is why one must insist on the indefinite article. For example, in origami one may speak of *a* folded piece of paper.

Yet the 'event' of origami, of folding, is neither the actual composition of relations that lends the apparent out-turn consistency – *a* hat, *a* plane, *a* rabbit, etcetera –, nor the virtual multiplicity of all singular compositions that are held in untimely abeyance on the plane of immanence. Deleuze (1997: 5) puts it beautifully: 'This indefinite life does not itself have moments, however close together they might be, but only meantimes (*des entre-temps*), between-moments.' An event passes between the folds: it is a becoming real without being an actual state of affairs; and it floats on the surface of things in a kind of suspended animation. Everything is (s)played out on the surface, on the plane of immanence and consistency. Space is always a real virtuality: it resists actualization, and is transformed on each occasion that it is actualized. 'It is the horizon itself that is in movement' (Deleuze and Guattari, 1994: 38).

In sum, space is a differential, and not a unifying, element. Paradoxically, spacing puts what it articulates out of joint. So far as it is spaced, either intensively or extensively, matter is always moving out of place. It never quite 'holds together', no matter how much force and binding are applied. Such is the 'world ... of folding and unfolding. The whole thing is a crossroads, a multiple connectedness' (Deleuze, 1995: 155). Little wonder, then, that an event is 'the part that eludes its own actualization in every-thing that happens' (Deleuze and Guattari, 1994: 156). It has 'kept the infinite movement to which it gives consistency;' it is a 'virtual that is real without being actual, ideal without being abstract.' So, Deleuze always sought to activate those halved-together point-folds that re-release varia-tion, those point-folds which, when activated, would begin to slide and make the entire constellation of apparently stable and sedimented forces slide. (This resonates with Derrida's rereading of Hegelianism on the back of Bataille: Derrida, 1978.) This required a 'philosophy of passage, and not of ground or of territory for'

> traversing the chaos: not explaining or interpreting it, but traversing it, all the way across, in a traverse which orders the planes, land-scapes, coordinates, but which leaves behind it the chaos, closing on itself like the sea on the wake of a ship.
>
> (Nancy, 1996: 112)

So:

> there are always many infinite movements caught within each other, each folded in the others, so that the return of one instan-taneously relaunches another in such a way that the plane of immanence is ceaselessly being woven. ... Diverse movements of the infinite are so mixed in with each other that, far from breaking up the One-All of the plane of immanence, they constitute its

variable curvature, its concavities and convexities, its fractal nature as it were. It is this fractal nature that makes the planomenon an infinite that is always different from any surface or volume determinable as a concept. Every movement passes through the whole of the plane by immediately turning back on and folding itself and also by folding other movements or allowing itself to be folded by them, giving rise to retroactions, connections, and proliferations in the fractalization of this infinitely folded up infinity.

(Deleuze and Guattari, 1994: 38–9)

Following in the wake of Deleuze, there is nothing left for the spatial scientist but the play of joints (and ... and ... and). Consistency remains when all apparent constancy has discharged. What remains is precisely that which maintains the different detached pieces in their incalculable disjointure – AND ... AND ... AND – : the interval takes all; the ontology of being gets carried away by the conjunctives. Moreover, the taking on of consistency requires one to 'discover the right running speed,' as Martin (1996: 19) so aptly puts it. As with the cinema, not any old differential calculus will do the trick: it is always a matter of spacing and pacing, of speed and slowness, of rhythm and expression.

Inscribed on the plane of consistency are *haecceities*, events, incorporeal transformations that are apprehended in themselves; *nomadic essences*, vague yet rigorous; *continuums of intensities* or continuous variations, which go beyond constants and variables; *becomings*, which have neither culmination nor subject, but draw one another into zones of proximity or undecidability; *smooth spaces*, composed from within striated space.

(Deleuze and Guattari, 1988: 507)

As this chapter begins to wind down, it may be worth differentiating between striated space and smooth space. This distinction – or more accurately: Möbius spiralling – takes us to the nub of spatial science in the wake of Dr Seuss' Cats in Hats and the Deleuzoguattarian storm. In a striated space pointillism prevails, which is invariably energized by a binary-machine that breaks up becomings into a distribution of sedentary and arboreal points. 'Points everywhere are offered as ends of trajectories, lines, and curves, and they, through a transcendental regulative principle, control the itinerary' (Martin, 1994: 269). In a smooth space, by contrast, pointillistic striation is deconstructed through a double invagination: points are returned to the manifold; and the reconstructed point-folds are dissimilated into an infinite, fractal and chaosmotic abyss. In other words, the smoothing of striation enables the interval to once again take all. Spatialization becomes substance itself. On this basis, a structural, genetic, molar, and arboreal *plane of organization*, transcendence, principle, and finality unfolds a decentred, mutational, molecular, and rhizomatic *plane of consistency*, immanence, contingency, and becoming. Constancy gives way to consistency; oneness gives way to multiplicity; and evolutionary hylomorphism gives way to 'viroid life'. Moreover, whilst State philosophy is arborescent (rooted, towering, branching, overhanging, phallogocentric, and transcendent), nomad thought is rhizomatic (adrift, flat, transversal, interleaved, invaginated, and immanent). Specifically, rhizomes fulfil two crucial functions: they strangle the roots and scramble the codes of all arboreal and sedentary thought; and they exemplify fractal surfaces that express the continuous variation of multiplicities without unity. By fixing and extending a central point-fold a root establishes an order, out of which emerges a pre-programmed, irreversible, and essentially hierarchical series of bifurcations (aborescence, structuralism, stricturalism). By contrast, every point-fold in a rhizome is connectable and disconnectable, reversible and displaceable, and everything can be either broken-off or set into play. And unlike arboreal–sedentary concepts, rhizomatic–nomadic ones are distributional rather than punctiform, articular rather than anchored, compositional rather than unifying, immanent rather than transcendent, contingent rather than essential, and spectral rather than given. In this way, nomad thought yields a geography that relays the twitching of the schizoanalytic switchboard and the stuttering of difference-producing repetition. Hereinafter:

> the fissure has become primary, and as such grows larger. It is not a matter of following a chain ... even across voids, but of getting out of the chain or the association. ... It is the method of BETWEEN ... which does away with all ... of the One. It is the method of AND, 'this and then that,' which does away with all ... of Being. ... The whole undergoes a mutation, because it has ceased to be the One-Being, in order to become the constitutive between-two. ... The whole thus merges with what Blanchot calls the force of 'dispersal of the Outside' or the 'vertigo of spacing'.
>
> (Deleuze, 1994: 179–180)

To round things off, let me simply note the fact that the chaosmos can be folded in many ways. And while it is indeed the case that many folded figures are lent to us as ready-mades there is nevertheless no manner or style of folding, unfolding, and refolding that is set in advance. So it remains for the geographer to (s)play along with the folds and to become swept up by the variable consistency of a certain context. Such would be the ethics of the event. Deleuze and Guattari (1988: 482) put it nicely: 'Voyage in place. ... Voyage smoothly or in striation, and think the same way.' In effect, the trick of poststructuralist geography is not merely to try to 'Never miss a twist or a fold' (Derrida, 1989: 10), but to *open up* the givens; to *open up* and dissimilate the events; to *open up* the chaosmotic singularities and multiplicities; and to thereby enable something other to happen. Letting space take place. That's all.

Letting space take place:

it happens – to have gone.

Presence absents: absence presents.

NowHere: NoWhere.

Becoming-only-to-fade.

Hauntology, vibratology.

Ontological flicker.

[re.] ~~~~~~~~~~~~~~~~~~~~~~~~~~~~ [mix]

Spatial science un-glunked
(or, Gilles Deleuze plays 'pop geography')

... and ... and ... and ... and ... and ... and ... and ...

... geo ... graphein ... geo ... graphein ... geo ... graphein

... writing ... earth ... writing ... earth ... writing ... earth

... rhizome ... nomad ... schizo ... swarm ... storm ... vapour
... chaosmos ... consistency ...

[... and also the trees lent support to the cure.]

[Mad.Deleuzion] ~~~~~ (...) ~~~~~ (χ) ~~~~~ [3.59]

What, then, is the status of this chapter? I have repeatedly insisted on the illusory nature of pointillism: that there are no constants in this or any other universe. Rather, there are only degrees of consistency that will never cease being worked over by the disarranging force of chaosmotic variation. No beings (*iss*), just becomings (*ands*). Yet this chapter seems to be strewn with countless '*iss*' and other such constants, with insistent statements and fast-and-frozen results. It has its own array of un-thunkable Glunks. Where is the consistency in that? Some take it for granted that the '*is*', as the exemplary motif of the metaphysics of presence and the ontology of immutable being, has a certain phallic value: it stands tall and erect; alone and self-assured; without any recourse to the hand of another.

The phallic One has only itself: pure positivity at the source. Given time it would be worth pursuing this phallic structuration as an inaugural gesture of Western metaphysics: how a certain characterization of sexual difference animates ontology (cf. Derrida, 1995; Krell, 1997). Yet it should be self-evident that an erection is neither given nor constant. It is established in specific contexts, according to particular relations of affectation, and endures only so long as that which flows through it is constricted and held under pressure. Indeed, there is no phallogocentrism that is not worked over by fluidity and vapory: a gust of fresh air; a taste of the outside. ... Moreover, the phallic term is energized not by auto-affection or autoeroticism, but by way of this constitutive outside. A battery of flows animate and energize everything that appears phallic, so that every constant moves and trembles according to the rhythm of the flows that surge through it: rhythmanalysis. This is why the phallic 'is' vibrates. Ontology gives way to vibratology.

So, with regard to all of those 'iss', it is not a question of expulsion or exorcism – as if one could have done with the spectre of hauntology! It is not that constancy is bad or wrong. It is simply unbecoming and ill-mannered. Constancy is bereft of consistency: that's all. Ontology should be placed under erasure, to be sure; but real consistency takes hold when 'is' becomes a deterritorialized term, wrenched from its usual phallogocentric context in order to be (s)played out – once again – as a vibrator, according to the ebb and flow of incalculable 'ands' that are forever coming and soliciting our affection. For there are multiple becomings in every body: this is what moves the Earth and shakes the World. In this way, rhythmanalysis gets swept up by schizoanalysis. Such is the art of spatial science when it truly expresses the differentials that space makes. Even the ink on this page vibrates.

References

Ansell Pearson K (1997) *Deleuze and Philosophy: The Difference Engineer* (Routledge, London)

Boundas C V, Olkowski D (eds) (1994) *Gilles Deleuze and the Theatre of Philosophy* (Routledge, London)

Clarke D B, Doel M A, McDonough F X (1996) 'Holocaust topologies: singularity, politics, space' *Political Geography* **15** (6 and 7): 457–89

Deleuze G (1983) 'Politics,' in *On the Line*, Deleuze and Guattari, trans. J Johnston (Semiotext(e), New York): 69–115

Deleuze G (1990) *The Logic of Sense*, trans. M Lester, ed. C V Boundas (Columbia University Press, New York)

Deleuze G (1993a) *The Deleuze Reader*, ed. C V Boundas (Columbia University Press, New York)

Deleuze G (1993b) *The Fold: Leibniz and the Baroque*, trans. T Conley (University of Minnesota Press, Minneapolis)

Deleuze G (1994) *Difference and Repetition*, trans. P Patton (Athlone, London)

Deleuze G (1995) *Negotiations, 1972–1990*, trans. M Joughin (Columbia University Press, New York)

Deleuze G (1997) 'Immanence: a life . . .' *Theory, Culture and Society* **14** (2): 3–7

Deleuze G, Guattari F (1984) *Anti-Oedipus: Capitalism and Schizophrenia*, trans. R Hurley, M Seem, H R Lane (Athlone, London)

Deleuze G, Guattari F (1988) *A Thousand Plateaus: Capitalism and Schizophrenia*, trans. B Massumi (Athlone, London)

Deleuze G, Guattari F (1994) *What is Philosophy?*, trans. G Burchell, H Tomlinson (Verso, London)

Deleuze G, Parnet C (1987) *Dialogues*, trans. H Tomlinson, B Habberjam (Athlone, London)

Derrida J (1978) 'From restricted to general economy: A Hegelianism without reserve,' in *Writing and Difference*, trans. A Bass (University of Chicago Press, Chicago): 251–77

Derrida J (1988) *Limited Inc*, ed. G Graff, trans. S Weber, J Mehlman (Northwestern University Press, Evanston)

Derrida J (1989) 'Introduction: desistance,' in *Typography: Mimesis, Philosophy, Politics* P Lacoue-Labarthe, ed. C Fynsk (Harvard University Press, Cambridge): 1–42

Derrida J (1994) *Specters of Marx: The State of the Debt, the Work of Mourning, and the New International* trans. P Kamuf (Routledge, London)

Doel M A (1994) 'Deconstruction on the move: from libidinal economy to liminal materialism' *Environment and Planning A* **26**: 1041–59

Doel M A (1995) 'Bodies without organs: deconstruction and schizoanalysis,' in *Mapping the Subject: Geographies of Cultural Transformation*, eds S Pile, N Thrift (Routledge, London): 227–41

Doel M A (1996) 'A hundred thousand lines of flight: a machinic introduction to the nomad thought and scrumpled geography of Gilles Deleuze and Félix Guattari' *Environment and Planning D: Society and Space* **14** (4): 421–39

Doel M A (1999) *Poststructuralist Geographies: The Diabolical Art of Spatial Science* (Edinburgh University Press, Edinburgh)

Doel M A, Clarke D B (1999) 'Virtual worlds: simulation, suppletion, s(ed)uction, and simulacra,' in *Virtual Geographies: Bodies, Spaces and Relations*, eds M Crang, P Crang, J May (Routledge, London): 261–83

Dr Seuss (1957) *The Cat in the Hat* (Random House, New York)

Dr Seuss (1958) *The Cat in the Hat Comes Back* (Random House, New York)

Dr Seuss (1969) *I Can Lick Thirty Tigers Today! and Other Stories* (Random House, New York)

Foucault M (1970) *The Order of Things: An Archaeology of the Human Sciences* (Tavistock, Andover)

Guattari F (1992) *Chaosmosis: An Ethico-Aesthetic Paradigm*, trans. P Bains, J Pefanis (Power Publications, Sydney)

Irigaray L (1991) *The Irigaray Reader*, ed. M Whitford (Blackwell, Oxford)

Krell D F (1997) *Archeticture: Ecstasies of Space, Time, and the Human Body* (SUNY, Albany)

Lyotard J-F (1990) *Duchamp's TRANS/formers*, trans. I McLeod (Lapis, Venice)

Malabou C (1996) 'Who's afraid of Hegelian wolves?,' in *Deleuze: A Critical Reader*, ed. P Patton (Blackwell, Oxford): 114–38

Martin J-C (1994) 'Cartography of the year 1000: variations on A Thousand Plateaus,' in *Gilles Deleuze and the Theatre of Philosophy*, eds C V Boundas, D Olkowski (Routledge, London): 265–88

Martin J-C (1996) 'The eye of the outside,' in *Deleuze: A Critical Reader*, ed. P Patton (Blackwell, Oxford): 18–28

Massumi B (1996) 'Becoming-deleuzian' *Environment and Planning D: Society and Space* **14** (4): 395–406

Nancy J-L (1996) 'The Deleuzian fold of thought,' in *Deleuze: A Critical Reader*, ed. P Patton (Blackwell, Oxford): 107–13

Tschumi B (1994) *Architecture and Disjunction* (MIT Press, London)

6

RELICS, PLACES AND
UNWRITTEN GEOGRAPHIES
IN THE WORK OF MICHEL
DE CERTEAU (1925–86)

Mike Crang

Placing de Certeau in geography

Michel de Certeau has lately become a small-scale mantra in geographical writings. His name is recited to authorise three points. The first from urban theory, relates to planning and views from on high:

> Seeing Manhattan from the 110th floor of the World Trade Centre. Beneath the haze stirred up by the winds, the urban island, a sea in the middle of the sea, lifts up the skyscrapers. . . . A wave of verticals. Its agitation momentarily arrested by vision. The gigantic mass is immobilized before the eyes. It is transformed into a texturology. . . . To what erotics of knowledge does the ecstasy of reading such a cosmos belong? Having taken voluptuous pleasure in it, I wonder what is the source of this pleasure of 'seeing the whole', of looking down on, totalizing the most immoderate of human texts.
>
> (1984: 91–2; reprised 1985a, 1980a)

This comes from his essay 'Walking in the City' and connects with a series of critiques of the viewpoint of planners, the panoptic disciplining of space and the pretensions of social theory. For many geographers this is the only point of contact with his work. In this first invocation, de Certeau becomes the champion of the common folk and street level social theory. It is in this guise that he has become a darling to some, as a counterpoint to stratospheric theory, and villain to others, as an example of micro-theory romanticising the popular.

A second invocation is as a theorist of consumption. De Certeau extends his vision of walking to a more general metaphor of reading. His critique

136

of texturology from on high leads to a poetics of reading. These are his most celebratory passages about the activity of 'making do'.[1] In this case he uses the city as metaphor for modern life.

> Cut loose from the traditional communities that circumscribed their functioning, they have begun to wander everywhere in a space which is becoming at once more homogeneous and more extensive. Consumers are transformed into immigrants. The system in which they move is too vast to be able to fix them in one place, but too constraining for them to ever be able to escape from it and go into exile elsewhere. There is no longer an elsewhere. And because of this, the strategic model is also transformed, as if defeated by its own success: it was by definition based on the definition of a 'proper' distinct from everything else; but now that 'proper' has become the whole. It could be that, little by little, it will exhaust its capacity to transform itself and constitute the only space ... the scene of Brownian movements of invisible and innumerable tactics. One would thus have a proliferation of aleatory and indeterminable manipulations within an immense framework of socioeconomic constraints and securities: myriads of almost invisible movements, playing on the more and more refined texture of a place that is even, continuous, and constitutes a proper place for all people. Is this already the present or the future of the great city?
>
> (1984: 40–1)

There is much here that will be developed later on. This is a figure for consumption situated in the world depicted by the Frankfurt school – a world of mass availability and control at the same time (1997b: 91–2, 107–10; cf. Frow 1991). De Certeau wants to open a space to say that our studies of technologies of power have often led us to believe their own statements of efficacy. We have missed the 'nocturnal' and hidden realm of use (1997b: 138). In this he suggests we have privileged writing over reading, production over consumption. A concern with the unrecognised and hidden activity of ordinary folk is his hallmark.

This cut brings us to the final general invocation of de Certeau, the power relations he analyses in terms of realms of strategy and tactics. Strategy he sees as the imposition of power through the disciplining and organisation of space. Tactics are the 'ruses' that take the predisposition of the world and make it over, that convert it to the purposes of ordinary people. I shall develop this later but here is de Certeau's interest in talking about practices and events. All these invocations suggest work that begins a programme for attending to the spatial practices of people. In the language he applies to both intellectual and popular fields, he is interested in the relationships of place as a fixed position and space as a realm of practices – counterposing the fixity of the map to the practice of travelling.

Intellectual travails and travels

The above invocations serve as a jumping-off point, but as they are currently recited they do not fully capture the breadth and richness of writings that range from sixteenth-century theology to Latin American ethnology, to literary theory, to psychoanalysis to urban living. Geography has fastened rather too eagerly on only the last element of his work. I shall take this opportunity then to sketch his intellectual itinerary – or at least the traces of his passage.

Michel de Certeau was born in Chambéry in 1925. He joined the Jesuits in 1950 and was ordained in 1956 receiving a doctorate in religious science in 1960 from the Sorbonne. He became a specialist in early modern religious history and began his studies of mystics. In response to events of 1968, he changed course (de Certeau 1997b). His work began to address the problems and issues of contemporary society and theory moving to deal with heterogeneous issues, and audiences. In his footnotes there is a staggering array of anthropologists, historians, writers and even a reference to *Environment and Planning A*. One might add his membership of Jacques Lacan's *École Freudienne* from its inception to its demise. The diversity of his writings was both product and key to his work. As Ahearne writes:

> The extraordinary intelligence at work in his thought . . . is the product of this untiring textual, cultural and interlocutory 'travel', coupled with a form of interior distancing or 'quiet' born of a life long immersion in the demanding texts of the Christian mystics.
> (1995: 2)

He worked full-time in California from 1978 to 1984, before returning to the Écoles des Hautes Études Sciences Sociales. He died in January 1986, receiving eulogies from historians like Roger Chartier, anthropologists like Marc Augé, a tribute in *Liberation* from Julia Kristeva and has been the subject of conferences and special issues of journals such as *Diacritics*.

My purpose in outlining this itinerary is that it is emblematic of his project as a whole. His peripatetic intellectual wanderings need to be read as a refusal of disciplinary authority, by displacing disciplines. Placement he saw as about 'proper' knowledge, as an orchestrated gathering of topics, sanctioned and limited by its point of speech (1997b: 123). There is a hint of how he saw his intellectual practice as an evasion of the current order of knowledge:

working with its machines and making use of its scraps, diverting time owed to the institution; we can make textual objects that signify an art and solidarities; we can play the game of free exchange, even if it is penalized by bosses and colleagues when they are not willing to 'turn a blind eye' on it; and, in these ways, we can subvert the law that, in the scientific factory ... progressively destroys the requirement of creation and the 'obligation to give'. ... Realizing no profit (profit is produced by work done for the factory), and often at a loss, they take something from the order of knowledge in order to inscribe 'artistic achievements' on it and to carve on it the graffiti of their debts of honor.

(1984: 28)

In reaction to disciplinary boundaries we find instead a variety of spatial practices and a language of thought that is truly thinking through space. I will thus follow his work from a perspective perhaps close to that of Godzich (1986: vii) in linking de Certeau with a tradition of philosophy that is sceptical to 'the identity of thought and being' (and by implication dialectics). I want to suggest that de Certeau's work is an attempt to approach, or circle round 'being' – without reducing it to the categories of thought.

This chapter will follow this thread through his works on an ethics of knowledge, the spatiality of knowledge about the Americas, practices of knowledge and, to finish where we entered, ordinary practices in the city. Some words of caution are needed about this route. First, de Certeau himself reminds us that such theoretical narratives risk a totalising gesture that stockpiles the prior ideas – an 'Occidental capitalization of knowledge' (1986: 146). Instead this narrative is put together in what he calls a piling up of insufficiencies – a continual feeling 'that's not quite it', a gesture which propels the story outwards. Second, de Certeau's writing is very much the performance of such a dispersion (Conley 1992). As François Hartog summarised it:

He discovered, but without measuring, he travelled through, but without inhabiting, this heterological space of which he was, in a certain way, the inventor and historian, but a historian without territory, the instigator of a proceeding rather than the founder of a new discipline.

(cited Giard 1991: 219)

It is almost to do a violence to his work to draw it together under a few themes. As Hartog suggests de Certeau was 'inscribing his work in the space of the other' only then 'to disappear in the rumour of the crowd' (cited Giard 1991: 219). Or as de Certeau put it:

> When someone departs the security of being *there* together . . .
> another time begins, made of other sorts of excursions – more
> secret, more abstract, or 'intellectual' as one might say. These are
> traces of the things which we learn to seek through rational and
> 'academic' paths, but in fact they cannot be separated from chance,
> from fortuitous encounters, from a kind of knowing astonishment.
>
> (in Terdiman 1992: 2)

De Certeau conceptualised his practice (and that of others) through a
spatialised vocabulary. This was no mere affectation but an alertness to
epistemic practices. As his early mentor Dupront put it, in his *Espace et
Humanisme* (1946), possession through vision 'provides the definition of
modern knowledge, whose progress is made in the reading of space . . .
[Modern knowledge] is expressed in . . . a shifting from space traversed
to space that is read' (cited Giard 1991: 215). My aim is to keep this piece
traversing rather than try and render visible de Certeau's work as though
on a slab.

Ethics of the Other

Godzich (1986) suggests links with the work of Emmanuel Levinas who
saw thought as a response to a truth that was alien to the subject. Levinas
(1989) critiqued dialectics that interiorised, colonised and mastered the
unknown by placing the terrain of knowledge under the law of the Same.
Alternately Buchanan (1997) links de Certeau's notion of practice to the
non-representational logics of Deleuze's transcendent empiricism. Both see
de Certeau's work as opposing Hegelian dialectics of idea and object, and
taking up ideas where truth is not mastery to explore the relationship of
the representable and unsayable – a zone of indeterminacy. De Certeau
(1983) likens this to a philosophy sinking into the world rather than trying
to dominate it so that, as in Wittgenstein, knowledge is of a kind with
ordinary speech acts and language games rather than standing over them.
Thus philosophical discourse is haunted by an exteriority that has a (necessary and fortunate) quality of being unfinished leaving wounds in the text.
The philosophy does not dominate its object. It is seized by what it speaks
of (1983: 26). So, for instance, vision becomes an opening to being Other,
a general field of desire and differentiation in which twining and folding
relationships between objects are played out.

The relationship to the Other is thus a central concern in his
thought as well as his substantive topics. 'It is not certain that de Certeau
thought such a "science of the Other" to be constructible; rather it
constituted a horizon of intelligibility toward which his work addressed
itself in its entirety' (Giard 1991: 217). Louis Marin put it more poetically
that 'the other is always subject to a kind of Cheshire cat devolution or

disappearance ... this necessary horizon of communicative loss' (in Terdiman 1992: 4). The mark of a perturbation or rupture as a necessary part of thinking about practices, as a mark of heterology, is a theme in his work. He explored what has to be forgotten to make things intelligible, the survivals of other ways of thinking that creep in as 'lapses in the syntax created by the law of place' where 'they symbolize a return of the repressed, that is, a return of what, at a given moment, has *become* unthinkable in order for a new identity to *become* thinkable' (1988: 94). He saw the 'place of knowledge' as under the logic of the Same, where things were rendered transparent, intelligible and visible (1988: 333) – a place, where everything is spread out before the gaze of theory, and differentiated only by location relative to each other (1987; Castoriadis 1987: 201). Wary that science, 'by substituting its own places for the complex geography of social ruses and its "artificial" languages for ordinary language, has allowed and even required reason to adopt a logic of mastery and transparency' (1984: 22) he wrote so as to interrogate each figure without creating a stable centre. Thus we have a thought never in repose but constructed through itinerancy; a Freudian reading of texts as what they bring forth (Giard 1991: 218).

In this light he likened sixteenth- and seventeenth-century mystics to the statues on the boundary of an ancient Greek city, marking what is unknowable (1992b: 2). Mysticism was the visible performance of contact with the unspeakable, inconceivable Absolute (1992a: 14). Mysticism opens up a sense of a beyond (in this case the unfathomable essence of the universe); it opens up an itinerary (1992a: 19). Mystical knowledge offers a way of thinking that in place of dominating physically welcomes the repressed element of the other (1992b: 47). A parallel is Bosch's painting *The Garden of Earthly Delights*, where '[t]he secret of *the Garden* is to *make you believe* that it possesses some sayable secret' (1992b: 52). Bosch's figures suggest visible but unreadable stories. It plays on our need to decipher, to find hidden schemes; the placing of elements provokes a speculation on the logic linking them (Conley 1992).

Spatial encounter, spatial history

A concern with how the unrepresentable punctures symbolic systems is not surprising given de Certeau's Lacanian background. Yet this is a profoundly historicised and spatialised version of Lacan. I shall outline the historical inflection before looking at the geographical imagination. The historicisation is a twofold inflection, not just the chronology expected of a writer who ranges from mediaeval to contemporary. The first inflection is the articulation of the difference of psychoanalytic and historiographic temporalities. In historiography time is a sequence in which the past is absolutely and irrevocably distinct. In the Freudian imaginary, by contrast, it is as though each of Rome's spatial configurations remained intact and interpenetrated the others (1992b: 65). If for Lacan the unconscious is structured like a language, for de Certeau language is structured like a city. The second inflection comes out of the historian acting in the present, through specific present things (archives, monuments), to reconstruct a past time leaving a constitutive split. History is poised between fictions of the past and claims of an authority based on present practice (Poster 1992).

Spatial practices are equally bound up in an economy of representation and difference. This comes through most obviously at two levels: first, the substantive focus on encounters between spatially separated cultures; second, the economy of representation itself is worked through in spatial terms. In fact a bold reading would suggest that the encounters with the new world were merely one part of this wider economy. Certainly one could find statements to support that, yet I think it misses the way de Certeau worked – historically embedded examples are chosen as prototypical points in the evolution of practices. De Certeau was no idealist. For him '[t]he world of objects is there, terribly "real" in resisting human modification' (Conley 1988: xvii). But contrary to many materialisms, the objects are quite the opposite of empirically self-evident. Their irreducible 'thinginess' renders them resistant to representation (1997b: 141). They can never be fully captured, and appear as limits or gaps that elicit ever more erudition. Echoing Lacan, the Real is what is problematic in discourse, not what is self-evident. There is no invariate model. Choosing moments of crisis and change was a matter of highlighting the changing representational systems – and the practices through which we now view past worlds in their full strangeness. Each stands thus at the tension of current systems of intelligibility and those which sustained them then/there.

The *Writing of History* starts with the allegorical etching of Amerigo Vespucci encountering America (1625). A spatial juxtaposition where:

> Amerigo Vespucci the voyager arrives from the sea. A crusader standing erect, his body in armor, he bears the European weapons of meaning. Behind him are the vessels that will bring back to the European West the spoils of a paradise. Before him is the Indian 'America', a nude woman reclining in her hammock, an unnamed presence of difference, a body which awakens within a space of exotic flora and fauna. An inaugural scene: after a moment of stupor, on this threshold dotted with colonnades of trees, the conqueror will write the body of the other and trace there his own history.
>
> (1988: xxv)

This space and time of contact should not be taken as a generic model. The scene is counterpointed in his reading of another frontispiece etching of a century later – this time on the works of Lafitau (1980b) – of a study in the west. The 'muse' once more is female, the scattered artefacts of ancient places and new worlds lie about the writer. Here the broken images form a landscape of ruins. The study is a thinking laboratory where the ruins are teased into timeless ethnological theory. Time disappears from theory yet peeks through in these artefacts. The time of contact forms the repressed context of theory. 'The law of producing a text on the site of ruins imposes itself. Henceforth it will be necessary to create writing with the debris of the Other' (1980b: 50).

Between these two allegorical places, there is a tension. The time of co-living with hosts tends to sink into oblivion in the time of writing and intellectual profit (de Certeau 1986: 25). Yet not total oblivion. Neither space can be sealed from the other, the place of writing is 'constantly altered by the inaccessible (t)exterior [hors-texte] which authorizes that writing' (1986: 69) the trace of each inside the other creates a gap. A gap that calls forth a narrative. The texts create a series of surprises and inter-vals (classically voyages) that substantiate the alterity of the savage and thus authorise speech. The ahistorical descriptions of peoples are framed by these meta-discursive spatial practices. For instance Montaigne's texts, like the *ars memoriae*, work by placing things with the assumption that there is a stable topography with a place for every figure (1986: 70; cf. Yates 1968, Carruthers 1990). Yet this is a precarious achievement, since the tale is structured around three sorts of accounts, each of which shows the gaps in the others – the common-sense reaction lacks reasoning, the ancients lacked knowledge, and contemporary accounts were too often untrustworthy. It creates a negativity at its centre.

The ways difference fractured the space of western thought was crucial to de Certeau. It takes us beyond binary inversions or stable oppositions, like primitive versus civilised. The topology is more complex than this. So while the western text manufactures time and reason, in contrast to a space 'over there' that forms a place for pleasure, the effect of this is to make pleasure the unsayable remainder of the inexpressible primitive (1988: 227). The figure of leisure and desire leaves a trace of pleasure, a profit brought back from the Other. These practices form an unsayable, unlocalisable kernel that is written around – a process of circum-scription.

The writing of Jean de Lery (1578) serves to spatialise cultures as a synchronic picture (1988: 205), beginning a form of knowledge that creates the appearance of coherence (1997b: 150). The description of the Other is characterised 'not by localizations or geographical routes . . . but by a taxonomy of living beings' (1988: 226). The appearance of a spatial taxonomy is however belied by the way the circulatory practices of travel and study intertwine the supposedly separate cultures. It creates the appear-ance of bounded, discrete places while turning them into unstable meeting points (1997a: 91). However, the structure of the account works to divide

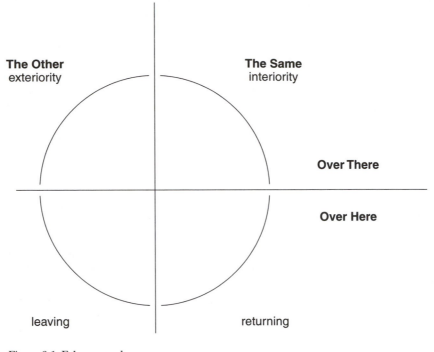

Figure 6.1 Ethno-graphy
Source: de Certeau 1988: 221

Alterity and the Same and also 'over there' and 'over here' as spatial categories (Figure 6.1). This forms a double location in theory and in practice
since here/there and Same/Other form different axes. Thus the space of
'over here' anticipates the Other as the ethnographer sets out , while 'over
there' is marked by introflection and a created sense of interiority. Narrative
happens 'back here', while 'over there' is characterised by static description; the practice of the travel, represented in the detailed log, opposes a
space of objects.

The voyage is transformed into a cycle which creates a centre of knowledge, a place of accumulation which fixes the other in place. The writings
form a technology invading space and capitalising time (1988: 216), an
accumulation premised on speaking in the name of the Other, turning
the motion around the circuit of Figure 6.1 into a stockpile. And yet, the
circle never quite closes, rather the transformations of the encounter mean
change is introduced, and there is a gap between starting and finishing
point. This gap, the space created through the Other, drives narratives
onwards to try and fill it, yet they only reinscribe the loss. In later accounts
we still see this process, so in Jules Verne's travel story, *20,000 Leagues Under*

the Sea, the narrator works in the library of the fabulous submarine. In so doing Verne narrates, in a displaced form, the researcher, Marcel, he employed to scour libraries for eighteenth-century sources (1986: 140). The effect is an unfurling sequence of writing and voyage where both Marcel and Verne labour on other texts then bury them in their own. The effect is of fragments piled up, of citations of citations and ruins of ruins, so that:

> the narrative displays a multiplication of trajectories, which unfurl an earlier writing in space, and of documents, which bury the past beneath displacements of location. But all of this occurs in the same place, in a book, or rather collection of books, each of which, due to its particular geography, is different from the preceding one, in other words stands *beside* the other, yet nevertheless repeats the same depth effect by placing itself *above* or *below* the other.
>
> (1986: 143)

This depth effect is the accumulatory economy. Through the citations and endless erudition of naming places Verne is working to textualise space, to make 'spatial history' (Carter 1987). Different epochs are marked by different relationships of time and space in texts and their practices – a double modality of textuality and geography (Giard 1991: 213). So mediaeval hagiographies, the lives of saints, provide a geography of the sacred, with the places where events occur but not the time – since their temporality is that of the cycle of calendar and festival. The itinerant reader is led to the named place of the saint (1988: 280–2), whereas cartographic and literary practices in early modern France charted the changing shape of the individual. Literature appealed to mapping to create new forms for new ideas of self and subject. The 'view shifts from one of the microcosmic self as mirror of the macrocosmic world to one in which both the reader and the characters discover that every figure counts as an insular entity among thousands of others' (Conley 1996: 177). This latter singularisation goes with a depiction of self against unknown and fragmented descriptions with travels between instances. The total scheme of the global cosmography is replaced by knowledge organised around the places of islands. This moment is the start of a split of map and travel stories (Conley 1996: 193–7).

The relationship of map to story is a key part of this analysis. The map and the list make knowledge as a field of equivalent points (1987). By contrast narratives are about motion. The fractures in narratives create routes from idea to episode in a spatio-temporal practice (cf. Sieburth 1987). This suggests a role for stories that is not about emplacing things but rather creating a theatre of frontiers and interactions, about the deformation rather than topical definition of places. Narrative is a relationship between structure and events comprising a topography and its alteration by otherness. Otherness introduces temporality so that '[e]very play or story is the progressive transformation of a spatial order into a temporal series' (1986: 22–3).

Otherness puncturing narrative wholeness is a modified working of Lacanian ideas of desire and lack – a performativity structured around absences – while the spatialised approach leads back to Lacan's seminars on a 'general theory of space for thinking out language' (1986: 49). So writing

> with no other ties than these forests of signs, these symbols of absence and, as the kaballah put it, these letters, initiating other encounters and other spaces. These are icons of what one begins to understand might still be said resembling the 'angels' which have become nothing more than a manner of speaking.
>
> (de Certeau in Terdiman 1992: 2)

These ideas seem prescient of the imaginative geographies depicted by Said (1978). However, they take writing and knowledge as spatial practices rather than representational systems. This echoes work currently being done by writers such as Michel Serres (1982, 1995). The idea of displacement, and working through materials as properties that may be rearranged but not appropriated, has already found a ready audience in post-colonial writing (Dhareshwar 1989: 153). As we shall see this makes de Certeau both cautious of representational knowledge, where practices are made to denote a shift in some system to an observer conceived as outside it (Hetherington 1998), but moreover makes him eschew the idea of localised and resistant communities and places. Instead he suggests a way to 'think space' in a consciousness that 'binds local space to the world beyond national boundaries' (1997a: 109).

The practices and places of knowledge

I have thus far introduced some ideas of de Certeau's practice as a mobile engagement itself. This section will clarify how he saw science as spatial practice. Like many current commentators he reinserted academic studies in their social context – including the position of enunciation as well as the enunciated. This opens up fairly obvious avenues to look at the authority mechanisms through which speech is credentialised (1985b). Taking this further, the reality effects are often necessarily dependent on obscuring the practices that created them, so that '[r]epresentations are authorized to speak in the name of the "real" only if they are successful in obliterating any memory of the conditions under which they were produced' (1988:

208). As a practice organising notes, producing sources and making them into a collection exiles them from practice and makes them objects of representational knowledge. The best research now is about producing 'useful lacunae' (1988: 72–8). So science works on practices by anticipating them 'within that grid of hypotheses and models which will "make them speak", its battery of questions, like so many hunters' traps, transforming the silence of things into answers, into language' (1980a: 22).

Thus he suggested Foucault too often set up traps for the world in advance, so after initial surprise, a stimulus of heteronomy, the world becomes remarkably ordered again – leaving practices 'the black sun of theory'. A rhetoric of clarity, where one thing is cut out and turned upside down to reveal everything else, makes the theoretical stance almost panoptical in itself (1984: 44–8, 1986: 187–91). The choice of structures focuses on regularities still, not the 'scattered polytheism' of relic systems of thought, on dominant models of the episteme not the continued performance of other practices. This awareness of how the tools of theory operate marked his understanding of all cultural studies:

> to outline the functioning of a cultural aggregate, to make its laws
> visible, to hear its silences, to structure a landscape that is nothing
> if it is not more than a simple reflection. But it would be wrong
> to think that these tools are neutral, or their gaze inert: nothing
> gives itself up, everything has to be seized, and the same inter-
> pretive violence can either create or destroy.
>
> (1986: 135)

We must recognise the guiles of theory without then moving from seeing them as transparent to seeing them totally obviating the outside world. De Certeau related the fields of practices and theories while, like Wittgenstein, admitting 'we are foreigners on the inside – but there is no outside' (1984: 13–14).

Practices are often treated as either inert contents or empty structures (Brammer 1992). This reveals only the 'Beauty of the dead' where theory communicates with cadavers (1986: 20–1). This stance de Certeau linked with the creation of *locis proprii* or *propres* of knowledge. These are places that allow vision to present a localizable object. 'Our society is characterized by a cancerous growth of vision, measuring everything by its ability to show or be shown and transmuting communication into a visual journey. It is a sort of epic of the eye and the impulse to read' (1984: xxi). The *propre* creates objects through transforming the uncertainties of history into readable spaces (1984: 36). The propre is the 'mastery of time through the foundation of an autonomous place'; place is antagonistic to time. It appears autonomous of the practices that create it, and thus atemporal. If we return to the opening analysis of the city, representational art and science 'immobilize its opaque mobility in a transparent text' through a 'no-when' synchronic system flattening all readings into one plane of data (1984: 94) and the 'empire of the evident' (1984: 204). Science is in these terms the creation of representational knowledge.

This proper vision of the 'concept city' is in decay, de Certeau is relieved to note – though he remained concerned about the tendency for the technologisation of the economic occurring alongside the relegation of culture to folkloric idiosyncrasy (1997b: 134). He refused to be worried about the decline of grand visions, suggesting '[t]he ministers of knowledge have always assumed that the whole universe was threatened by the very changes that affected their ideologies and positions. They transmute the misfortune of their theories into theories of misfortune' (1984: 95). The inability of these theoretical fictions to apprehend practices he likens to 'a comedy of mourning in the tomb of the absent' (1984: 157) – like time geographies of routes which refer to what has gone by not the act itself of passing by. The agenda he sets emphasises the plurality of practices – not comprising a series but an innumerable mass of singularities. These are the ruins of non-hegemonic systems, that form the raw material worked on by theories. The dispersed knowledges of practices elude the gaze of theory. The city is not univocal, and he sought to replace the homogeneity of single rationales (1997b: 116). Nor would (or could) science eventually make princesses of all these Cinderellas (1984: 67) for that would necessarily reduce them to representations rather than practices. Instead he suggested a mode of knowledge through travel to open space to difference, since stories about places are makeshift things, composed of the world's debris (1984: 107). Practices have no place of their own but move in the territory of the other (1986: 202). We can note then that de Certeau has a tension between a knowledge that wishes to make things legible by placing them and illegible practices that move over those places with a relationship of absolute difference between them. These are very fixed positions with a monolithic view of knowledge rather like Deleuze and Guattari's (1987) 'Royal Science'. It might be argued it leaves the popular only as existing as it is marked through its exclusion – without a positivity of its own. The popular is only visible through the lens of theory which cannot see the life but only the dead objects ripped out of a living culture (1986, 1997b: 41, Frow 1991, Morris 1988, Schirato 1993).

Practices and tactics

De Certeau comes dangerously close to seeing the popular as a remainder defined by exclusion from proper knowledge, even if it is so enormous that it is larger than the authorised practices of knowledge (1997b: 134). There are some defences to this charge, for instance by separating the philosophical and sociological moments of his work (Buchanan 1997), but principally I think we can reply that his work is a political exhortation not simply an empirical statement (Kinser 1992). His is not a model for all time but concerns how science is being applied to the world – a concern that carries echoes of Weber, the Frankfurt school and humanistic geography. De Certeau sketches the fear of a system of instrumental knowledge expanding all the time. His concern is to regain a sense of doing and knowing without being a means (1997b: 118). His answer to the advancing 'rationalisation of society' is that the ensuing giant order is both vast yet also strangely tenuous when set against the maritime immensity of scattered practices – the city is an 'order-sieve' (1984: 143). The gaze of power transfixes objects but also thus becomes blind to a vast array of things that do not fit its categories (1997b: 138). We might note that there is an almost nostalgic feel for a western European mode of urbanity in his writing (Kinser 1992 but see de Certeau 1997a: 91–2).[2] Equally the focus on grand power, on totalisation, means a view of power as singular with no mediating levels, where power is a one-way flow from a central source (Frow 1991: 57–8). It is true that writers like Fiske (1989) have taken this to suggest a romanticised popular culture, but, as Buchanan (1997) argues, this ontologises de Certeau's ideas, translating his concern for indeterminate practice (tactics) into resistant people and entities (the popular). He does not simply celebrate the people nor all their actions. His allusive chapters in *The Practice Of Everyday Life* lack historical specificities, indeed as Bennett (1998: 174) argues, they tend to somewhat dangerously homogenise all 'resistances' under a mythologised generic popular, 'a unifying myth of common otherness' as Morris put it (in Bennett 1998: 174).

De Certeau can seem to unproblematically valorise tactics though. We might ask whether tactical transgressions change anything (Frow 1991). To which his reply might be rather that they mean that things are not what they seem, not changing possible futures but our notion of what currently is. I would suggest the vision of power is not meant to be directly transposed onto institutions. The vision of power is related to Lacanian ideas of the Law of the Name of the Father as controlling and classifying. The idea of lack, absence, the role of vision and non-place offer a spatial and urban metaphor for psychoanalytic ideas of language. Language becomes city. But not simply landscape becoming text – de Certeau saw texturology and reading the city as complicit in the strategies of power. Rather the city becomes an arena of stories – where narrative offers the presence but not absorption of alterity. A theory of narration is indissoluble from practices, where narration proceeds by way of coups and detours by way of the past and quotation in the sudden opportunistic connections from memories (1984: 78–9; 1986: 192). 'The verbal relics of which narrative is made up (fragments of forgotten stories and opaque gestures) are juxtaposed in a collage in which their relationships are not thought out and therefore form a symbolic whole. They are articulated by lacunae' (1984: 143). Stories are not *about* movement, but *make* movements, not objects but effects, they transform, they do exactly what they say they do. In other words, his interest in practice is as a process without product, where open-ended communication is about action (1997a: 56). This is very much the Greek *metis* as a logic immersed in practices (1984). These stories mobilise memories that profit from the order without creating their own; they use occasions not create them; they bring invisible geographies into contact with the ordered realm of the rational.

Strategic power works by controlling and organising space to construct proper knowledge. In contrast, tactics – the arts of making do, like reading, or cooking – use what is there in multiple permutations. In de Certeau's terms they pass without occupying space. Unlike the formulations of the Birmingham school, de Certeau does not offer a metaphorical image of subcultures as resistant spaces, besieged by hegemonic powers (Bennett 1998: 176). Instead they are a bet on time, as an adaptive process based not on a balance of power (domination against resistance, local cultures against dominant global ones and so forth) but an absence of power – tactics are the weapons of the weak. This sort of practical knowledge of the city transforms and crosses spaces, creates new links (metonymy rather than metaphor), comprising a mobile geography of looks and glances. A crucial well spring is memory. It forms an anti-museum, which does not catalogue and place events, but takes fragments and propels them into the present. Time introduces alterity to space through the sudden deployment of memories. This is ' "Memory", in the ancient sense of the term, which designates a presence to the plurality of times and is thus not limited to the past' (1984: 82, n7). The alterity is that these memories do not just contain events, but still carry the remains of different conceptual systems from whence they came. These then are the ghosts in the machine. They bring the immediate and millenary, the novel and the permanent into contact (1997b: 137). Walking is thus to create non-sites and haunted geographies.

The model is not then grammar but reading, narrating and speaking. Where 'pedestrian utterances' speak the city, through metonymic tricks such as synecdoche and asyndeton the space of the city expands and contracts. Through the way thinking of one place brings in another, speaking the city can make wild temporal and spatial leaps – sudden connections and shifts. De Certeau used the term trajectory (1984: 33, 1997b: 145) to suggest that no matter what place we were in our space pointed to and from elsewhere – not as points on a sequence but immanent in each point. Trajectory suggests a temporal movement through space, that is, the unity of a diachronic succession of points through which it passes. In this sense comes the famous tag-line that space is practised place (1984: 117). Spatial transformation is mediated via memory.

For one who is often labelled as a high priest of agency-centred accounts of the city this leaves an unusual vision of the subject and agent. First, it is an agent constituted against a monolithic vision of power – a socio-psychological version of modernity. De Certeau believed totalities exist as real tendencies and projects, even if he did not think they would ever succeed totally (1997b: 136). His vision is replete with structures. It is not an agency free for all. But second, his subject is informed by psycho-analysis. The role of alterity and the idea of heterology have been carried from encountering the other to encountering daily life. The subject calling up memories to act is 'constructed as the stratification of heterogeneous moments' and this form of time results in 'the impossibility of an iden-tity fixed by place' (1986: 218). It is his insistence on the roles of particular places and paths which gives the spatial grid defining people's memories and imagination an independent heuristic value, as a topic in its own right, beyond the abstractions of social systems on the one hand and the empiri-cism of tracing individual behaviour on the other (Kinser 1992). Poster (1992: 102) comments that it offers a vision of a subject that neither recre-ates the unity of liberal theories nor evaporates the possibility of agency. The work on the city suggests instead:

> A piling up of heterogenous places. Each one, like the deterio-rating page of a book, refers to a different mode of territorial unity, of socioeconomic distribution, of political conflicts and of identifying symbolism . . . The whole, made up of pieces that are not contemporary and still linked to totalities that have fallen into ruins . . .
>
> (1984: 201)

This then corresponds to a vision of subjectivity that is equally open, where 'what is memorable is what we can dream about a site. In any palimpsestic site, subjectivity is already articulated on the absence that structures it like existence.' (1984: 144).

A letter returned

Michel de Certeau was a complex and fascinating thinker. His work looks to the proliferation of meanings but through his spatialised language he manages to keep hold of a context of power that is too easy to lose. His focus on modalities of use offers a sense of agency which has chimed with, and contributed to, rethinking from historical studies (Chartier 1987), to cultural studies (Frow 1991, Poster 1992) and geography. The agent is linked to an aporia or limit to knowledge; a non-discursive kernel not symbolised in language. Knowledge of practices is 'thus a mark in place of acts, a relic in place of performances: it is only their remainder, the

151

sign of their erasure' (1984: 35). He begins to provide a theory of practice where alterity is a pluralising element, that gets away from notions of authenticity (Buchanan 1997: 175). His emphasis on the role of narration and on the spatiality of that process, his spatial stories, remains a novel contribution. It certainly takes us beyond the reductions of his work to an account of agency or tactical use of the city. In reading his work, I am reminded most often of the way he described Michel Foucault, suggesting that he writes brilliantly – a little too brilliantly, leaving the reader trying to recall what it was that won them over. Of his academic practice and the assembly of material the comment that he was a 'dancer disguised as an archivist' (1984: 80) also seems to say a lot about de Certeau.

Notes

1   The subtitle of this book is French is *L'art de faire*.
2   While criticising ideas of information as implying a transparency inimicable to the resistant opacity of inhabiting the local and particular, he nonetheless attacks 'the nostalgias of an illusory utopia decorated with all the charms of social conviviality and threatened by the dark clouds of technology' (1997a: 92).

References

Ahearne, J. (1995) *Michel de Certeau: Interpretation and Its Other*. Polity Press, Cambridge.

Bennett, T. (1998) *Culture: a Reformer's Science*. London, Sage.

Brammer, M. (1992) Thinking Practice: Michel de Certeau and the Theorization of Mysticism. *Diacritics* 22 (2): 26–37.

Buchanan, I (1997) de Certeau and Cultural Studies. *New Formations* 31 Summer: 175–88.

Carruthers, M. (1990) *The Book of Memory: A Study of Memory in Mediaeval Cultures*. Cambridge University Press, Cambridge.

Carter, P. (1987) *The Road to Botany Bay: An Essay in Spatial History*. Faber & Faber, London.

Castoriadis, C. (1987) *The Imaginary Institution of Society*. Polity Press, Cambridge

Certeau, M. De (1980a) On the Oppositional Practices of Everyday Life. *Social Text* 1 (3): 3–43.

Certeau, M. De (1980b) Writing vs. Time: History and Anthropology in the Works of Lafitau. *Yale French Studies* 59: 37–64.

Certeau, M. De (1983) The Madness of Vision. *Enclitic* 7 (1): 24–31.

Certeau, M. De (1984) *The Practice of Everyday Life*. California University Press, Berkeley. trans S Rendall.

Certeau, M. De (1985a) The Practices of Space in Blonsky, M. (ed.) *On Signs*. Blackwell, Oxford: 122–45.

Certeau, M. De (1985b) The Jabbering of Social Life in Blonsky, M. (ed.) *On Signs*. Blackwell, Oxford: 146–54.

Certeau, M. De (1986) *Heterologies: Discourse on the Other*. Manchester University Press, Manchester. trans B Massumi.

Certeau, M. De (1987) The Gaze: Nicholas of Cusa.(trans C Porter) *Diacritics* 17 (3): 2–38.

Certeau, M. De (1988) *The Writing of History*. Columbia University Press, New York. trans T Conley.

Certeau, M. De (1992a) Mysticism. (trans M Brammer) *Diacritics* 22 (2): 11–25.

Certeau, M. De (1992b) *The Mystic Fable (Vol 1): The Sixteenth and Seventeenth Centuries.* University of Chicago Press, Chicago. trans M Smith.

Certeau, M. De (1997a) *The Capture of Speech and Other Political Writings.* University of Minnesota Press, Minneapolis, trans T Conley.

Certeau, M. De (1997b) *Culture in the Plural.* University of Minnesota Press, Minneapolis, trans T Conley.

Chartier, R. (1987) *The Cultural Uses of Print in Early Modern France.* Princeton University Press, NJ.

Conley, T. (1988) Translator's introduction to *The Writing of History*, M. De Certeau. Columbia University Press, New York.

Conley, T. (1992) Michel de Certeau and the Textual Icon. *Diacritics* 22 (2): 38–49.

Conley, T. (1996) *The Self-Made Map: Cartographic Writing in Early Modern France.* University of Minnesota Press, Minneapolis.

Deleuze, G. and Guattari, F. (1987) *A Thousand Plateaux*, University of Minnesota Press, Minneapolis.

Dhareshwar, V. (1989) Toward a Narrative Epistemology of the Postcolonial Predicament. *Inscriptions* 5 (*Traveling Theories, Traveling Theorists*): 135–58.

Fiske, J. (1989) *Reading the Popular.* Unwin Hyman, London.

Frow, J. (1991) Michel de Certeau and the Practice of Representation. *Cultural Studies* 5 (1): 52–60.

Giard, L. (1991) Michel de Certeau's Heterology and the New World. *Representations* 33: 212–221.

Godzich, W. (1986) The Further Possibilities of Knowledge, Foreword to Certeau, M. De *Heterologies: Discourse on the Other.* Manchester University Press, Manchester.

Hetherington, K. (1998) *Expressions of Identity: Space, Performance and Politics.* Sage, London.

Kinser, S. (1992) Everyday Ordinary. *Diacritics* 22 (2): 70–82.

Levinas, E. (1989) *The Levinas Reader.* Blackwell, Oxford.

Morris, M. (1988) Banality in Cultural Studies. *Discourse* 10 (2): 3–29.

Poster, M. (1992) The Question of Agency: Michel de Certeau and the History of Consumerism. *Diacritics* 22 (2): 94–107.

Said, E. (1978) *Orientalism.* Routledge and Kegan Paul, London.

Schirato (1993) My Space or Yours?: de Certeau, Frow and the Meanings of Popular Culture. *Cultural Studies* 7 (2): 282–91.

Serres, M. (1982) *Hermes: Literature, Science, Philosophy.* Johns Hopkins University Press, Baltimore.

Serres, M. with Latour, B. (1995) *Conversations on Science, Culture and Time.* University of Michigan Press, Ann Arbor.

Sieburth, R. (1987) Sentimental Travelling: On the Road (and Off the Wall) with Lawrence Sterne. *Scripsi* 4 (3):196–211.

Terdiman, R. (1992) The Response of the Other. *Diacritics* 22 (2): 2–10.

Yates, F. (1968) *The Art of Memory.* Pimlico, London.

7

HÉLÈNE CIXOUS

Pam Shurmer-Smith

Of all the lands a geographer may visit, there is, perhaps, none so strange as Academia. Academics the world over recognise one another and, even if they do not speak the same surface language (French, Polish, Hindi) or even the same middle language (Physics, History, Lit. Crit.) they usually think, feel and communicate in the same deep academic language. All of the residents of Academia are naturalised citizens, none was born there, all are refugees from somewhere else, and, like most refugees, they have a simultaneous love and loathing for their new home, a place where their foreignness can suddenly confront them, just as they were feeling settled.

Hélène Cixous has been a resident and a foreigner in Academia since the late fifties. Had she wished to, she could have had a conventional academic career – multilingual, erudite, well connected, she could have been an establishment figure in literary studies or critical theory and yet she has mapped out a life which emphasises her temporary resident status. However, she certainly has not been sidelined either – she took her first university appointment (aged twenty-five) at the University of Bordeaux in 1962, and in 1968, along with Foucault, Deleuze and Serres, was a founder of the new University at Vincennes, where she still is Professor of Literature. In spite of this, is still hard to say what Hélène Cixous *is* other than Hélène Cixous. Cixous has written political tracts, essays, novels, plays, literary criticism, but she rarely does so using the conventions and manners of Academia. Cixous believes in stretching language to breaking point, and this cannot be done within formal conventions of genre – her novels neglect to have 'plot', her academic writing is under-referenced and defies conventional structure. The commentaries on her (including this one) tell one more about the commentator's reading of Cixous than about Cixous herself, for she is a thinker and writer (writer and thinker) who constantly eludes those who would set themselves up as 'experts' on her thought. The best way to know Cixous is to read Cixous and to have faith in one's own reading. People differ radically on what they think she is saying. These are not just minor differences of interpretation (the sort of thing that the residents of Academia love to converse about for hours and years on end). There are major gaps in understanding of what, concerning anyone but Cixous, one would think of as fundamentals.

Reading Susan Sellers I experienced a sensation of a chasm opening under my feet; Sellers is regarded as a major authority on Cixous, someone one revises one's own inexpert reading in the light of. When she stated quite incontrovertibly that 'It is [the] double role of the mother in initiating separation and safeguarding life that is the concern of *Angst*' (Sellers 1996: 40) I went blank. I'd published a paper (Shurmer-Smith 1994) about Cixous' terrifying ability to write placeless space, focusing on *Angst* (Cixous 1977b) which I was convinced was about a woman separating herself from her dying (male) lover. I first read *Angst* while my own mother was dying – how could I not have noticed that the book was about mothers? How could it have dragged me back to something I thought I had carefully packed away in my past, the lingering death of a lover, if it had been 'about' what was going on before my eyes? How could two readers have such different views of the subject of a book? Which of us was wrong? How could one ever know? Why should geographers even care to struggle with a writer whose meaning is so clouded?

Coming cold to Cixous in the late nineties, one might be forgiven for wondering what variety of madness is spread across page after page, for it is virtually impossible to read Cixous without understanding the source of her desires, the nature of her frustrations and the extent to which these have been shared with other women. Perversely, as her project becomes progressively incorporated into mainstream thinking, so her strategy appears increasingly bizarre and one wonders at her deliberate strangeness. Why does she distance herself in a language of her own making? Why does this language seem more alien now that it has been 'canonised' than it did when she first used it? How can Cixous be read today?

Cixous is invariably cited as one of the 'big three' French feminist theorists, alongside Kristeva and Irigaray. In writing this sentence I unwittingly stumbled straight into the sort of problem that Cixous has addressed throughout her whole career. I initially wrote 'triumvirate', but deleted it straight away, since that means 'three ruling *men*'; then I *wrote* 'big three' – a more colloquial expression. I did not want to be colloquial, but there was no available word, just as there are no dignified words or constructions for many of the things that Cixous wants to write. This, a thousand times over, is what is meant when she talks of writing from the margins, or against the walls, or with her body.

To start again, though Cixous is regarded as a major French feminist theorist, this is not how she sees herself, complaining that, particularly in English-speaking countries, she is known for her early political writings (*Le Rire de la Méduse*, 1975a [tr. K. and P. Cohen 1976 as *The Laugh of the Medusa*]; *La Jeune Née*, 1975 [tr. B. Wing 1986 as *The Newly Born Woman*] and *La Venue à l'écriture*, 1977a [tr. S. Cornell *et al.* 1991 as *Coming to Writing*]) rather than her poetic creations. She even rejects the term 'feminist' because she sees feminism as having been constructed out of a false opposition which can be thought only through the distorted language of masculine domination. She has, however, continued to publish with the feminist publishing house, *Des Femmes*, and to extol the virtues of its head, Antoinette Fouque, whom she sees as taking up the activist role she does not herself fulfil. Whatever one decides to label her, Cixous has been very important in the reconceptualising of both women and the feminine but this reconceptualisation has not been through any variety of conventional social science. Cixous does not address the social construction of roles and statuses, does not indulge in debates about whether patriarchy is grounded in modes of production or reproduction, indeed she seems supremely unimpressed by social and historical 'facts'. She takes a far broader agenda, that of articulating how one can live; live the body, live the spirit, for she believes that it is in giving voice that life acquires meaning.

For Cixous, writing is not a mere matter of representation; she does not just write to communicate what she experiences, what she believes or what she has thought. She writes as her mode of being, believing and thinking. She writes endlessly (only a fraction is published), the writing itself is the communication and this is far more important than its content. Though this realisation invariably comes as a shock to her readers, it should not, for it is arguable that, outside Academia, the only communication which exists for the sake of its content comes in the form of instructions or demands, whereas sociable communication uses content as its vehicle (we talk about *something* so that we can talk, we write letters so that we can stay in contact, we watch the television so that we do not feel cut off). Like most other readers, I often found myself wondering, 'what is she going on about now?' until I realised that Cixous writes in an endless dialogue with herself and also with language. She muses over images; she meditates upon mantras, emptying her mind of normal constructions; she plays with words, allowing one word to suggest others that sound like it in a slipperiness we enjoy in poetry but are confounded by in academic writing.

Cixous writes in French, but she is not French. Who and what she is a major part of the Cixous project; a project which can make concerns with 'positionality' seem unsubtle. For Cixous, a position is a moment on a voyage, it is not a standpoint, but a fleeting glimpse of the world, and she writes by flashing images one after another. Though she writes a great deal about herself there is never a complete revelation, but, somehow, this partial exposure gives more sense of intimacy than many complete autobiographies. Cixous was born in Algeria into a Jewish family, with a father who spoke French and a mother and grandmother who spoke German. Outside there was Arabic. Also outside was French colonial society.

The landscape of my childhood was double. On one hand there was North Africa, a powerful sensual body that I shared, bread, fruit, odours, with my brother. On the other hand existed a landscape with the snow of my mother. And above the countries, the always present History of wars.

(Cixous and Calle-Gruber 1994. tr. Prenowitz 1997: 196)

Her father was a renowned doctor in Oran who died of tuberculosis when Cixous was young; his death haunts Cixous' writing right up to the present, but so does his being and his language:

Humour was a second language for him. He played on everything, members of the family, situations, and above all, signifiers. He was the enchanter. The universe was slightly translated. He had married a German woman and he had a house where we spoke German because my grandmother had arrived and spoke almost no French. So my father had forged, in a Joycian way, an entire system of jokes on the German language that became part of the family idiom. We all juggled.

Perhaps the verbal virtuosity or versatility that there is in my writing comes to me from my father: as if he had made to me a gift of keys or of linguistics.

(Cixous and Calle-Gruber 1994. tr. Prenowitz 1997: 196)

In *Dedans* (1969) Cixous writes of her father's death as the pervasive theme of her childhood, establishing the absence, lack, which will prove to be the focus for her search for herself. In her reflexive book *Photos de racines* (1994) (translated as *Rootprints* (1997)), written in conversation with Mireille Calle-Gruber, Cixous prompts Calle-Gruber to begin her *Générique* (curriculum vitae) not with her birth date but, '12 fev 1948 – mort du père à l'âge de 39 ans' (p. 209). The household of Cixous' childhood was warm, inward-looking and woman-dominated. Her widowed mother trained as a midwife and the young Hélène attended births with her – death and the births of the poor were early experiences which made it impossible for her ever to be able to separate the bodily from the intellectual, perhaps a strange migrant to the land of Academia; but she also came from a background of books and words, words in many languages.

At her secondary school, which had a restriction on the number of Jews admitted, she was the only Jewish child in her class; after that she attended a *lycée* for *boys*. She experienced school always from the margins, eccentrically. At university, too, she was marginalised by the fact that she had married young (aged only eighteen) and her student days in Paris had to accommodate not only a husband and the births of children, but also her sense of isolation from the land of her birth. No one who has not grown up taking scorching sunlight for granted can ever understand the home-sickness for the sun that preoccupies Cixous, Camus and millions of less known refugees. '*J'aime le passé passé. J'aime avoir perdu. J'aime Oran que je n'ai jamais revue et ne reverrai jamais*' (*Ibid.* (1994): 106) massacred in its trans-lation as 'I like the past. I like to have lost. I like Oran which I never saw again and will never see again . . .' (*Ibid.* (1997): 97) – Surely here *aimer* is loving not liking, *le passé passé* is a passion not for the simple past but the past which has passed, the grammatical plus-perfect, the past which is out of fashion, the unrecoverable which will not just not be seen again, but . . . never revered, never dreamed (never returned, like a genie, to a glass?)

The story of Hélène's family is part of the story of her philosophy – a mother throughout her academic life, joined by her own mother and grandmother, who were rendered destitute and stateless when they were expelled from Algeria, the country whose Independence her whole family had actively supported. Her beloved younger brother had been sentenced to death in Algeria by the OAS, but escaped to Bordeaux; on Independence he intended to become an Algerian citizen, but then came the pain of his imprisonment by the new Algerian government. The family's experience of multiple exclusion was acute. Cixous' sense of marginality, difference, otherness, homelessness is deeply ingrained; for her, foreignness is heaped upon foreignness many times over, a foreignness impossible to escape, wher-ever she goes, whatever she does. It was almost inevitable that she should have been drawn to poststructuralist thought and that, for so many of her readers, her most valuable contribution was the problematisation of oppo-sitional categories (1975, tr. 1986). 'Deconstruction, the style of thought which lets us retrieve the living element of life from where it has been walled up' (1994: 91 my translation).

Most famously, Cixous is responsible for what is known as *l'écriture femi-nine*, which does not translate easily to 'feminine writing'. *L'écriture feminine* starts from the assumption that feminine thoughts are shackled by mascu-line language, that it is only through experimentation that this language can be stretched to accommodate the expression of the feminine. (I have deliberately avoided using the terms *women* and *men* here, for Cixous sees masculine and feminine going beyond this categorisation and draws some of her major examples of *l'écriture feminine* from Genet and Kafka.) What we know of her special feelings about her father makes easy sloganising about patriarchy difficult and she is more concerned with Lacanian phal-logocentrism. Cixous claimed early on (1977 tr. 1991) that she was writing to demolish the walls constructed by language, the 'little cages of meaning assigned, as you know, to keep us from getting mixed up with each other' (tr. 1991: 49). Reading this experiment is pleasurable primarily for the excite-ment of witnessing the collapse, but this excitement has inevitably dulled with time, particularly in the light of the incorporation of the problema-tisation of language into the mainstream, as, for example, the requirement in many academic journals and elsewhere that non-sexist, non-racist language be used.

Cixous' version of *l'écriture féminine* is full of puns and other plays on words; this makes what she is doing difficult to translate (as I hope I have already demonstrated) Much of the time translators fix a word with one meaning, where more are lying beneath the surface or sidling up. They are forced to do this or to interrupt the flow with too many brackets or footnotes. Like so many other people who experienced the British education system, my command of French is far from confident, but the audacity of her project has forced me to the originals, dictionary to hand, using translations as a guide; I want to know what she is *writing*, not just what it can mean. Cixous would probably rather like this process of groping and guessing one's way through her writing. One is supposed to labour over Cixous, for meaning is only deceptively apparent. After the labouring, enlightenment sometimes comes in a flash. Her texts are readerly, the reader must contribute to the reading. In French, punning is not primarily for the sake of groan-inducing humour, it is to unsettle meaning, to make one realise that one can think two or more things almost simultaneously and that meanings can infect or subvert one another. So, for example, one is supposed to read her admiration for waves (*vagues*) and her frequent watery imagery as including her preference for indeterminacy (vagueness). In the title of her polemical work *La Jeune Née* (1975 tr. 1986) she calls up *Genet* (the writer and the plant) and *la je nais* (the thing I give birth to), *la journée* (the day's course, from which the English journey) all of which has to be forfeited when the title is rendered into English as *The Newly Born Woman* (tr. Betsy Wing 1986) (see Conley 1992: 54) Perhaps her most famous punning is in *Vivre l'Orange* – a small book in which she meditates upon a stream of ideas triggered by hearing news from Iran, from Iran to Oran, the place of her childhood, to herself then as Orange, a product of Oran, but also oranges (not the only fruit); herself as an orange, juicy, contained in a skin ... (logic does not help one understand, but poetry does) the self as exotic and familiar, natural and constructed.

> Organised discourse is no use to me. Of course, what I do is nonetheless grammatical, but everyday language is no good for this [subtlety]. It's even bad for it. Indeed because there is this everyday language, which is useful, too often one goes no further than everyday – when one must go to eternity.
> (Cixous and Calle-Gruber 1994. tr. Prenowitz 1997: 196)

Cixous today does not repudiate her early writings, but she does regret that in the English-speaking world her contributions to feminist polemic obscure her later work. She claims that in the seventies she:

... planted those essays *deliberately*, at a very dated, entirely historical moment, to mark off a field; so that we would not lose
sight of it entirely – to have done something deliberately: that
already tells you what it is! 'The Laugh of the Medusa' and other
texts of this type were a conscious, pedagogic, didactic effort on
my part to class, to organise certain reflections, to emphasise
a minimum of sense. Of common sense ... I was inspired to
write those texts by the urgency of a moment in the general
discourse concerning 'sexual difference'. Which appeared to me
to be confused and to be producing repression and loss of life and
sense.

(*Ibid.*)

Of course, that moment has passed; the discourse of sexual difference is
routine, it sits in respectable journals and on approved curricula. No one
wants to be frozen in her youth, her words of more than twenty years
ago quoted in the present tense. Cixous has grown and changed, but
certain core characteristics have remained constant, notably the preoccupation with the problem of how to be in time and space.

Cixous has written plays since the seventies, but, gradually, her enthusiasm for the theatre has overtaken other media of expression. Here she
parallels (but in my opinion outstrips) the career of Marguerite Duras, for
whom she has considerable admiration. Theatre is a vehicle for communication in which an author can take on many different voices, allowing her
multiple creations of herself and against herself to speak. 'The whole
body, the whole being is a theatre' (*Ibid*: 103). In theatre the voices can
speak at the same time, or they can remain silent; space and time can
be brought into play; staging, lighting, sound, all speak and can be apprehended simultaneously. In theatre voices cutting in, drowning out,
dwindling away, do not seem as difficult to cope with as they can on the
page. Theatre was clearly an ideal medium for someone like Cixous who
wanted always to let her writing flow unbounded, never to let her meanings become too fixed, always to allow them to take on a life of their own.
In theatre it is not only a matter of collaboration between actors, director,
writer and a host of technicians, there is also a tangible and interactive
encounter with the audience, but in theatre a writer can also become her
own audience:

I remember extraordinary experiences (in the theatre); being
doubled over with laughter seeing them play a scene I had written
thinking I had made an extremely tragic scene. Hearing this interpretation, I was captivated: what a lesson! We do not know what
we do. It was very beautiful: to have the experience of point of
view, of point of hearing ...

In the theatrical text the audience is implicated, it is actively present *in* the space of language. Would speak to himself, to herself. The audience is the reflexive *Self* of all the characters.

(*Ibid*: 101–2. Emphasis in original)

In the theatre Cixous favours epic subjects but, whilst retaining the large-scale and the grand sweep, she simultaneously focuses in on tiny inter-personal encounters. She has a considerable interest in opera and one can see the operatic emerging in her plays. Even the play *Voile Noire, Voile Blanche* (Black Sail, White Sail) (1994b), which is predominantly set in the poet Anna Akhmatova's apartment, avoids the domestic scale by allowing the whole of Stalin's Russia constantly to buffet against the characters, who can find no refuge. Akhmatova is portrayed as a mythical tragic heroine, a vulnerable tyrant balancing on the edge of destruction, a poet who cannot dare to write down her words but must commit them to the mind of a companion whose memory is as malleable as all memories. Something as tiny as the precise word in a poem takes on the tragedy of forgetting and remembering of an entire era.

Cixous uses theatre to think her own relationship with important events which are removed in time or space – those things which involve everyone in their moral magnitude but from which one is also remote. She has set her plays in Cambodia (*L'histoire terrible mais inachevée de Norodom Sihanouk, roi du Cambodge*, 1985) in India (*La Prise de l'école du Madhubai*, 1984 and *L'Indiade ou l'Inde de leurs Rêves*, 1987) and in the mythic past (the libretto for the opera *Le nom d'Oedipe: Chant du corps interdit*, 1978, *Les Eumenides*, 1992, *L'histoire (qu'on ne connaitra jamais* 1994a).)

Few of Cixous' plays have been published in English and opportunities for seeing them performed, even in France, are rare. For English readers, Morag Shiagh's (1991) treatment of Cixous' early performances is often as close as they will be able to get, but Shiagh is a wonderfully empathic commentator who captures not only the substance of the plays but also the ambiance of the event and the sheer audacity of Cixous' use of theatre. Shiagh's account of *L'Indiade ou l'Inde de leurs Rêves*, performed in 1987 at the *Théatre du Soleil*, is so persuasive that I merge it into my reading of the play, my readings on Indian Independence and my multiple experiences of India, and create a false memory that I attended this spectacular performance. The play was embedded into a simulacrum of an exotic India of sumptuous costumes and drapes, swords and turbans. The whole theatre was Indianised, Indian food was on sale, scents and sounds evoked India – but *whose* India? The India of *their* dreams. The India of *Others* and of othering, but also an attempt to catch the dreams of Nehru, Gandhi and a host of freedom fighters. *L'Indiade*, calling up the *Iliad*, catching at an Indian modern epic in the way in which western Europe had long since appropriated Ancient Greece as part of its own mythic past. For me,

161

Gayatri Spivak's difficulty with Cixous' tendency towards the exotic in the theatre misses the central point of Cixous' attempt to engage with European Orientalism:

> When she writes her Indian and Indonesian plays, her take on the complexity and hybridity of so-called post-colonial nations is shaky. Her work with the *Theatre of the Sun* can unfortunately be seen as perpetuating a kind of inspired, too admiring ethnography and a romanticising historiography.
>
> (Spivak 1993: 159)

Remembering Cixous' background, what better way could there be of exposing the impossibility of an insider view when one is situated in the crack between nations, cultures? Cixous is neither 'genuine' as colonised nor as colonial and I see her romanticising of India as prompting the question how one might more authentically engage in the dreams of an unspecified other. However, I can also see that it is Spivak's place to bristle at the use of what she regards as her own history.

Perhaps Cixous is on safer ground when she looks to European myth. In 1994 *L'histoire (qu'on ne connaitra jamais)* ['The Story (that one will never know')] was performed at the Theatre de la Ville in Paris and, having spent much of the previous year worrying about how to understand Cixous, I went to Paris for a weekend to experience her play at first hand. I am writing this in the personal because it is impossible for me to conceive of this play other than as a bodily sensation. It was hot, humid and the air was heavy; the rather grand theatre was packed with the sort of Parisian intelligentsia that strikes terror into the provincial British heart. The stage was entirely mirror – one easily became confused by what was material and what was reflection, which reflections were reflections of reflections and what was horizontal and what vertical surface. Costumes flowed in sumptuous silk, characters emerged through mirrored traps – an element of the story would reach a denouement and then start again. And it went on and on and on. After two hours, an interval. The Parisian intelligentsia looked less blasé and I realised that it wasn't just my shaky French that was leaving me disoriented, people were consulting their programmes to check that it really was an interval, there had been no obvious clue as to whether the play had finished or paused. The performance continued for another two hours and, after midnight the audience stumbled into a street washed by a thunder-storm we had missed, incarcerated in the theatre. Space, stars, velvety sky, clean cool air, everything sharply defined. Obviously not even Cixous was capable of summoning up the elements to lay further weight upon her drama, but it was (and is) impossible for me to separate out the component parts of the experience; the night outside and the release from the performance remain the final act.

The play itself is a working and reworking of the myth of the Niebelung, but, as the title suggests, the subject is the familiarity but unknowable nature of myth, the claim of myth to be able to catch essences. The play was about the telling of tales and the impossibility of either faithful representing or true understanding; it was about looking at things from all angles and yet still not knowing.

It would be unrealistic to claim that Cixous has had a significant direct impact on geography as an academic discipline (her status is even less in the geography generated in France than in English-speaking countries) but I do believe that she can make a major contribution to the way in which geographers think about categories, boundaries, ethnicities, about embodied spaces, about senses and abstractions. I certainly do not think that she has a contribution to make only to feminist geographies, though she has valuably problematised gender and installed a maternal perspective. Cixous' importance is in terms of experimentation, rather than the setting up of theoretical positions or the presentation of fact. She has few answers but poses devastating questions about the nature of being. Dealing as she does in abstractions, she draws upon space not just as a metaphor but also as a tangible ingredient:

> His command of walls: he built them. Each sentence constructed like a wall to surround his people; called to me. I listened. I came: there was a wall, that he had just made. I could hear him panting on the other side of the thought that he'd had difficulty getting over; which I had neither the strength nor desire to get over; but which rose up in front of me and communicated something of his excessive nature.
>
> (Cixous 1977b tr. Levy 1985: 132)

There is still much research to be done on the way in which people feel space(s), research which needs to go beyond mere socially constructed fears and apprehensions, though Horner and Zlosnik (1990) have ably examined generic landscapes in women's writing, and I have attempted to address the issue of sensuous space (Shurmer-Smith 1994). Here Cixous' utilisation of the idea of economies of pleasure and the notion of *jouissance*, an emphasis on the maternal gift rather than patriarchal profit, can offer much to a retheorisation of space and power which can accommodate gentler relationships with the environment and with a spectrum of difference. New(ish) enthusiasms for networks rather than structures, for unbounded notions of agency, draw freely upon the ideas which Cixous did so much to popularise.

Verena Andermatt Conley has been important in explaining Cixous' work to English speakers (Conley 1992) and has celebrated the project of *l'écriture feminine* as a feminist strategy. Much of her academic career has been devoted to the voluminous work of Cixous, but in her recent book *Ecopolitics: The environment in poststructuralist thought* (1997) we find a disillusionment, almost a sense of resentment at all that time wasted reading impenetrable writing instead of living. Always suspicious of Cixous' starquality, as in her description of her imperious performances at Vincennes (Conley 1984: 80), *Ecopolitics* finally turns away from Cixous:

> A voluble utopianism of May 1968 seems to go unchecked, without any qualification or adaptation in view of the unparalleled violence enacted on the globe ... that had taken place over the passage of only thirty years. The 'self-discovery' that *l'écriture féminine* brought to the female writer in the 1970s stays at the threshold of a productive narcissism once the writing self is unveiled, it has to 'fragment' or 'disembody' itself ... in order not to be tempted into personifying its new being as a variant of a Goddess Natura, Ceres, Athena, Demeter, or other benevolent deities that would essentialise the mother. In an ecofeminist politics, writing has to work tirelessly in dialogue with specific issues that are not just human-centred but that mobilise a vision of and an attention to specifically connected elements in given environments.
>
> (Conley 1997: 138–9)

Am I right in reading this as 'we've got to get down to some commonsense issues and start clearing up, rather than listening to a lot of airy-fairy essentialising'? Am I also right to assume that all that has happened is that Cixous and Conley have aged in different ways, have different new enthusiasms? That Cixous uses ideas about the environment to feel her way through the problem of being, whilst Conley, along with so many others has shifted to a prioritisation of the environment in itself, an activism and a neo-puritan denial of self?

> Cixous now seem(s) mired in an egocentric politics ...
>   Thirty or more plays and novels have appeared, and in such febrile frenzy, it would be impossible for any ecologist or feminist to divide attention between activism and careful assimilation of Cixous writing.
>   When Cixous is left to the devices of her own *écriture féminine*, the consciousness gets attenuated.
>
> (Conley 1997: 139)

With a weariness, I read the sad old struggle of the student to demolish her teacher, the devotee to outgrow her heroine, the pretender to kill the queen – the old myth of the Golden Bough, this time told in the feminine. Surely Cixous has shown us that there must be much more to a tale told in the feminine?

I sense in some of the critical geography of today an impatience with the intellectual problematisations of the recent past, a desire to return to moral certainties (and moral high-grounds), to issues and action. It would seem that Cixous with her preoccupation with sensation (sensuousness, sensationalism) does not chime with the times. Unlike Conley, I do not see Cixous as 'mired in egocentric politics', because I believe that if we suppress the experiencing self there is no reference point from which we can guard against totalising views. It is hard to imagine Cixous in an activist role – though she writes 'with' her body, she is pure intellect. She operates in the realm of the experimental, pushing to the limits her attempt to merge experience with representation. Now that she is over sixty, it is arguable that neither her young self nor her present self speaks easily to younger women today who now take for granted so much that she has accomplished.

Though I mentioned her earlier as part of the French Feminist 'big three', Cixous is in fact personally and intellectually closer to Deleuze, Foucault and, particularly, Derrida ('whom I have always considered to be my "other"' 1997: 80), all of whom have been more readily incorporated into the geographical imagination than she. And yet it is with a geographical description that Cixous remembers her first sight of Derrida:

> he was walking on the crest of a mountain . . . from where I was I saw him clearly advancing black on the light sky, feet on the edge, the crest was blade thin . . . his progression on the limit between the mountain and the sky melted into one another.
>
> ('Quelle heure est-il', undated, quoted in Cixous and Calle-Gruber 1994 tr. Prenowitz 1997: 79)

In her notebooks she works up this memory into a metaphor about deconstruction:

> he is situated at the point of contact between two slopes, versants, inclines, sides – at the reversal point of climb and descent, of desire into mourning, or mourning into burst of life, of you into me, of he into she . . .
>   J.D.
> Could only have inhabited language, place where the two sides can co-exist with their in, their between, their exchange, space of amphibologies. Language (the) only medium that gives the time at once stopped and mobile to describe the interstitial.
>   The *interre*stitial.
>
> (Cixous and Calle-Gruber 1994 tr. Prenowitz 1997: 80)

Cixous and Derrida, children of the margin, have known one another since they started to write and much of their writing and thinking has been in dialogue. This dialogue has produced a shared passion for the deconstruction of categories, oppositions, spaces of exchange and between-ness, a passion which an engaged geography cannot afford to ignore.

References

Cixous, H (1969) *Dedans* Paris. Grasset.

Cixous, H (1975a) 'Le Rire de la Méduse' *L'Arc* 61 39–54 (Tr. K and P Cohen [1976] 'The Laugh of the Medusa' *Signs* 1 (4) 875–93).

Cixous, H (1975b) 'Sorties' In Cixous, H and Clement, C (1975).

Cixous, H (1977a) 'La venue à l'écriture' in Cixous, H with Gagnon M and Leclerc A. (1977) *'La Venue à l'écriture'* Paris Union General d'Editions. (Tr S Cornell *et al.* [1991] 'Coming to Writing': 1–58. In (ed.) D Jensen *Coming to Writing and Other Essays.* Cambridge, Massachusetts. Harvard University Press.)

Cixous, H (1977b) *Angst* Paris. Des Femmes. (Tr. J Levy [1985] *Angst* London. John Calder.)

Cixous, H (1979) *Vivre l'orange* (Tr. A Liddle and S Cornell – Bilingual Publication) Paris. Des Femmes.

Cixous, H (1985) *L'histoire terrible mais inachevée de Norodom Sihanouk, roi du Cambodge.* Paris. Théâtre du Soleil.

Cixous, H (1987) *L'Indiade ou l'Inde de leurs Rêves et Quelques Écrits sur le Théâtre.* Paris. Théâtre du Soleil.

Cixous, H (1994a) *L'histoire (qu'on ne connaitra jamais)* Paris. Des Femmes, Antoinette Fouque.

Cixous, H (1994b) *Voile Noire Voile Blanche,* Paris. Théâtre du Soleil.

Cixous, H and Clement, C (1975 *La Jeune Née* Paris. *Collection 10/18* (Tr. B Wing [1986] *The Newly Born Woman* Minneapolis. University of Minnesota Press.

Cixous, H and Calle-Gruber, M (1994) *Hélène Cixous, Photos de Racines.* Paris. Des Femmes, Antoinette Fouque. (Tr. E Prenowitz [1997] *Hélène Cixous, Rootprints: Memory and Life Writing.* London. Routledge.

Conley, V (1984) *Writing the Feminine: Hélène Cixous.* Lincoln. University of Nebraska Press.

Conley, V (1992) *Hélène Cixous.* London. Harvester Wheatsheaf.

Conley, V (1997) *Ecopolitics: The Environment in Poststructuralist Thought.* London. Routledge.

Horner, A and Zloznik, S (1990) *Landscape of Desire: Metaphors in Modern Women's Fiction.* London. Harvester.

Sellers, S (1996) *Hélène Cixous: Authorship, Autobiography and Love.* London. Routledge.

Shiagh, M (1991) *Hélène Cixous: A Politics of Writing.* London. Routledge.

Shurmer-Smith, P (1994) 'Cixous' Spaces: Sensuous Space in Women's Writing' *Ecumene* 1 (4) 152–170.

Spivak, G (1993) 'French Feminism Revisited' in *Outside in the Teaching Machine.* 141–172. London. Routledge.

8

HENRI LEFEBVRE

A socialist in space

Andy Merrifield

I

Henri Lefebvre was on British television a little while ago. The show, 'The Spirit of Freedom,' was strictly for insomniacs and appeared in the wee hours on Channel 4. There were four programmes in all and each one tried to reassess the legacy of Left French intellectuals during the twentieth century. The tone was cynical and pejorative throughout, which wasn't surprising given that the series was written and narrated by one of France's more recent philosophical bad boys, Bernard-Henri Levy.[1] In front of the camera the night I watched sat an old white-haired man, dressed in a shabby jacket and blue denim work shirt. It was obvious to viewers that the nonagenarian hadn't long left to live. Even Levy described his interviewee as

> tired that afternoon. His face was pallid, his eyes bloodshot. I felt he was overwhelmed from the start and clearly bored at having to answer my questions. He spoke with difficulty, and when the memories were painful, it was sometimes hard for him to mention certain people I got him to recall. He told me several times he would rather talk about the present and the future, about things going on around him in the world.[2]

'We exchanged questions and answers,' Levy says a little later, 'arguments and clarifications. I'd come hoping he would play a certain role, and this he did with a show of goodwill I hadn't expected. I have to admit he also did it with skill and style.'

But what was the 'certain role' Levy wanted Henri Lefebvre to play? I want to begin this chapter with such a question because the answer reveals much about the role Lefebvre himself has played in France's twentieth-century intellectual history. For one thing, he's been around for almost its entirety. The title of his biography confirms as much: *Henri Lefebvre et*

L'Aventure du Siècle.[3] During that time, he had lived through two World Wars, drunk wine and coffee with the Surrealists, joined and left and joined again the French Communist Party, fought for the Resistance Movement in the early 40s, driven a cab in Paris, taught sociology and philosophy at numerous French universities, been one of the intellectual godfathers of the 1968 generation. Meanwhile, he'd authored and introduced into France a whole body of Marxism, and written prolifically on urbanism, on everyday life, and on space. Throughout the twentieth century, clearly, Henri Lefebvre has done and seen and heard a lot.

And yet it wasn't Lefebvre's own work that concerned Levy. Levy was much more interested in other figures from France's past: Paul Nizan, Georges Politzer and Alexandre Kojève. Lefebvre knew all three, and that's what Levy wanted him to talk about. 'The astonishing thing was,' Levy admitted, 'he understood what I wanted and went along with it.' So that was Lefebvre's 'role': he was, for Levy, an indispensable observer and reporter. He was somebody who'd been there and had befriended other French intellectuals and who'd outlived them all. Now he could recount old tales and re-live the life and death struggles of bygone days. In the illustrious company of Sartre or Camus or Breton or Nizan or Malraux or Althusser, Lefebvre was a minor and relatively unknown figure. (Even the on-screen caption for the programme got it wrong: it introduced Lefebvre as a 'Historian'!)

II

Given this minority status in France, why then has Lefebvre become such a cult figure in Anglo-American intellectual circles today? Did his work on urbanism and space initially lead to bad press in France? Maybe it did. For when he began writing about both later in his career, orthodox Marxists couldn't figure either out. So maybe his spatial turn sounded the death knell to his stardom? Even his great spatial book, *The Production of Space* – regarded by many contemporary geographers and urbanists as his magnum opus – was misunderstood and overlooked when it hit the French bookshelves in 1974. The timing couldn't have been worse: Althusser's reputation was formidable then and his structural Marxism was *de rigueur*. And if you didn't agree with Althusser and you were still a Marxist, it was to Garaudy's humanism you'd turn, not Lefebvre's. And a book about space? Well, that's what most socialist radicals seemed to need like a hole in the head! When things did assume a spatial turn after the late 60s urban riots and student and worker protests, Althusser still snuck in ahead of Lefebvre. It was the former's Marxism, after all, which underwrote Manuel Castells's highly influential sociological research on urbanization. And Castells's *La Question Urbaine* – replete with attacks on former mentor Lefebvre – made it to press two years before *The Production of Space*.

In *The Urban Question*, Castells at once undercut his senior's humanist predilections and the intellectual credibility of Lefebvre's object of analysis. Castells boldly asked whether the 'urban' was a legitimate object of enquiry at all. The 'urban question' for him was above all a question of how an urbanizing *capitalist mode of production* functioned. In Castells's spatial universe, the city was indeed a *container* of social and class relationships. But it was these social relations which had primacy over any explicit 'urban' or 'spatial' category. Lefebvre, for Castells, was just a little too lax in his reification of space. Castells even caught a whiff of spatial fetishism going on. He wasn't impressed. Lefebvre had strayed irrevocably. From trying to develop a 'Marxist analysis of the urban phenomenon,' he, Castells suggested, 'comes closer and closer, through a rather curious intellectual evolution, to an *urbanistic theorization of the Marxist problematic.*'[4] No compliment intended: This was a stinging criticism which probably helped assure the relative neglect of Lefebvre's work during the 1970s.[5]

While Lefebvre's rejoinder maintained that Castells didn't understand space – 'He sets aside space,' Lefebvre said. 'His is still a simplistic Marxist schema'[6] – it was David Harvey in 1973 who first brought Lefebvre to the attention of Anglophone audiences. In *Social Justice and the City*, however, Harvey's Lefebvre was Lefebvre-*lite*: the French Marxist played only a cameo part in the 'Conclusions and Reflections' chapter. But Lefebvre's idea that a distinctively 'urban revolution' was supplanting an 'industrial revolution', and that this urban revolution was somehow a spatial revolution as well, had a deep and lasting resonance in critical urbanism and geography – longer-lasting, it seems, than Castells's own urban research, which was reaching its sell-by-date as early as the mid-80s. Soon Harvey was to deepen his appropriation of Lefebvre. For instance, in a brilliant essay called 'Class-Monopoly Rent, Finance Capital and the Urban Revolution', published in 1974, Harvey used Lefebvre to shed light on how Baltimore's spatial organization and housing markets got structured by financial institutions. Nevertheless, he also warned readers that Lefebvre's thesis was 'startling in its implications and obviously requires careful consideration before being accepted or rejected.'[7] Over the next few years, a Lefebvrian cottage industry began to spawn; and some of it took up Harvey's challenge.

In this context I'd like to suggest that rather than Lefebvre influencing Anglo-American geography and urbanism, it is perhaps the other way around: Maybe it's been Anglo-American geography and urbanism that has resuscitated Lefebvre's flagging spatial career and prompted his more recent claim to fame. One wonders how well-known his work would have become without the dedicated mediation of David Harvey, Ed Soja, Fredric Jameson, Mark Gottdiener, Derek Gregory *et al*. One wonders, too, whether we would have ever seen *The Production of Space* appear in English. God knows, seventeen years is long enough anyway. (A far cry from Althusser's

For Marx, published in France in 1965 and making it to English book-shelves a couple of years later.) No surprises, then, that interest in Lefebvre proliferated most of all in Anglo-American radical geography.[8] And with this proliferation, the more Lefebvre's name and work became known, and the more he himself, ironically, became a marketable publishing commodity. Hence the spate of translated works over recent years.

The Production of Space, published in 1991 by Basil Blackwell (Harvey's old publisher), and diligently translated by one-time Situationist Donald Nicholson-Smith, has been the biggest catalyst here. In a way, its appear-ance has been *the* event within critical human geography over the 1990s. No more vicarious appropriation now; no more do we only have to listen to big boys and other French-speakers citing Lefebvre. English-speaking readers and lesser spatial thinkers can now have their say, form their own opinions, debate Lefebvre's tantalizingly loose, prolix and episodic style. So, after a very long wait, everybody has been given access to one of the most original Marxist thinkers of the twentieth-century. *The Production of Space* is here, and it has sparked a thorough reevaluation of social and spatial theory on both sides of the Atlantic. At last, Lefebvre, while not yet a household name, has achieved posthumous notoriety. But what's all the fuss been about?

III

The explorations in *The Production of Space* (POS) are the explorations of an extraordinarily protean intellectual. But this protean intellectual was also a seventy-year-old *French Marxist*. The strengths and failings of the book should be considered in this light. Of course, there's much more going on than plain old-fashioned Marxism: Hegel crops up often; Nietzsche's spirit, as I've argued elsewhere, is palpable;[9] Lefebvre's inti-mate grasp of romantic poetry and of modern art and architecture is demonstrable; meanwhile, Lefebvre breezes through the history of Western philosophy as if it's kids' stuff. Nonetheless, POS is a text which is somehow quintessentially Marxist, socialist and modernist, and that, I think, shouldn't be forgotten.

The book begins with a 'Plan of the Present Work.'[10] This opening gambit is surprising in its coherence, and the argument proceeds with considerable analytical consistency. Immediately we get a compressed account of the concept of space, listen to how it has been denigrated in Western thought, and hear how Lefebvre himself aims to work through this motley state of affairs. On the face of it, this all sounds like a tame philosophical dilemma, hardly one to change the world. But as we follow Lefebvre onwards through POS we soon see its radical import. After a while his pursuit for a 'unitary theory of space' unfolds – critically and flamboyantly. The project he coins is *spatiology*, and it involves, amongst other things, a rapprochement between *physical* space (nature), *mental* space (formal abstractions about space), and *social* space (the space of human action and conflict and 'sensory phenomena'). These different 'fields' of space have, Lefebvre thinks, suffered at the hands of many philosophers, scientists, and social scientists, not least because they've been apprehended as separate domains.

POS seeks to 'detonate' everything here. For Lefebvre sees fragmentation and conceptual dislocation as serving distinctively ideological purposes. Separation ensures consent, perpetuates misunderstanding, and worse: it reproduces the status quo. By bringing these different modalities of space together within a single theory, therefore, Lefebvre seeks to *expose* and *decode* space, and thereby empower socialists everywhere in their analysis of, and struggle against, an urbanizing modern capitalism. The key concept to contend with, however, is *production*.

The emphasis on production, of course, chimes with the radical manner in which Marx himself emphasized it. Marx, remember, suggested that to be radical meant 'going to the root of things'. And his obsession with production was designed to do just that: to get to the root of capitalist society, to delve into its 'hidden abode', to go beyond the fetishisms of observable appearance, and to trace out its 'inner dynamics' holistically, in all its gory horror. Lefebvre, correspondingly, tries to demystify capitalist social space by tracing out its inner dynamics and *generative moments* in all their various guises and obfuscations. Here, generative means 'active' and 'creative', and creation, says Lefebvre, 'is, in fact, a *process*' (POS: 34) (original emphasis). Thus getting at this generative aspect of space necessitates exploring how space gets *actively produced*. Again, like Marx, Lefebvre makes political use of process thinking in his theoretical quest for explanation.[11] Now, in Lefebvre's hands, space becomes redescribed not as a dead, inert thing or object, but as organic and fluid and alive; it has a pulse, it palpitates, it flows and collides with other spaces. And these interpenetrations – many with different temporalities – get superimposed upon one another to create a *present* space. As such, each present space is 'the outcome of a process with many aspects and many contributing currents' (POS: 110). But all this presents certain problems. The biggest, says Lefebvre, is that it's 'never easy to get back from the object [the present space] to the activity that produced and/or created it' (POS: 113). Because once the 'construction is completed, the scaffolding is taken down; likewise, the fate of an author's rough draft is to be torn up and tossed away' (*ibid.*). So what needs to be done is to 'reconstitute the process of its genesis and the development of its meaning'.

What we have here is a spatialized rendering of Marx's famous analysis on the *fetishism of commodities* from Volume One of *Capital*. This rested on the recognition that commodities assume a strange 'thing-like' character once they get exchanged at the marketplace. There, what are fundamentally inter-subjective relations become, Marx says, perceived by people as objective. There, in Marx's words, 'it is a definite social relation between men, that assumes, in their eyes, the fantastic form of a relation between things.'[12] Marx calls this masking effect 'fetishism' and it gives commodities a special 'mystical' and 'mist-enveloped' quality. At the level of exchange – the traditional focus of analysis for bourgeois economists – it is nigh impossible, Marx says, fully to apprehend the social relations, activities and exploitations occurring in the productive labour process. Lefebvre's shift, accordingly, from conceiving 'things in space' to that of the actual 'production of space' itself, is the same conceptual and political shift that Marx made from 'things in exchange' to 'social relations of production'. Let's hear the former expound further:

> instead of uncovering the social relationships (including class relationships) that are latent in spaces, instead of concentrating our attention on the production of space and the social relationships inherent to it – relationships which introduce specific contradictions into production, so echoing the contradiction between private ownership of the means of production and the social character of the productive forces – we fall into the trap of treating space 'in itself,' as space as such. We come to think in terms of spatiality, and so fetishize space in a way reminiscent of the old fetishism of commodities, where the trap lay in exchange, and the error was to consider 'things' in isolation, as 'things in themselves'.
>
> (POS: 90)

Now, the production of space can be likened to the production of any other sort of merchandise, to any other sort of commodity. Now, too, we can perhaps begin to see how Lefebvre's ideas diverge from those of Manuel Castells. Recall how the urban question for Castells was a question of *reproduction*; the urban crisis for him was a structural crisis of *consumption*. All the action in Castells's drama got foisted into the reproductive rather than productive realm. Given his intellectual debt to Althusser, this is hardly surprising.[13] Lefebvre, on the other hand, assumes a much more active understanding of space. For him, space isn't just a passive surface for reproductive activity. Of course, spaces do permit commodity transactions and the reproduction of labour-power to all 'take place.' Castells's work demonstrated this tellingly enough. But to leave it at only that would, Lefebvre insists, miss much, would fall into the trap of treating space 'in itself'. Because now, he says, space is itself actively produced as part of capitalist accumulation strategies. And, importantly, space gets produced before it is reproduced – even though reproduction is obviously a necessary condition for further production.

So space – urban space, social space, physical space, experiential space – isn't just the staging of reproductive requirements, but part of the cast, and a vital, productive member of the cast at that. Space, in the apt words of David Harvey, is an 'active moment' in expansion and reproduction of capitalism. It is a phenomenon which is colonized and commodified, bought and sold, created and torn down, used and abused, speculated on and fought over. It all comes together in space: space *internalizes* the contradictions of modern capitalism; capitalist contradictions are contradictions *of* space. Here Harvey and Lefebvre find broad agreement. To know how and what space internalizes is to learn how to produce something better, is to learn how to produce another city, another space, another space for and of socialism. To change life is to change space; to change space is to change life. Architecture or revolution? Neither can be avoided. This is Lefebvre's radiant dream, his great vision of a concrete utopia. It's a dream that underwrites POS.

IV

Critical knowledge has to capture in thought the actual process of production of space. This is the gist of Lefebvre's message. Theory must render intelligible qualities of space which are at once perceptible and imperceptible to the senses. It is a task that necessitates both empirical and theoretical research, and it's destined to be difficult. It will doubtless involve careful excavation and reconstruction, necessitate both induction and deduction, journey between the concrete and the abstract, between the local and the global, between self and society, between what's possible and what's impossible. Theory must somehow trace out the actual dynamic and complex interplay of space itself – of buildings, monuments, neighbourhoods, whole cities, the world – exposing and decoding those multitudinous imperceptible processes involved in production. So far so good. But how can this be done?

Lefebvre works through these dilemmas himself by constructing a complex heuristic device: he calls it a 'spatial triad,' and it forms the central epistemological pillar of POS. Unfortunately – or fortunately – he sketches this out only in preliminary fashion; he leaves us to add our own flesh and to re-write it as part of our own chapter or research agenda. What's more, while Lefebvre suggests that the triad is something we will encounter 'over and over again' in POS, its appearance beyond the initial chapter is more implicit than explicit, assumed rather than affirmed. Why? Because it's not a mechanical framework or typology he's bequeathed us here, but a dialectical simplification, fluid and alive, and each moment messily blurs into other moments in the real life contexts. Notwithstanding, three moments are identified: representations of space, representational space, and spatial practices. Let's look more closely at each in turn.

(a) *Representations of space* refers to conceptualized space, to the space constructed by assorted professionals and technocrats. The list might include planners, engineers, developers, architects, urbanists, geographers, and others of a scientific bent. This space comprises the various arcane signs, jargon, codifications and objectified representations used and produced by these agents and actors. Lefebvre says that it's always a space which is *conceived*, and invariably ideology, power and knowledge are embedded in this representation. It's the dominant space of any society because it is intimately 'tied to the relations of production and to the "order" which those relations impose, and hence to knowledge, to signs, to codes, and to "frontal" relations' (p33). Since Lefebvre believes this space to be the space of capital, conceived representations of space play a 'substantial role and a specific influence in the production of space' (p42), finding 'objective expression' in monuments, towers, factories, office blocks, and the 'bureaucratic and political authoritarianism immanent to a repressive space' (p49).

(b) *Representational space* is directly *lived* space, the space of everyday experience. It is space experienced through complex symbols and images of its 'inhabitants' and 'users,' and 'overlays physical space, making symbolic use of its objects' (p39). Representational space may be linked to underground and clandestine sides of social life and doesn't obey rules of consistency or cohesiveness, neither does it involve too much 'head': it's rather felt more than thought. It is simply *alive*. In lived representational space, there's more *there* there:

> it speaks. It has an affective kernel or centre: Ego, bed, bedroom, dwelling, house; or: square, church, graveyard. It embraces the loci of passion, of action and of lived situations, and thus immediately implies time. Consequently it may be qualified in various ways: it may be directional, situational or relational, because it is essentially qualitative, fluid and dynamic.
>
> (POS: 42)

Lived space is an elusive space, so elusive in fact that thought and conception usually seek to appropriate and dominate it. Lived space is the experiential realm that conceived and ordered space will try to intervene in, rationalize, and ultimately usurp. On the whole, architects, planners, developers and others, are, willy-nilly, active in this very pursuit.

(c) *Spatial practices* are practices which Lefebvre says 'secrete' society's space; they propound and presuppose it, in a dialectical interaction. Spatial practices can be revealed by 'deciphering' space and have close affinities with *perceived* space, to people's perceptions of the world, of their world, particularly with respect to their everyday world and its space. Thus spatial practices structure everyday reality and broader social and urban reality, and include routes and networks and patterns of interaction that link places set aside for work, play and leisure. Such practices embrace both production and reproduction, conception and execution, the conceived and the lived, and somehow ensure societal cohesion, continuity, and what Lefebvre calls a 'spatial competence' (p33). Still, cohesiveness doesn't imply coherence, and Lefebvre is vague about the precise manner in which spatial practices mediate between the conceived and the lived, about how spatial practices keep representations of space and representational space together, yet apart. One thing he's more sure of, though, is that there are 'three elements' here not two. It's not, he says, about a simple binary between lived and conceived, but a 'triple determination': each instance internalizes and takes on meaning through other instances.

Relations between the conceived–perceived–lived aren't ever stable and exhibit historically defined attributes and content. So it follows that Lefebvre's triad loses its political and analytical resonance if it gets treated merely in the abstract: it needs to be *embodied* with actual flesh and blood and culture, with real life relationships and events. But Lefebvre has experienced a lot, in life and as a researcher at the Centre National de la Recherche Scientifique (CNRS) and at the Institute de Sociologie Urbaine, to know that an unrestrained capitalism always and everywhere gives primacy to the conceived realm. Lefebvre knows too well, for example, that the social space of lived experience gets crushed and vanquished by an *abstract* conceived space. In our society, in other words, what is lived and perceived is of secondary importance compared to what is conceived. And what is conceived is usually an *objective abstraction*, an oppressive objective abstraction, which renders less significant both conscious and unconscious levels of lived experience. Conceptions, it seems, rule our lives, sometimes for the good, but more often – given the structure of society – to our detriment.

It ought to be pointed out here that Lefebvre's emphasis on 'abstract' has clear Marxian overtones: *abstract space* bears close resemblance to Marx's notion of *abstract labour*. But Lefebvre goes a lot further than Marx, for whom 'abstract' still operated mainly as a temporal phenomenon. Marx held that qualitatively different (concrete) labour activities under the bourgeois system got reduced to one quantitative measure: money. This standard becomes the common denominator for all things as commodity relations colonize everywhere and everybody; Marx coined this kind of labour, *abstract labour*, labour in general, and it is intimately tied to the law of value, to socially necessary labour *time*. Of course, in no way does 'abstract' imply a mental abstraction: it has a very real *social existence*, just as exchange value and the value form themselves have.

In like vein, abstract space has a very real social existence. It gains objective expression in different buildings, places, activities, and modes of social intercourse over and through space. But its underlying dynamic is conditioned by a logic which has no *real* interest in qualitative difference. Its ultimate arbiter is none other than value. Value, money (the universal measure of value), and exchange value (price) all, by hook or by crook, set the tone of the structural conception of abstract space. Thus value dictates underwrite conceived space. Here exigencies of banks, business centres, productive agglomerations, information networks, law and order, all reign supreme – or try to. Just as abstract labour denies true concrete labour, abstract space likewise denies true concrete qualitative space: it denies the generalization of what Lefebvre calls *differential space*: a space which doesn't look superficially different, but is different, different to its very core. It's different because it celebrates particularly – both bodily and experiential. Hence abstract space isn't just the repressive economic and political space of the bourgeoisie; it's also, Lefebvre suggests, a repressive male space which finds its representation in the 'phallic erectility' of towers and skyscrapers, symbols of force, of male fertility, and of masculine violence. Insofar as abstract space is formal, homogeneous and quantitative, it erases all differences that originate in the body (like sex and ethnicity) or else reifies them for its own quantitative ends. True differential space is a burden. It cannot, must not, be allowed to flourish by the powers that be. It places unacceptable demands on accumulation and growth.

V

In response, Lefebvre invokes the lived and perceived over the conceived. Or, perhaps more accurately, he seeks to transcend their factitious separation under modern capitalism. Here Lefebvre's earlier invectives on alienation and everyday life, first expounded in *Critique of Everyday life – Volume One* (written in 1947 and published in English in 1991),[14] enter the fray. There, Lefebvre stressed the *dialectical* nature of everyday life. It is the realm, he said, which is colonized by the commodity and so is shrouded in all manners of mystification. At the same time, it remains a primal site of meaningful social resistance. Everyday life thus becomes the 'inevitable starting point for the realization of the possible.'[15] Everyday life, in other words, internalizes all three moments of Lefebvre's spatial triad; it's a space – the only space – which brings 'wisdom, knowledge and power (*la sagesse, le savoir, le pouvoir*) to judgment.'[16]

The compartmentalization of different spheres of human practice has led to what Lefebvre calls the 'despoliation' of everyday life. What this begets, in turn, are human beings who experience, in Marx's words from

The Economic and Philosophical Manuscripts, 'one-sided individuality'. Overcoming one-sidedness, for Lefebvre, means recovering a 'genuine humanism' ('*véritable humanisme*'). Implied herein is a more wholesome personhood and spatial organization. Crucial therein would be a reconciliation between thinking and living, between the head with the heart, between theory and practice, between what Lefebvre sees with what he wants. The reassertion of the *spatialized body* in critical thought is a first step towards this reconciliation. So like the young Marx, Lefebvre affirms a humanist-naturalism: 'space', he says, 'does not consist in the projection of an intellectual representation, does not arise from the visible–readable realm, but it is first of all *heard* (listened to) and *enacted* (through physical gestures and movements)' (POS: 200).

Descartes and the Cartesian tradition began this severing, first carved out this debilitating disjunction between body and mind. And the shortcomings of the Cartesian Logos revealed themselves to Lefebvre in the growing technocratization and bureaucratization of social life. This programming has continued apace in both Europe and the United States since the late 50s to the degree that now all of us 'are being looked after, cared for, told how to live better, how to dress fashionably, how to decorate [our] house, in short, how to exist; [we] are totally and thoroughly programmed.'[17] The spatial embodiment of Logos is immortalized in the modern planning and New-Town movement, and Lefebvre witnessed the French version first-hand at Moureux, a New Town near his home of Navarreux, in Southwest France. There, he claims, 'modernity opened its pages to me'. 'Whenever I set foot in Moureux,' he laments, 'I am filled with dread.'[18] Here, in Moureux, as in other New Towns and suburban developments, Lefebvre believes that *ennui* has set in long ago. Here spontaneous vitality and creativity has been wrung out of its inhabitants and its spaces. Moureux's desert spaces perpetuated deserts of the mind. Here, in this ordered, enclosed and controlled world, Lefebvre felt that people are crushed by routine. No adventure or thrill now: everything gets dictated by the predictable mathematical exactitude of the Cartesian 'masterplan'. This is the world satirized so magnificently in Jean-Luc Godard's film *Alphaville* (1965). In such spaces, Lefebvre witnessed the end of romance and uncertainty. He heard, too, the death knell of the spirit.

Not so in Nararrenx. That is a picturesque medieval town, and Lefebvre can't hide his fondness for it. But his nostalgia here isn't backward looking. His is no Heideggerian atavistic model of authenticity and the Good Life. Lefebvre's nostalgia is firmly for the future and he uses the past only as a vehicle for going forwards and onwards, towards a higher plane of critical thinking and awareness. So Lefebvre's philosophy is no ordinary philosophy and he no ordinary philosopher. His is a *meta*-philosophy, he is a *meta*-philosopher. Such people, he claims, don't build abstract systems but instead 'aim to take from philosophy those ideas which are capable of arousing critical consciousness, ideas that are destined for a higher and at the same time more profound consciousness of the world in which we live.'[19] The goal of meta-philosophy is to 'uncover the characteristics of the philosophy that used to be, its language and its goals, to demonstrate their limits and to transcend them' (POS: 405). Meta-philosophy is an antidote: it attempts to surmount separations and sunderings, tries to unite speculative philosophy and critical theory with political action. It seeks critical and self-critical knowledge. It alone can expose phony transcendence in the name of real transcendence. Meta-philosophy can only be sanctioned in revolt, in individual and collective revolt – a revolt inside one's head and out on the street with others.

VI

Lefebvre is such a good Marxist here because his Marxism is so bad, is so heterodox. Marx's cult-hero was Prometheus. It was Prometheus, remember, who suffered because he stole fire from the Gods. It was he who appeared in *Capital* in the noble guise of the proletariat chained to capital. The Promethean principle is one of daring, inventiveness, and productivity. Marx appropriated it, was inspired by it. But Lefebvre is no Promethean. His ideals seem more akin to an Orpheus, maybe even to a Narcissus. Neither toiled or commanded but stood back, were unproductive, sang and listened to music. Lefebvre's radicalism revels in this and in 'Dionysiac life', a world of drink and feast, mockery and irony. This line tows no Party line. It moves in the shadows and remains on the outside. Nobody ever knew what Lefebvre was going to do next because Lefebvre never knew what he was going to do next either. This made for a reluctant and problematical Party man: he couldn't be trusted. His agenda rallies around erotic not rational knowledge; his Marxism is more about love and life than Five Year Plans. His Marxism sounds more like libertarian anarchism.

His is an ambiguous, festive, urban Marxism. Alongside Marx, we find Hegel; alongside Hegel, we find Freud; alongside Freud, we find Nietzsche. In Freud, Lefebvre found the unconscious; in Hegel, consciousness; in Marx, practical conscious activity; in Nietzsche, language and power. In the city, Lefebvre made space for all four. But there, in the city, unconscious desires and passions lay dormant, dormant beneath the surface of the real, within the *sur*real. There, Lefebvre reckons, they are waiting for judgment day, for the day when they can be realized in actual conscious life. And Marx is right: political–economic forces both shape and constrain these unconscious desires and passions. Economic forces inevitably suppress passion or else create new false passions – ones enveloped in all manner of mystifications and fetishisms.

But instead of mystification Lefebvre wants cities to release repression. He wants them to provide the means for 'free associative' expression, be arenas of *jouissance*, of intense sensual and sexual pleasure and excitement. He wants everyday life and everyday space – urban representational space – to be reclaimed for itself, reclaimed as a decisive 'lived moment'. Lived moments somehow have to *dis*alienate the everyday. They involve collective and individual rituals of resistance; they would be both serious – sometimes deadly serious – and playful; indeed, they should be luminous 'festivals of the people'.

Festivals are the veritable antithesis of bureaucratic domination and ordering. Festival day, Lefebvre says, 'is a day of excess (*le jour de la démesure*). Anything goes. This exuberance, this enormous orgy of eating and drinking – [has] no limits, no rules.'[20] Of course, it is rural festivals he's evoking here. These are, he thinks, associated with 'human joyfulness' (*réjouissances humaines*). They clearly left a lasting impression on him. Doubtless they activated involuntary memory, aroused childhood visions of paradise, tasted a bit like that Proustian *madeleine* dipped in tea. But the mature Lefebvre says festivals also 'tighten social links and at the same time give rein to all desires which have been pent up by collective discipline and the necessities of everyday work.'[21] True, he says, they always 'contrasted violently with everyday life.' But, and this is an important but, *'they were not separate from it.'*[22] On the contrary, festivals 'differed from everyday life only in the explosion of forces which had been slowly accumulated in and via everyday life itself.'

From the standpoint of classical Marxism, all this sounds pretty weird stuff. Yet Lefebvre sees no necessary contradiction between his ideas on festival and Marx's and Lenin's ideas on workers' self-management. Besides, 'revolutions of the past', Lefebvre claims, 'were festivals – cruel, yes, but then is there not always something cruel, wild and violent in festivals?'[23] Now, though, Lefebvre wants to project these ideas into a modern urban context, while giving them a few added twists to boot. Now, his vision posits the street as a kind of stage. The drama here might be epic or absurd or both, scripted by Brecht or Artaud or Chaplin or even Rabelais – who could tell? It's intended to be spontaneous, after all. In any event, street actions and demonstrations would become festivals of the city's citizens, and they'd try to forge together reproduction and production, residence and workplace, blend rent strikes with a general strike, all the time keeping hold of – but only just – a rambunctious carnivalesque spirit.

These ideas formed the lifeblood of the May '68 protests, and Lefebvre lectured to many protagonists, including Daniel Cohn-Bendit. At the same time, his ideas complemented the subversive radicalism of the Situationists. For a while, Lefebvre taught Guy Debord and worked with various members of the movement until 1963, when an acrimonious squabble caused a split. Then Lefebvre was accused – probably falsely – of plagiarism, of ripping off the Situationists' ideas about urban revolution and festival, specifically their interpretation of the 1871 Paris Commune. Both he and the Situationists had celebrated the Commune as an incomparable 'spatial revolution'; Lefebvre called it 'the only realization of revolutionary urbanism to date'. Its issues, he said, were territorial and urban; the Communards, he said, spoke the language of the everyday, demanded freedom and self-determination, destroyed symbols of bourgeois power and authority, occupied the streets and shouted and sang and died for their 'right to the city'. This, Lefebvre thought, was 'the city's grand and supreme attempt to construct itself as the measure and norm of human reality'. Ninety-seven years later, this manifesto had been re-enacted on Parisian streets.

The Commune prefigured the heady days of May. But the French Communist Party had denounced the May street actions; Lefebvre praised them, criticized them, tried to understand them.[24] For him, it was a momentary realization of the possible: 'imagination had seized power'. For a while, Paris existed as an island of liberated differential space in a sea of abstract space. Therein lay its strength as well as its weakness. Lefebvre's Marxism and anarchism scuffed up against each other in a creative, though problematical, tension. In both 1871 and 1968, we had a new kind of upheaval: an urban revolution, a reclamation of space for itself, a space and time for human development. It didn't last long. The protagonists were famous for fifteen minutes. In 1968, they were the children of Marx and Coca-Cola. Both upheavals provided a glimmer of vindication: Lefebvre, the socialist in space, was clearly on to something.

VII

Nowadays, we can still bring Lefebvre into our own cities and into its spaces. His urban visions still have a lot to tell us. His ideas remain a vital point of reference for any contemporary discussion on the future of the city. His thoughts about festival equally have a surprising resonance. One only has to glance around the political landscape in Britain now to witness groups such as 'Reclaim the Streets' blending direct action with quasi-anarchistic carnival. Over the last couple of years, such activity has occupied public spaces in north, south and central London, danced and shouted in the street, united men and women and children from all backgrounds, brought traffic to a standstill, and demanded pedestrians' right to the city. Like Lefebvre, these people have a keen sense that cities should not only be fun places to live in, but that urban politics can be a whole lot of fun as well. Much the way Lefebvre did with their 1968 forebears, the imaginative power of these kinds of protest needs to be understood, harnessed and channelled into a meaningful and coherent radical politics – especially given their amazing capacity to politicize young people alienated from ballot-box Parliamentary politics. Clearly, academics and urbanists and everybody else concerned about the fate of our cities can help out here. There is plainly still plenty to do to make our cities exciting as well as liveable, aesthetical as well as ethical, ordered as well as disordered, managed yet somehow spontaneous. But for us in the academy, for us who write about cities and space in the public realm, in books like this one, we can help a lot. But only if we bring Lefebvre's ideas a little closer to home.

For we scholars and intellectuals who operate in the academy now find our own space and lives increasingly under assault from the same commodification Lefebvre tried to demystify years ago. Our space – our academic space, in our department, on paper – is itself becoming (has become?) yet another *abstract space* of capitalism, and we ourselves are the perpetrators, are the formulators of new kinds of representations that are inexorably tied to relations of production and to the 'order' they impose. In our own daily practice, we deal more and more with abstract representations and codifications of society which are wrenched out of the lived experience of both ourselves and others outside the academy. Thus, when we write about daily life now, we should think very carefully about whose daily life we are talking about. When we write about space, we should likewise think about whose space we mean. When we write about radical intellectuals like Henri Lefebvre *et al.*, we should think about our own role as radical intellectuals.

Of course, we need Lefebvrian criticism and self-criticism. But now we should turn it on ourselves too, analyse our own daily lives and spaces. Better to bite the hand that feeds us than to remain toothless academic hacks. Lefebvre's maverick free spirit can still inspire us in our work and in our lives. And we need inspiration if we are to resist the growing rationalization and professionalization of university life, together with the lures of the academic marketplace, where promotion seduces commotion and where lies supplant in-your-face truth. So we might want to reclaim our own space at the same time as we help to reclaim, for its citizens, the space of our cities. Yet before imagination can seize power once again we firstly need to develop some imagination. We need to imagine a space that can free ourselves and our thought and our cities. That, for me anyway, has to be what 'thinking space' is really all about.

Notes

1 The whole series was transcribed and translated and later published in book form under the English title, *Adventures on the Freedom Road* (Harvill, London, 1995).
2 '"A Group of Young Philosophers": A Conversation with Henri Lefebvre,' in Levy, *Adventures on the Freedom Road*: 131.
3 See Remi Hess, *Henri Lefebvre et L'aventure du Siècle*, A. M. Métailié, Paris, 1988.

4 Castells, *The Urban Question*, Edward Arnold, London, 1977: 87 (original emphasis).
5 In a personal communication, Marshall Berman told me that throughout the 1970s he tried to no avail to talk various publishers into translating Lefebvre. 'Hopeless!' was how Berman described it.
6 Cited in Gailia Burgel *et al.*, 'An Interview with Henri Lefebvre,' *Environment & Planning D: Society and Space*, 5: 27–38. Castells's inert rendering of space in his research on urbanization and urban social movements also formed the bone of contention at his doctoral defence in Paris in 1976. Lefebvre was one of the examiners and apparently gave the young(ish) pretender a grilling. (I am grateful to David Harvey – who was present at the formalities – for sharing this tale with me.)
7 See *Regional Studies*, 1974, 8: 239–55.
8 Lefebvre received vital airtime in the journal of radical geography, *Antipode* which, like anglophone radical geography more generally, had been thriving since 1969. In 1976, for example, *Antipode* published Lefebvre's 'Reflections on the Politics of Space' (8, 2: 30–7). The essay later reappeared in Richard Peet's edited collection *Radical Geography* (Maaroufa Press, New York, 1977).
9 See Andy Merrifield, 'Lefebvre, Anti-Logos and Nietzsche: An Alternative Reading of "The Production of Space",' *Antipode*, 27, 3 (1995): 294–303.
10 The page references that follow refer to Donald Nicholson-Smith's English translation (e.g. Basil Blackwell, Oxford, 1991).
11 For a more in-depth discussion on the pluses and pitfalls of this, see my recent essay 'Between Process and Individuation: Translating Metaphors and Narratives of Urban Space, *Antipode*, 29 (October, 1997).
12 Karl Marx, *Capital I*, Chapt.1, Sect.4: 72 (International Publishers, New York, 1967).
13 Cf. Althusser's dictum: 'The ultimate condition of production is therefore the reproduction of the conditions of production' (see 'Ideology and Ideological State Apparatuses' in *Lenin and Philosophy*, New Left Books, London, 1971). In *The Urban Question*, Castells admitted that '[b]y urban system, I mean the specific articulation of the instances of a social structure within a (spatial) unit of the *reproduction* of labour power' (p237) (emphasis added).
14 *Critique of Everyday Life – Volume One*, Verso, London, 1991.
15 *Everyday Life in the Modern World*, Penguin, Harmondsworth, 1971: 35.
16 *Critique*: 6. See, too, *Critique de la Vie Quotidienne I*, L'Arche Editeur, Paris, 1958: 13.
17 *Everyday Life in the Modern World*: 107.
18 *Introduction to Modernity*, Verso, London, 1995: 119: 118.
19 Cited in 'Leszek Kolakowski and Henri Lefebvre – Evolution or Revolution (Interview)' in A. Naess (ed.) *Reflexive Waters – Basic Concerns of Mankind*, Condor Books, London: 202.
20 *Critique*: 202; and *Critique de la Vie Quotidienne*: 216.
21 *Ibid.*
22 *Critique*, 207 (emphasis in original).
23 *Everyday Life in the Modern World*: 36.
24 For specific details, see Lefebvre's short book, *The Explosion: Marxism and the French Revolution of May 1968* (Monthly Review Press, New York, 1969).

9

JACQUES LACAN'S TWO-DIMENSIONAL SUBJECTIVITY

Virginia Blum and Heidi Nast

Jacques Lacan is the twentieth-century psychoanalyst who, more than any other, has stressed the formative role of visual identification in human subjectivity. The very process of becoming what we call human, he theorized, happens in relation to images on which we model ourselves as though in a mirror. Certainly, his insights are critical to a twentieth-century subject of an image-centred society. With so much of western society growing up in front of televisions, finding not only our relationship to the world but the very structure of our identities through visual media, Lacan's emphasis on the degree to which our identities are shaped in relation to a two-dimensional mirror has much to tell us about the production of a twentieth-century subjectivity (see Ewen and Ewen 1992, Boorstin 1961, Postman 1986). Yet, to privilege the visual is we argue, to collapse, the subject into two dimensions, a collapse that Lacan reads as *inevitable* to subject-formation in general. Through a spatial analysis of this theorist, we illustrate not only the degree to which he is embedded in a two-dimensional account of identity-formation, but also how his universalization of the two-dimensional subject guarantees for Lacan an implacable bourgeois order of the nuclear family. Furthermore, because Lacan's theories, particularly that of the 'mirror stage,' have proved enormously influential among contemporary critical discussions of subject-formation and identification, we want to interrogate the *spatial* limitations of his theory in order to consider what geographers might take from (or make of) Lacan.

We begin by explaining Lacan's three registers of subject-formation, the real, the imaginary, and the symbolic, discussing them in terms of their spatial consequences for Lacan's notion of the human subject. Lacan registers loosely equate the real with the body, the imaginary with the ego, and the symbolic with the linguistic and cultural order which organizes us into social subjects. We are particularly interested in the unstated but nevertheless implacable limitations placed upon subjectivity – especially as the subject emerges in and through the spatial. Because the French sociologist, Henri Lefebvre, has criticized at length Lacan's spatial reduction of the subject, we draw on his objections as a springboard for our own critique.

Lefebvre criticizes Lacan for what he considers his privileging of the visual over the spatial in signifying lived experience and subjectivity. Lacan's subject, Lefebvre avers, is produced exclusively in the arena of images and language; consequently, the body is reduced to two dimensions. Originally no more than the effect of a two-dimensional image (indeed the image of an image, as we discuss below in our section on the mirror stage), which is then processed through the realm of the signifier (language), the three-dimensional body is collapsed by Lacan. The only 'third dimension' theorized by Lacan, the dimension that founds and mediates alterity, is the phallus – the signifier without a signified – which produces and sustains all meaning in the world without itself being implicated in the meaning-making machinery. The phallus is never located. Indeed, as Lefebvre indicates, it is detached (both literally and figuratively) from the body that needs to be suppressed in order for the Lacanian phallic economy to function. The clearest illustration of this suppression of the body is, as we show, in Lacan's insistence upon the distinction between the symbolic phallus and the bodily penis. We now turn to an exploration of Lacan's theory of the 'real' in order to understand how crucial it is for his system of signification to disavow the body.

The real mother

The 'real' is among Lacan's most elusive concepts.[1] While in some ways similar to the Freudian world of drives, Lacan's account of the real emphasizes the difference between subjects prior to and within the symbolic order. Thus, the real is more radically the register of the body in contrast to the symbolic order of mind. Lacan's implicit dualism 'The lack of the lack makes the real,' writes Lacan, 'which emerges only there, as a cork. This cork is supported by the term of the impossible – and the little we know about the real shows its antinomy to all verisimilitude' (1978, page ix). Bruce Fink's discussion of the role of the real in Lacanian register theory is possibly the best to date. Fink makes a distinction between two forms of the real, the first of which is presymbolic – in other words, the world of the prelinguistic infant. This is the order of the real experienced prior to subject-formation. For Lacan, becoming a subject means integration in the symbolic order. It means creating distinctions, differences, hierarchizing experience and phenomena in relation to which the subject is always carving a *separate identity*. 'Remember this, regarding externality and internality – this distinction makes no sense at all at the level of the real. The real is without fissure' (Lacan 1988, page 97). The organism's conscious separation from what is at first an undisrupted merger with the environment is a necessary part of becoming a subject. This will become an important aspect of the real as we go on to consider the implied link of the maternal and the real.

The second order of the real is the real from the perspective of the postsymbolic subject. This is the real that remains unintegrated in one's symbolic system, that haunts one from the margins of subjectivity. Fink argues that Lacanian psychoanalysis is largely about bringing the symbolic to bear upon this unintegrated real, a process called subjectivization. While Lacan would claim that the real is anything *but* a naive notion of 'reality', there are times when this order strikingly resembles Nature in the way that it both opposes Culture and is connotatively linked to the maternal – specifically, an infantile perspective on the maternal. To the degree that the real is the order of the presymbolic, subjectivity is achieved through transcending this infantile link to what is necessarily a maternal world in which the infant's experience is governed by primary caretaking of bodily needs. Luce Irigaray has criticized Lacan at length for just this double bind: there is a presymbolic order but there is nothing to be said about it because it is prior to and outside the order of the subject.[2] The presymbolic, Irigaray points out, is tacitly associated with the feminine.[3]

While Irigaray suggests that Lacan merely discounts feminine sexuality, rather, Lacan represents the maternal as an enormous threat to the subject – as a threat to there even *being* a subject.[4] The 'lack of the lack' that Lacan calls the real is a lost (and clearly fantasmatic) plenitude, a perfect continuity between the world's (mother's) providing and the infant's need. This is the deception, for Lacan, at the heart of the maternal function – her invidious lure that threatens to make psychotic the subject who fails to separate. The psychotic, unmoved by the father's Law (the paternal metaphor, what Lacan terms the Name-of-the-Father) is perhaps the subject living most clearly in what would be the second order of the real, refusing the symbolic order from his or her position *within* the symbolic order. Abiding by laws of the paternal order, namely, the incest prohibition and the consequent punishment by castration, ensures the child's (always assumed to be male) entrance into the symbolic chain of signification; indeed, all human signifying systems (e.g., kinship, juridical) collapse without this sustaining function of the signifier to harness the infant's desire and divert it from the mother. Alterity, then, is founded through renouncing the mother's body.

Such renunciation of the maternal body leads, as Lefebvre maintains, to the suppression (prohibition/exile) of the body itself from Lacan's psychical economy, this body whose 'blueprint' is the body of the mother. In the following passage Lefebvre points to Lacan's need to abject the maternal through representing the incest taboo as the cornerstone of civilization. Lefebvre writes:

[T]he prohibition which separates the (male) child from his mother because incest is forbidden, and the prohibition which separates the child from its body because language in constituting conscious-ness breaks down the unmediated unity of the body – because, in other words, the (male) child suffers symbolic castration and his own phallus is objectified for him as part of outside reality. Hence the Mother, her sex and her blood, are relegated to the realm of the cursed and the sacred – along with sexual pleasure, which is thus rendered both fascinating and inaccessible.

The trouble with this thesis is that it assumes the logical, epis-temological, and anthropological priority of language over space. By the same token, it puts prohibitions – among them that against incest – and not productive activity at the origin of society.

(1991, pages 35–6)

Indeed, a social system that takes prohibition as its founding moment demands, as Lefebvre points out, the subjection of the body to what is imagined to be a 'higher' order – that of the *law* – based in *language*. In Lacan's seminar on psychosis, he emphatically privileges 'the intervention of the order of speech':

The Oedipus complex means that the imaginary, in itself an inces-tuous and conflictual relation, is doomed to conflict and ruin. In order for the human being to be able to establish the most natural of relations, that between male and female, a third party has to intervene, one that is the image of something successful, the model of some harmony. This does not go far enough – there has to be a law, a chain, a symbolic order, the intervention of the order of speech, that is, of the father. Not the natural father, but what is called the father. The order that prevents the collision and explo-sion of the situation as a whole is founded on the existence of this name of the father.

(1993, page 96)

The importance here of the 'father' is that he arrives on the scene as the intervening third party between the mother and the child (assumed to be male) – which retrospectively becomes the model of harmony. But it is the phallus, a cut-off bit of the father (this phallus that comes to distract the mother's desire) that takes the place of what was imagined to have obtained in the mother–child dyad. The father as a social entity, specifically the father's name, oversees the whole order inasmuch as this split between the natural body of the father and his social form suppresses the ungovernable impulses of a polymorphously (incestuously) desiring body.

Typically, Lacan links bodies to women and mind to men – with a twist. In order for Lacan's social order to function rationally, to sidestep the psychosis threatened by the mother's body, subjects must come into social being through identification with two-dimensional images, first through the mirror stage (the imaginary) and then through the order of language (the symbolic). One wonders, given the connection between the real, psychosis, and incest, if the real is where incest can happen? More importantly, if the prohibition placed on the mother's body is enforced through paternal law, is the real the place of the maternal? While Lacan's real certainly cannot be directly equated with the undifferentiated state of wholeness experienced by the infant in relation to the mother – a theory propounded by diverse psychoanalysts[5] – one wonders why it is that the failure of the incest taboo (the cornerstone of the symbolic order), winds up stranding the subject in the real – with his mother? Why is it that the mother (here presented as the dangerous lure into insanity) is aligned against the Law, against the Symbolic order, against Culture? Unless she is Nature? Might she even be – Death?[6]

Lacan's maternal constitutes the primary threat to the Symbolic order, the maternal that is the realm of the real. It is his feminization (and maternalization) of the real that ultimately leads to a dis-embodied account of the spatial and what he will formulate as the symbolic order.

Two-dimensionality and the 'threshold of the visible'

For both Lefebvre and Lacan, psychical and corporeal separations from the maternal realm are what become the basis for constructions of difference and subjectivity (see Blum and Nast 1996), both theorists depending on tropes of the 'mirror' to explain how this separation occurs. For Lacan, subjectivity is precipitated during 'the mirror stage', a transitional period that typically occurs when an infant is between six to eighteen months old. The characteristics of this stage are briefly as follows: a child recognizes and hence situates itself through identifying with an image outside itself, 'out there'. Lacan calls upon the trope of the mirror image to stand paradigmatically for any image with which the infant identifies. The point is not that it is *in* a mirror that the child finds the contours of its self, but rather that the process of subject-formation is a *mirroring* one. This is for Lacan the founding model for an illusory totality of a 'self'; the mirror-image is the *ideal* or totalized ego whom the infant longs to *become*. Lacan calls this internalized mirror image an *imago* to emphasize the fantasmatic quality of the relationship between a perceived image and a perceiving infant who is not yet a subject. It is only consequent to the child's psychical internalization of the image that the ego is founded. Subjectivity is spatially and ontologically *decentred*; the subject is shaped literally from the *outside in.* Crucially, both subjectivity and alterity, which are mutually constitutive, happen in the child's relationship with its *own* image.

The sense of alterity established through the mirror-image is, moreover, a complexly negotiated one. On the one hand, it is fundamentally based upon misrecognition: the mirror-image 'out there' is in one sense 'me'; on the other hand, it *is* 'out there' and therefore not-me. Of equal importance is the fact that the image 'out there' produces a Gestalt of wholeness (the image is a coherent unity) that exceeds the infant's feelings of bodily awkwardness and fragmentation. At the same time, this wholeness is what makes the infant aware that it is fragmented (it is not whole, yet) and thus in situational rivalry with its mirror-image.

> The mirror stage is a drama whose internal thrust is precipitated from insufficiency to anticipation – and which manufactures for the subject, caught up in the lure of spatial identification, the succession of phantasies that extends from a fragmented body-image to a form of its totality that I shall call orthopaedic.
>
> (Lacan 1977b, page 4)[7]

By 'insufficiency', Lacan is referring to the felt motor uncoordination of the newly crawling infant; by 'anticipation', he means the glimpse of the 'future' the infant gets in the coordinated fluid mirror-image with which it identifies. What is noted but nonetheless left underdeveloped by Lacan is the degree to which mirroring entails a number of *spatial* disjunctions: First, I (here) am *there* (in the mirror-image); *There* (the mirror-image) is *here* (ego). Second, the image itself is two-dimensional and accordingly founds a two-dimensional subject. Third, the image is a symmetrical inversion of the spectating body. That the mirror allows the spectating child to occupy both positions at once means that the distance, differences in dimensionality, and asymmetry between subject and image are fantasmatically collapsed. Connected to the mirror-stage spatial disruptions is the distinction Lacan makes between the eye and the gaze, which importantly structures gender identity as yet another spatial break.

This formation of the 'I' through the relationship with the mirror image is, as Lacan puts it, 'the threshold of the visible world'. Following what Freud (1923) calls the body-ego, Lacan agrees that the body is the blueprint for what comes to be the ego. This body-ego is then internalized as what is felt to be the 'whole self' but is really just, according to Lacan, the ego – a limited psychical agency forged through the various misrecognitions of the mirror stage. Prior to the emergence of the 'I' as body-ego and then ego, the visible world is undifferentiated. The visible world is then differentiated from the 'I' only as a secondary effect of mirroring – the narcissistic pattern of subject formation. In other words, the 'visible world' is constituted as such through its subordination to the subject's emergence qua subject through narcissistic forms of identification.

Does this mean, then, that the visible world, subordinated as it is to the originary process of misrecognition, is itself merely the screen for the subject's projections? Let us consider what Lacan has to say on the topic:

> What is the image in the mirror? The rays which return on to the mirror make us locate in an imaginary space the object which moreover is somewhere in reality. ... Suppose all men to have disappeared from the world. I say *men* on account of the high value which you attribute to consciousness. That is already enough to raise the question – *What is left in the mirror?* But let us take it to the point of supposing that all living beings have disappeared. There are only waterfalls and springs left – lightning and thunder too. The image in the mirror, the image in the lake – do they still exist?
>
> (Lacan 1988, page 46)

Lacan goes on to insist that they do exist and they owe their existence to optical devices; this insight will lead to his distinction between the eye and the gaze. He writes: 'Despite all living beings having disappeared, the camera can nonetheless record the image of the mountain in the lake, or that of the Café de Flore crumbling away in total solitude.' Even if no human subject sees any of this, Lacan asserts that it all 'exists' because of the invention of recording instruments – cameras. In his remarkable discussion of the proliferation of visual apparatuses in the nineteenth century, Jonathan Crary (1990, page 136) maintains that the 'new camera [was] an apparatus fundamentally independent of the spectator, yet which masqueraded as a transparent and incorporeal intermediary between observer and world'.

Confirming Lacan's insight, Crary points to the camera's deployment as an independent spectator. The camera pulls together the field of vision through its unifying eye. We ascribe to it consciousness. The bodies the camera admires become the bodies we want to be – the bodies with which we identify. Thus, we become, as Lacan puts it 'a picture' (1978, page 106). Subjectivity happens in relation to a gaze that emanates from this effect: 'It is through the gaze that I enter light and it is from the gaze that I receive its effects. Hence it comes about that the gaze is the instrument through which light is embodied and through which – if you will allow me to use a word, as I often do, in a fragmented form – I am *photo-graphed*' (1978, page 102)

Arguing that each subject perceives only part of the whole, Lacan makes a distinction between the eye and the gaze (1978). In the field of the visible, the eye is an organ that locates the specificity of each subject's anatomically restricted perspective in contrast to being *seen* (the gaze). Imagining oneself through others' field of vision is constitutive of the gaze. This distinction between what one can and cannot see of *oneself*, the experienced inadequacy of one's own limited visual field in relation to the entire field of vision where the remainder of one's identity *happens*, is increasingly important to a culture overly invested in control and empowerment through what is visually accessible. As Michel Foucault has observed (1980, page 153), there were significant transformations in patterns of social control in the seventeenth and eighteenth centuries, the emergence of systems of visual surveillance and increasing concern with opening up to inspection formerly dark places. The ability to 'identify' or *name* deviant members of the society is inextricably linked to identity-formation as itself a process of identification. Louis Althusser has decried at length the process whereby individuals become subjects of and to ideology. He calls this process 'interpellation'; individuals are in a sense 'hailed' or identified by the culture which both transforms us into subjects (by naming/locating us) and 'subjects' us to the practices of the culture that *can* name and locate us. To identify the self is to both identify and dis-identify with others (Althusser 1971).

This subject who is in a sense severed from his/her own identity (always behind or outside the visual field) is a subject conceived only in two dimensions. The eyes, then, become the organ of identification and the gaze becomes the process of subjectivization. Like the television screen that has no 'depth' or the mirror image that leaves any animal indifferent once it realizes there is nothing 'behind' it, Lacan's subject is always a two-dimensional screen in search of depth through identifications with other 'screens'.

While the eye is the organ of sight, the gaze emphasizes the process of making and becoming visible. The *not-seen* in psychoanalytic theory becomes inextricably bound up with its theory of castration because in the case of female genitals, what is not-seen is construed as absent or missing. According to Freudian theory, in reaction to seeing the penis, the little girl believes she *is* castrated and, similarly, as a consequence of seeing female genitals, little boys *fear castration*. From Freud (1923, 1931, 1933) on, it has been taken for granted by the psychoanalytic community that the little boy's horrific encounter with what is 'missing' from the body of the female is ever after internalized as castration anxiety. Subsequently, whatever is construed as missing from the visual field is metonymically linked to castration. In other words, the not-seen of every body is projected onto the female body alone and fetishized as her genitals. Lacan emphasizes the omission of female sexuality from the linguistic (in contrast to the anatomical) arena. Thus, in his famous *Encore Seminar* on female sexuality, he writes: 'There is a jouissance that is hers (*à elle*), that belongs to that 'she' (*elle*) that doesn't exist and doesn't signify anything' (1985a, page 74).

Because, in language woman only exists in relation to man without any primary sexual identity of her own, then she is not herself a linguistic entity. Thus the putative invisibility of her sex organs becomes the silence of her sexuality. It is through his emphasis on the linguistic that Lacan can pretend he is not making the same old assumptions about the visibility/absence of the penis. Importantly, it is through his professed indifference to the three-dimensional body that he freezes for all time Freud's masculinist representations of sexual difference.

Pile (1996, page 128), building upon Lacan's differentiation of the eye from the gaze points to the spatial consequences of the split and the spatial nature of castration anxiety and visual basis of sexual difference. As he writes:

> The split between the eye and the gaze is not achieved without cost, for it is instituted by an anxiety – the threat of castration (which both correlates with Freud's account of the child's understanding of the anatomical differences between the sexes and also explains the predominance of the phallus). The gaze slides over this anxiety and escapes consciousness. In this spatial topography of the mind, the gaze always lies behind or beyond understanding – once more evoking the idea that the subject's relationship to its specular image is founded by a profound failure-to-recognize its place.

Pile thus offers an important analysis of how the distinction between the eye and the gaze constitutes human subjectivity as spatially divided.

Nevertheless, in other psychoanalytic accounts of infancy,[8] one of the earliest forms of anxiety is separation anxiety from the mother – not castration anxiety. What is left out of Lacan's story of the mirror-stage child (and Pile's spatial elaboration of the gaze/eye split) is the mother's body as *the* place the child leaves behind, the place that made it whole before it became what Lacan calls 'a body in pieces', 'sunk in motor uncoordination' that looks to the mirror image to put it back together again. Thus, the original 'all-seeing' place is the *mother* – even though she may not be looking, even the very possibility of her not looking opens in the child the related possibility of the loss of her love. This is because the 'look' of the mother is what signifies her care and attention. Always having her look is necessarily called into question once the space between them is opened up. Not only is the mother's potential indifference spatially registered, if she stops looking, the child may cease to exist – as the object of her desire.

The child finds its own 'place', then, through bodily and visual displacement from the mother's body. Its movement away from the mother's body is what spatially constitutes it as not where it thought it was. In its progress away from the mother's body, she becomes one place among many at the same time that the spatial emerges as such for the infant.

We might say, then, that the subject emerges through a *spatial* 'fall' into difference and that it is this fall that allows a child to assume a personal identity. This 'fall', embodied in the child's crawl away from the mother, is the story *between* the separation from the mother and finding 'oneself' in the mirror image.

What Lacan omits from his story of the body-in-pieces that finds its antidote in the mirror Gestalt, is that body takes shape and finds itself *as* a body along the trail between the mother's body and its point of arrival – the end place it reaches before it returns, or before she rushes over to retrieve it. Each stage along the crawl away, then, founds a relationship to space that is at once formative (a new place 'discovered' by the body) and provisional (it will fall behind as the next step is achieved). It is not until the child discovers itself as a perceiving subject through inhabiting the 'I' that the mother's look is no longer required to sustain it. The child cannot truly separate from the mother until it takes over its own look.

Our discussion brings us to conclude that Lacan conflates castration anxiety (visually registered) with separation anxiety (spatially registered). It is precisely this conflation that allows Lacan to collapse the spatial trajectory of the crawling infant into the mirror: The collapse denies the fact that the child finds itself as much in the journey as it does in the culminating mirror image. To convert the bodily experience of separation into the visual register of absence and presence (of both the mother and the phallus) is not only to efface the traces of the child's route away from the mother's body and into the world, it is ultimately to pretend that there is no body to leave, no scene to be lamented.

We urge readers to rethink Lacan's developmental scenario in the light of an embodied and three-dimensional subjectivity. The spatial is founded in the very route from the mother's body to the mirror (to be as concrete as possible). This is a subject whose emergent identity exceeds the visual. Yet what happens in Lacan's mirror stage is that the experience of a body-in-the-world is supplanted by two-dimensional images of space and spatial relationships. Why might Lacan omit such crucial (and embodied) transition? To explain, we return to Lacan's account of the mother – but this time, the mother as she functions in the mirror stage.

Heterosexuality and tourism

For Lacan, the mother plays a key role in the mirror stage in that she is assumed to be the most consistently proximal caregiver of the child throughout infancy and therefore the child's primary or dominant 'other' (Lacan 1977c). In this sense, the child is said to first know or negotiate its world through its mother, a dyadic way of knowing that Lacan calls *imaginary*.[9] This 'other' is not, however, unmediated and therefore experienced naively outside paternal law and language. Rather the mother-figure is seen as already bound, defined and structured by the Law, for which reason she is also (m)Other (see Bowie 1991, page 138). As Lacan puts it:

> The fact that the Father may be regarded as the original repre-
> sentative of this authority of the Law requires us to specify by
> what privileged mode of presence he is sustained beyond the
> subject who is actually led to occupy the place of the Other,
> namely, the Mother
>
> (Lacan 1977a, page 311)

It is the (m)Other, then, who introduces the world of the Father to the child by bringing paternal rules and regulations to the child performatively.

There are several objections to be levelled against Lacan's account of child development. First, Lacan only draws upon bourgeois and hetero-sexualized positions of the nuclear family. That is, bourgeois heterosexuality (or at least that version that existed from the nineteenth to mid-twentieth centuries) is what sustains and informs Lacan's theorization of psychosexuality. In so restricting himself in time and class, Lacan theoretically suppresses questions about the social and political origins of masculinity. As John Brenkman (1993, page 57) puts it in his brilliant account of how psychoanalysis has suppressed the social origins of masculinity:

> The male child encounters the law limiting his desire in the voice
> of the command and the symbols designating the father as *castrator*
> and *law-giver*. But it is through the process of recognizing himself
> in the father that he learns masculinity and heterosexuality. His
> relation to his mother *becomes* Oedipal only as he is socialized into
> masculinity and heterosexuality.

Moreover, as Henri Lefebvre points out, Lacan's notion of mirroring is both aspatial and de-corporealized. Explicitly countering Lacan, Lefebvre claims that mirroring is fundamentally not about a disembodied ego passively locating itself in some two-dimensional, apolitical mirror-surface. Nor, Lefebvre claims, is it about a disembodied ego serving as a tabula rasa onto which image-ideals are introjected passively, narcissistically. Moreover it is not only about the human form and dyadic, specular relationships between two individuals or between individual and image.

Lefebvre suggests instead that mirroring is an active process that obscures the material and political world. Mirroring, for Lefebvre, is ultimately a social practice that requires substantial labour. Collapsing the world into a spectacle for the sake of uninterrupted viewing is only possible through a privileged disavowal of material distractions and engagements. In the end, mirroring is about 'self-deception', the word deception implicitly suggesting critique and some sort of accountability. Lefebvre lays out his arguments through calling upon Lacan's notion of the mirror, only to re-situate the mirroring process politically and spatially, re-constituting it in dimensionally and materially different ways. Lefebvre begins (like Lacan) by presenting us with a subject looking out upon a 'mirror'; unlike Lacan, though, the mirror is not a human form or image, but a complex phys-ical and social landscape. The subject, looking out upon the landscape, imagines that it is s/he who has created it, projecting onto the world her or his own fantasmatic coherence. As Lefebvre writes, a landscape 'presents any susceptible viewer with an image at once true and false of a creative capacity which the subject (Ego) is able, during *a moment of marvellous self-deception*, to claim as his own' (1991, page 189, our emphasis).

Contrary to Lacan, then, Lefebvre presents us with a subject in mate-rial relationship with a world; additionally, the subject is self-deceiving. In this sense, mirroring 'cannot be reduced solely to the surprise of the Ego contemplating itself in the glass, and either discovering itself or slipping into narcissism' (page 189). The passage quoted above also suggests different degrees of human susceptibility: not everyone is equally deluded.

What is important, then, is that Lefebvre privileges self-deception in his description of mirroring, later discussing how self-deception is socially and politically facilitated. He also identifies the psychical and material effects of such deception: upon specular introjection of the landscape the ego's sense of power and coherence is shored up. More importantly, having reducing the world to a 'picture', the subject is seduced into believing that it alone has created the landscape as its own 'work':

> A landscape also has the seductive power of all pictures, and this is especially true of an urban landscape . . . that can impose itself immediately as a work. Whence the archetypal touristic delusion of being a participant in such a work, and of understanding it completely, even though the tourist merely passes through a country or countryside and absorbs its image in a quite passive way. The work in its concrete reality, its products, and the produc-tive activity involved are all thus obscured and indeed consigned to oblivion.
>
> (*ibid.*, his emphasis)

Thus, Lefebvre makes it clear that a subjectivity formed through the continual effacement of the work (read labour, bodies, and places) is, first, achieved through disengaged viewing, and second, inherently violent. Moreover, by pointing to the labour of others, Lefebvre points to other subjectivities beyond the spectating subject which may, in the end, contest the fact that they are 'consigned to oblivion'.

In any case, Lefebvre metaphorically connects spectating to the work of tourists. Even so, complete reduction of the material world to spectacle is, ultimately, impossible: not all persons can afford to be tourists; and besides, complete bodily disengagement from, or 'passing through', a place can never be perfectly achieved. Bodies and landscapes remain and, as such, threaten to disrupt spectacular 'delusions'. Lefebvre draws particular attention to the body-in-excess when he tells a tale about 'Ego':

> When 'Ego' arrives in an unknown country or city, he [sic] first experiences it through every part of his body – through his senses of smell and taste as (*provided he does not limit this by remaining in his car*) through his legs and feet. His hearing picks up the noises and the quality of the voices; *his eyes are assailed* by new impressions. For it is by means of the body that space is perceived, lived – and produced.
>
> (1991, page 162; our emphasis)

Again, key phrases such as 'provided he does not limit this by remaining in his car' and 'his eyes are assailed' suggests a level of accountability and complexity in worldly engagements that Lacan's theorizing of the mirror cannot contain. Narcissism, which may work in some instances, is always assailable by the materialities in which we find ourselves and by other subjectivities. For Lefebvre, then, what is lacking in Lacan's analysis of mirroring is some recognition of material, political, and spatial forces that exceed the visual domain (1991, page 185).

False hope

Lacan attempts to depict a three-dimensional order beyond the mirror stage through his theory of the phallus. It is worth exploring at length Lacan's account of the phallic symbolic system in order to have a full sense of his gendered spatial order of human subjectivity. We argue that Lacan's two-dimensional account of subjectivity is intimately bound up with not only his representations of gender difference but also with his projection onto the maternal body of all the perils of a fall from the symbolic, rational, paternal order. Ultimately, we show both that the mother is in the place of the body (the real, the unmediated relation between the drives and the world's presenting) and the locus of desire. The confusion occurs over whether she is herself the scene, source and object of embodied desire or whether she functions merely as the pattern for the child's emergent desire. The question is, then, whether the mother and desire are in a metaphoric relationship or a metonymic one. Is she the model for desire or its avenue? How do we reconcile her desire for the phallus along with her body being the original object? Where Lacan diverges from Freud in this instance is especially noteworthy.

195

In Freud's version, the mother is the original object of desire. He, too, requires that this incestuous object be interdicted in order for human culture and civilization to thrive. Similarly, the drives (the body) endanger the subject's entry into culture. From a Freudian perspective, one must subordinate the pleasure principle to the reality principle. Moreover, the mother in her role as primary caregiver is the one responsible for activating the body's erogenous zones. While Freud insists that bodily drives cannot be altogether suppressed without risking severe neurosis (civilized morality and . . .), Lacan implies that the body itself is an ever-present risk to subject-formation. Indeed, much of Freud's work points to the dangers of repressing instinctual impulses, and his patients were more often than not the victims of the discrepancy between bodily drives and cultural imperatives (1905; Dora). Moreover, Freud represents the child's desire for the mother as a crucial stage en route to becoming a sexual subject. In marked contrast to Freud, Lacan positions the link to the mother as the most primitive layer of the inchoate subject. Lacan merges the originary (real) mother of the infant (associated with plenitude) with the later mother (imaginary) of emergent desire. It is this confusion that locates the mother as the primary threat to the subject.

Moreover, if, as Lacan says, the child progresses from wanting to be the phallus for the mother to wanting to have the phallus – is it not the case that the mother as well is shifted in the child's psychical economy from *having* the phallus (being uncastrated) to *being* the phallus – in the form of the female body that always masquerades as the phallus (Lacan 1977a)? The unmediated identification with the image/mother in the imaginary register ruptures with the intrusion of the third term, the phallus, signifier of the division and desire that gives rise to subjectivity. This is a long process that, as we will show, originates with the cognitive and spatial separation of the child from the mother, continues through the mirror stage, and is subsequently worked out substitutively through language. It is the entrance into language that founds the symbolic domain in the subject. Most importantly, the phallus as the signifier that both arises from and represents alterity (the separation from the mother who is then 'other' and the recognition of sexual difference) is incorporated into the very structure of signification for Lacan.

For Lacan, the phallus is the signifier of the desire for that which will put an end to desire – which is ultimately unattainable by the subject. It is because nothing will suffice that Lacan calls the phallus 'a signifier without a signified'. Desire for the phallus takes the human subject for a ride along the chain of linguistic displacement and substitution, all objects substituting for the originary and impossible one. He writes that the phallus 'designate[s] as a whole the effect of there being a signified inasmuch as it conditions any such effect by its presence as signifier' (1985b, page 80). It is because it is not real (Lacan claims that it cannot be identified with an organ) that the effects of the phallus are all the more profound and unshakable. As Jacqueline Rose (1985, page 49) writes,

> For Lacan, men and women are only ever in language. ... All speaking beings must line themselves up on one side or the other of this division [which occurs in relation to the phallus], but anyone can cross over and inscribe themselves on the opposite side from that to which they are anatomically destined. It is, we could say, an either/or situation, but one whose fantasmatic nature was endlessly reiterated by Lacan.

The bodily penis is irrelevant, Lacan tells us, to the overwhelming order of the symbolic phallus. While Freud seemed to locate all human impulses in the genitals, Lacan dissevers (literally and figuratively) the organ from the body – thereby paradoxically disembodying the very instinctual subject Freud had elaborated. This very split between the real and symbolic penis/phallus should warn us just how distasteful Lacan finds the 'real' body. How clever of him, ultimately, to disembody the most embodied aspect of Freudian theory.

At one level, then, the body seems to be incidental to, and at times altogether eliminated from, Lacan's account of the emergence of sexual difference and the Symbolic order. To some degree his locating of identity in language itself (away from the body) must be understood as standing in opposition to a flourishing mid-fifties post-Freudian biologism. Yet, Lacan's anti-biologism, his implicit condemnation of the prevailing insistence upon a corporeal innateness and inevitability of masculinity and femininity,[10] leads him to the opposite extreme: He locates subjectivity entirely in language – of which the body becomes merely an effect. Lacan's assertion that the Symbolic order precedes the human subject means, then, that subjectivity comes at the price of shedding the body altogether. Lacan's omission of the body is not altogether complete, however, in that he depends upon essentialized anatomies to sustain much of his theory. Although, as Jacqueline Rose avers, he would claim that we transcend our anatomical destiny through language, his language paradoxically seems chained to 'anatomical' difference. Lacan's famous statement (1985b, page 138) that sexual difference only takes place 'in the case of the speaking being' is thus countervailed by his investment in the same old phallocentric story of organ difference and bourgeois heteropatriarchal child-rearing practices.

Despite his expressed effort to disengage anatomically gendered bodies from their traditional familial roles, his theory depends on positioning women as primary caretakers and identifying men with the extra-domestic sphere. Moreover, the maternal is characterized as that which the child needs to escape in order to achieve subjectivity; escape is facilitated through pursuit of the phallus. The phallus, then, is what draws the child out of the maternal swamp to the loftiness of the 'outside' symbolic paternal order. Harkening back to a very literal psychoanalytic understanding of penis envy, Lacan casts the mother as the originary desiring subject (of the penis) in his economy of psycho-sexual development. The dynamics of desire in the mother–child relation underscore not only the mother's desiring subjectivity but her pivotal role in producing a desiring subject, the child. Lacan emphasizes that insofar as the order of desire is processed through her, the mother occupies the place of the Other:

> the fact that the Father may be regarded as the original repre-
> sentative of this authority of the Law requires us to specify by
> what privileged mode of presence he is sustained beyond the
> subject who is actually led to occupy the place of the Other,
> namely, the Mother.
>
> (Lacan 1977a, page 311)

The child is said to turn away from the mother when it sees that it is no longer the centre of her world, that it alone is incapable of completing the world of the mother. Lacan tells us that the child's relation to the mother, constituted as it is 'by the desire for her desire, identifies himself with the imaginary object of this desire in so far as the mother herself symbolizes it in the phallus' (1977c, page 198). Whether the mother's attention is directed to the father, another child, a job, it is the structure of the divergence of her desire from the child that is crucial; it precipitates the child into subjectivity at the same time that subjectivity is characterized by the very split from which it emerges. These object relations are illustrated by Lacan in Figure 9.1:

> Here's the situation as I sketch it out – here, the imaginary, that
> is, the desire of the phallus on the part of the mother, there, the

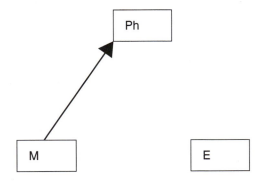

Figure 9.1 'La Nostalgie du Phallus', after Lacan
Note: M = Mother, E = Child; Ph = Phallus

198

child, our centre, who has to make the discovery of this beyond, the lack in the maternal object. This is at least one of the possible outcomes – from the time the child finds a way of saturating the situation and comes out conceiving the situation as possible, the situation turns around him.

What do we find effectively in the fantasy of the little girl, and also of the little boy? Inasmuch as the situation turns around the child, the little girl finds then the real penis there where it is, beyond, in he who can give her the child, identified, Freud tells us, in the father.

(Lacan 1994, page 202)[11]

The child's desire, then, is inaugurated by its recognition that it is not the sole object of the mother's desire, a recognition that produces a profound sense of loss. This 'something' other-than-the-child that the mother desires is what Lacan construes as the phallus – that object which (if the child possessed it) would restore the original relation of wholeness with the mother. What the mother's desire reveals to the child is that she is non-phallic (castrated), her 'lack' introducing the possibility of 'absence' into the child's psychical economy. The resulting body-image of the child is thereby constitutionally and continually threatened with fragmentation and loss. The child consequently strives to recover its central place in the mother's life and to overcome its loss by having what the mother wants, namely the phallus. As Lacan (1985a, page 83) writes,

If the desire of the mother *is* the phallus, then the child wishes to be the phallus so as to satisfy this desire. Thus the division immanent to desire already makes itself felt in the desire of the Other, since it stops the subject from being satisfied with presenting to the Other anything real it might *have* which corresponds to this phallus – what he has being worth no more than what he does not have as far as his demand for love is concerned, which requires that he *be* the phallus.

The phallus is at once the term of division and transcendence: it is that which leads the mother's desire away from the child and that which holds out the possibility of a reforged connection with the maternal.[12] The connection must be pursued indirectly through substitution, through the order of language. Significantly, it is only the boy's body which is indelibly marked as capable of delivering or giving up a phallus-gift, the penis.[13] This corporeal positioning of the gift once again undermines Lacan's putative distinction between corporeality and language, the penis and the phallus. As we can see, his psycho-sexual framework is structurally grounded in the heterosexual paradigm of the mother–son dyad, a potentially transgressive unit that is contained only by the incest taboo. This is because it is only the boy who is capable of action (in the form of giving), impressing (both in the sexual act and more generally impressing himself upon the world in the form of production), and signifying. Such reasoning suggests that only subjects who occupy the masculine position are capable of primary acts of signification.

At the same time, the feminine subject is informed through paternal law that she is 'not-all' in relation to the phallic function – she has no object to give, nothing to impress upon the world, nothing to inscribe. She enters the symbolic domain, but as a subject of the law, able only to recycle the symbolic domains created by men. She is not capable of primary acts of signification.

In the end, Lacan's universalizing, decorporealized, and culturally decontextualized account of psycho-sexual development forestalls political change through denying the political embeddedness of sexual identity. Not only does he theorize desire from within the bourgeois nuclear family, thus limiting familial and social diversity, he also grounds alterity and desire within naturalized anatomical difference. He consequently binds our options to functionalist (structuralist) reiterations of the patriarchal same. It seems that the only difference Lacan can recognize is that which distinguishes the child from the mother. Forcing upon us a script that casts the phallus as the only means of escape from the maternal, his 'family romance' celebrates the Culture-father releasing children from the predatory grasp of the Nature-mother. Indeed, we might argue that inasmuch as Lacan's phallus stands for an escape from the maternal, the mother is implicitly equated with the materiality of the body itself: she becomes 'body'. As a result, all acts of signification in the Lacanian schema reproduce the originary separation from the maternal/body; we speak our way out of the body. Language is thus simultaneously a release from the imperatives of the body and the grasp of the mother, which are metaphorically tethered through Lacanian theory.

Conclusions

Jacques Lacan's analysis of human subjectivity is flawed in that it, first, depends (like psychoanalysis generally) upon normative heterosexuality for it to make sense and, second, because it spatially reduces the world to two-dimensional signs of itself. Nonetheless, Lacan's work does describe the effects of modernity: commodification; the disembodiment and spatial reduction of life to images and language (two-dimensional signs); and the social prescription of normative, heterosexual, nuclear family life. As such, his work usefully describes how mirroring and the phallus structure sex and 'space'. Yet it is only Lefebvre who locates and describes the spatial violence behind modernity's oppressions. Even though Lacan might disavow his part in reproducing patriarchal relations, claiming that he is merely describing what exists, his account from the beginning depends upon a closed system of mirror and phallus. As such, his analytical system merely reproduces more of the structural same.

Nevertheless, Lacan presents geographers with important theoretical challenges. On the one hand, he describes a process which is ongoing: the structural reproduction of heteropatriarchy and the reduction of people and things to signs of themselves. One question might be then: How might geographers seize his analysis to find some crack within it, using his insights to show how the mirror and reductionism might be re-negotiated in ways that disrupt heteropatriarchy and spatial oppressions? Are we to look, as Lefebvre's (1991) work suggests, for a giant revolutionary step forward, or are there other kinds of spatial practices already in existence or waiting to be enacted that might work on numerous cultural fronts to gnaw from within. Our experiences of the body are treated as though merely an effect of the signifier, subordinating the corporeal to the linguistic. Several levels of spatial collapse occur simultaneously, as a result of this act of subordination; first, space itself is reduced to an effect of the signifier instead of having a role in creating meaning, in itself being meaningful. Space, in other words, becomes part of the illusion created through interpersonal relations in which the spaces between bodies are no more than projections of internal experience. Ultimately, negating the spatial as a primary signifying event leads to transforming bodies as well into reflections or imitations of a privileged linguistic *surface* reality.

Notes

1 This elusiveness is evident in the different ways in which the real has been interpreted by theorists of Lacan. Malcolm Bowie (1991, page 106), for example, asks if the real is 'outside or inside? Is it a vacuum or a plenum?'. By this he is referring both to the distinction Freud articulated between material reality [the outside] and psychical reality [the inside] as well as whether the real is a rupture in the otherwise intact fabric of the symbolic order of Western culture or a full and unmediated connection with the material world.
2 Julia Kristeva has argued that the presymbolic (what she calls the semiotic) is always imperilling the symbolic order in productive ways.
3 This is why Irigaray makes such a point of censuring Lacan for his account of female sexuality. She focuses on the Encore Seminar where he insists on the impossibility of understanding female sexuality because it cannot be processed through the symbolic order. 'They have nothing to say about it,' Lacan asserts.
4 Understanding Lacan's panic over the maternal realm does much to explain his criticism of those forms of psychoanalysis that encourage specifically maternal transferences onto the analyst, for example the 'holding environment' of British object-relations analyst, D. W. Winnicott (1986).
5 For example, Daniel Stern in *The Interpersonal World of the Infant* (1985).
6 Elsewhere, Lacan suggests that the real inevitably confronts us, regardless of our machinations, our symbolizations, and our inventions, in the form of chance,

the unassimilable (that is, trauma) and death (Lacan 1978, page 55). Chance events, for example, can destabilize (even momentarily) the symbolic (Cultural) edifice of subjectivity.

7 Importantly, this fragmentation is something experienced retroactively, that is, in relation to an anticipation of becoming whole, like the coherent unified image. As such, mirroring carries with it profound implications for cultural constructions of time. In particular, through the mirror, the infant's past is retroactively represented as a 'body-in-pieces' *at the same time* that the future is framed as a process of becoming the idealized image. It is this simultaneity that Lacan locates as the primordial 'violation of chronology' for the ego wherein 'both future and past are ... rooted in an illusion' (Gallop 1985, page 81).

8 Beginning with John Bowlby in the 50s, attachment theory has been central to psychoanalyses of the child.

9 Lacan uses the term 'imaginary' to refer to the psychical register where images are mistaken for reality.

10 See, for example, Helene Deutsch's (1930) work on female masochism and the maternal destiny of women.

11 This passage has been translated by the authors. The original text reads

> Voyez la position telle que je la dessine – ici l'imaginaire, c'est-à-dire le désir du phallus chez la mère, là l'enfant, notre centre, qui a à faire la découverte de cet au-delà, le manque dans l'objet maternel. C'est au moins une des issues possible – à partir du moment où l'enfant trouve à saturer la situation et à en sortir en la concevant elle-même comme possible, la situation pivote autour de lui.
>
> Que trouvons-nous effectivement dans le fantasme de la petite fille, et aussi du petit garçon? Pour autant que la situation pivote autour de l'enfant, la petite fille trouve alors le pénis réel là où il est, au-delà, dans celui qui peut lui donner l'enfant, à savoir, nous dit Freud, dans le père.

12 Having what the mother desires is negotiated differently for girls and boys. In identifying with the paternal order, the male child wants to have the phallus that will reconnect him with the mother. The penis of little boys is identified as that which might satisfy the mother, but this is coupled with an incest taboo. The little boy therefore decides to become just like his father and to wait to find a mother-substitute later in life to whom he can deliver his phallus-gift and make her complete. The little girl, in contrast, lacking on her body the signifier associated with the mother's desire, instead fantasmatically *becomes* the link that will restore her to a mythic prediscursive integrity.

13 For an excellent account of the distinction between the Freudian penis and the Lacanian phallus, see Jean-Joseph Goux (1992).

References

Althusser L, 1971 *Lenin and Philosophy and Other Essays* Trans. Ben Brewster (Monthly Review P, New York).

Blum V and Nast H, 1996 'Where's the Difference?: The Heterosexualization of Alterity in Henri Lefebvre and Jacques Lacan.' *Environment and Planning D: Society and Space* **14**: 559–580.

Boorstin D, 1987 *The Image: A Guide to Pseudo-Events in America* (Atheneum, New York). Rpt. 1961.

Bowie M, 1991 *Lacan* (Harvard UP, Cambridge).

Brenkman J, 1993 *Straight Male Modern: A Cultural Critique of Psychoanalysis* (Routledge, New York).

Crary J, 1990 *Techniques of the Observer: On Vision and Modernity in the Nineteenth Century* (MIT Press, Cambridge, MA).

Deutsch H, 1930, 'The significance of masochism in the mental life of women' *International Journal of Psycho-Analysis* **11:** 48–60.

Ewen S and Ewen E, 1992 *Channels of Desire: Mass Images and the Shaping of American Consciousness* (University of Minnesota Press, Minneapolis). Rpt. 1982.

Fink B, 1995 *The Lacanian Subject: Between Language and Jouissance* (Princeton UP, Princeton).

Foucault, M, 1980 *Power/Knowledge: Selected Interviews and Other Writings, 1972–1977* ed. Colin Gordon. Trans. Colin Gordon, Leo Marshall, John Mepham, Kate Soper (Pantheon Books, New York).

Freud, S 1905 'Fragment of an Analysis of a Case of Hysteria' *Standard Edition of the Complete Works of Sigmund Freud* ed. James Strachey 7: 7–122 (Hogarth Press, London).

Freud, 1918 'From the History of an Infantile Neurosis' *Standard Edition of the Complete Works of Sigmund Freud* ed. James Strachey 17: 1–71 (Hogarth Press, London).

Freud, S 1923 'The Ego and the Id' *Standard Edition of the Complete Works of Sigmund Freud* ed. James Strachey 19: 13–66 (Hogarth Press, London).

Freud, S 1931 'Female Sexuality' *Standard Edition of the Complete Works of Sigmund Freud* ed. James Strachey 21: 225–43 (Hogarth Press, London).

Freud, S 1933 'Feminity' *Standard Edition of the Complete Works of Sigmund Freud* ed. James Strachey 22: 112–35 (Hogarth Press, London).

Gallop J 1985 *Reading Lacon* (Cornell UP, Ithaca).

Goux Jean-Joseph 1992 'The phallus: masculine identity and the "exchange of women"' *Differences: A Journal of Feminist Cultural Studies* **4:** 40–75.

Irigaray L, 1985 *This Sex Which Is Not One* (Cornell University Press, Ithaca, New York).

Kristeva J, 1984 *Revolution in Poetic Language* Trans. Margaret Waller (Columbia UP, New York).

Lacan J, 1977a, 'The Subversion of the Subject and the Dialectic of Desire in the Freudian Unconscious', in *Ecrits: A Selection* Trans. Alan Sheridan (Norton, New York): 292–325.

Lacan J, 1977b, 'The Mirror Stage as Formative of the Function of the I as Revealed in Psychoanalytic Experience', in *Ecrits: A Selection* Trans. Alan Sheridan (Norton, New York): 1–7.

Lacan J, 1977c, 'On a Question Preliminary to any Possible Treatment of Psychosis', in *Ecrits: A Selection* Trans. Alan Sheridan (Norton, New York): 179–225.

Lacan J, 1978 *The Four Fundamental Concepts of Psycho-analysis* ed. J Alain-Miller. Trans. A Sheridan (Norton, New York).

Lacan J, 1985a, 'The Meaning of the Phallus', in *Feminine Sexuality* eds J Mitchell and J Rose (Norton, New York): 74–85.

Lacan J, 1985b, 'God and the "Jouissance" of Woman' in *Feminine Sexuality* eds J Mitchell and J Rose (Norton, New York): 137–148.

Lacan J, 1988 *The Seminar of Jacques Lacan: Book I, Freud's Papers on Technique 1953–1954* ed. Jacques-Alain Miller. Trans. John Forrester (Norton, New York).

Lacan J, 1993 *The Seminar of Jacques Lacan: Book III, The Psychoses 1955–1956* ed. Jacques-Alain Miller. Trans. Russell Grigg (Norton, New York).

Lacan J, 1994 *Le Seminar: La Relation d'Objet* Livre 4. ed. J Alain-Miller (Editions du seuil, Paris).

Lacan J, 1998 *The Seminar of Jacques Lacan: Book XX, Encore 1972–1973; On Feminine Sexuality, The Limits of Love and Knowledge,* ed. Jacques-Alain Miller. Trans. Bruce Fink (Norton, New York).
Lefebvre H, 1991 *The Production of Space* (Basil Blackwell, Oxford).
Pile S, 1996 *The Body and the City: Psychoanalysis, Space and Subjectivity* (Routledge, London).
Postman N, 1986 *Amusing Ourselves To Death: Public Discourse in the Age of Show Business* (Penguin, New York).
Rose J, 1985, 'Introduction – II', in *Feminine Sexuality* (Pantheon Books, New York).
Stern, D 1985 *The Interpersonal World of the Infant: A View from Psychoanalysis and Developmental Psychology* (Basic Books, New York).
Winnicott D W, 1986 *Holding and Interpretation: Fragment of an Analysis* ed. M. Masud R. Khan (Grove, New York).

10

FOUCAULT'S GEOGRAPHY

Chris Philo

1 Foucault, history, geography

But this inexhaustible wealth of visible things has the property (which both correlates and contradicts) of parading in an endless line; what is wholly visible is never seen in its entirety. It always shows something else asking to be seen; there's no end to it. Perhaps the essential has never been shown, or, rather, there's no knowing whether it has been seen or if it's still to come in this never-ending proliferation.

(Foucault, 1986a, page 110)

Roussel keeps them on their own level, in a way, and starting from the rabbit beating drums, makes the machine increasingly complex, but always remaining the same without ever passing to another register or level.

(Foucault, 1986a, page 179,
in an interview with Charles Ruas)

The purpose in this paper is to consider a few aspects of what I call Michel Foucault's 'geography', and in so doing I wish to sketch the outlines of an argument about how Foucault's vision of history (regarded both as what happened in the past and as attempts by intellectuals to inquire into these happenings) necessarily opens up a heightened sensitivity to the way in which *space* and *place* are inextricably bound up in this history. My ambitions here are indeed modest ones – notwithstanding Bonnett's (1990) remarks about 'mock humility' – for it will not be possible to create some impressive Foucauldian theory of how space enters into the constitution of social life, nor to synthesise Foucault's claims so that he appears to be saying much the same things as (supposedly) do other grand theorists of society and space. And neither do I want to suggest that this partic-ular luminary of social thought holds keys to an understanding unattainable elsewhere, for I suspect that not entirely dissimilar conclusions about the

205

'geography' of the world can be distilled from 'other' sources as diverse as Baudrillard (about whom, a little more later), certain existentialist traditions, the Geertzian anthropologists, and various feminist writers who have located in feminist thought many themes currently debated under headings such as 'postmodernism' and 'poststructuralism'. It is not a matter of 'creating yet another intellectual base to defend',[1] then, but of stirring another voice into the richness of recent 'geographical' debates: one that I continue to find inspirational both for theoretical reflection and substantive endeavour, and one that I also feel to have been somewhat misrepresented in the existing literature of human geography. Foucault's texts are obviously at liberty to escape their author as they are read and written about by geographers, but it seems to me that the 'story' to be told about Foucault and geography has been agreed upon all too quickly and simplistically.

It might be thought that Foucault has long been the focus of discussion in the hallways of human geography, and Gregson (1989, page 236, footnote 6) goes so far as to assert that 'Giddens and Foucault are the only modern social theorists to have attracted prolonged recent interest in human geography'. It is true that a number of geographers are beginning to embark upon substantive inquiries where Foucauldian concerns are vital – and here I am thinking of those dealing with the role of institutional and settlement spaces in the management of difficult or 'other' human populations (for example, see Driver, 1985a, 1985b, 1990; Ogborn, 1990; Philo, 1989a; Robinson, 1990) – and it is also the case that Foucault gets mentioned in more overtly theoretical accounts of space, territoriality, and social reproduction (for example, see passages throughout Sack, 1986; Wolch and Dear, 1989). There is also some indication that the historical critique of rationality embedded in much of Foucault's work, along with his sophisticated treatment of the 'power–knowledge' couplet, will *soon* become significant to the sorts of debates being pursued in this journal. However, I doubt if Foucault has yet attracted anything like as many column inches of geographical journal space as has Giddens, and I would also want to argue that what is surprising is the *absence* to date of any sustained theoretical engagement with Foucault on the part of theoretically minded geographers. There is certainly one exception in this respect, though, and this is to be found in the opening chapter of Soja's important text *Postmodern Geographies* (1989) where he examines Foucault's 'ambivalent spatiality' as a straw in the wind of '*re*asserting space in critical social theory'. The result is a thoughtful and intriguing introduction to Foucault's geography:

> The contributions of Foucault to the development of critical human geography must be drawn out archaeologically, for he buried his precursory spatial turn in brilliant wheels of historical

insight. He would no doubt have resisted being called a post-modern geographer, but he was one, *malgré lui*, from *Madness and Civilization* [1967] to his last works on *The History of Sexuality* [1979].

(Soja, 1989, page 16)

The line of Soja's reasoning here – and it parallels what he has to say about various other thinkers from Lefebvre to Berger – is to tease out Foucault's attentiveness to spatial relations, and then to claim Foucault as a 'postmodern geographer': a claim that Soja makes because for him the 'essence' of postmodernism (and I *deliberately* use this contradictory phrase) lies precisely in its rediscovery of space as something out of which human society is unavoidably fashioned. I stand here on the brink of a much broader critique of Soja that questions whether it is appropriate for him to reason as he does – to offer sweeping generalisations which seem to make of space an 'essence' present at the heart of any individual's social life and of any capitalist society's functioning – from a position that he describes as postmodernist (see also Gregory, 1990), and in what follows I will intimate that Foucault's own attack on 'totalising' theoretical endeavours could easily be turned back upon the still-very-grand ambitions of Soja in his book *Postmodern Geographies*.

To be more precise, though, what I will argue is that Foucault's geography is actually *far more* postmodern than Soja allows, and is hence far more fitting of the description 'postmodern geography' than is Soja's own version of human geography. And the hinge for my argument here is to spell out in some detail the way in which Foucault's attack on what he terms 'total history' calls forth a geographical way of looking at the world in which one sees *only* 'spaces of dispersion': spaces where things proliferate in a jumbled-up manner on the same 'level' as one another – on the one level where advanced capitalism and the toy rabbit beating a drum no longer exist in any hierarchical relation of the one being considered more important or fundamental than the other – and on which it can never be decided if 'the essential' has been sighted (because there simply is *no* 'essential' to be sighted or because, even if there is one, we can never know whether it has revealed itself). This may be an unorthodox reading of Foucault's geography, and it arises chiefly out of a confrontation of materials from the early pages of *The Archaeology of Knowledge* (1972) with Foucault's exegesis of the literature of Raymond Roussel in *Death and the Labyrinth* (1986a), two texts little consulted by geographers and other social scientists. An obvious objection to my reading is that the passages drawn upon may amount to nothing more than him using spatial *metaphors* to portray a new kind of intellectual route into the study of history, literature, or whatever, but my own view is that his thinking in this connection spills over from the realm of metaphor to embrace the *empirical* spaces and places existing in such messy abundance in, through, and around the substantive matters tackled in his historical works from *Madness and Civilization* (1967) onwards.

One of Soja's weaknesses is his failure to look much more closely at these historical treatments of substantive matters, and up to a point this is a failing in this paper as well (where such a failing is all the more acute, given my insistence on following the attention to detail demanded by Foucault's 'general history'), but I do make a few observations about these treatments and I can point to a sustained inspection of Foucault's 'spatial history' of madness conducted elsewhere (Philo, 1992b). Where Soja falls down further is in effectively projecting the account of spatial relations contained in *Discipline and Punish* (Foucault, 1977) – what Soja (1989, page 21) terms Foucault's 'provocative spatialisation of power' – on to the rest of Foucault's (theoretical and substantive) encounter with space, and it is through this manoeuvre that he misleadingly presents Foucault as someone concerned primarily with the way in which the operation of power seeps through the time–space *geometry* of how institutions, settlements, and (by implication) whole societies are arranged 'on the ground'. This is certainly not to deny this crucial dimension to Foucault's geography: rather, it is to suggest that by claiming to find here the 'essence' of Foucault's postmodern geography, and by in effect hooking up *his* own version of Foucault's geography to *his* own version of postmodernism, Soja ends up giving a somewhat misleading statement of Foucault's distinctive position relative to *both* geography *and* postmodernism. What I will hence attempt, through the abovementioned discussion of how Foucault's history shades into being a geography, is to provide an alternative account of Foucault as flagging (though in no sense exhausting the possibilities of) a 'truly' postmodern geography. We might not like this geography, but it seems to me that we ought to pause for a moment in our projects of combining Foucault with Giddens, Lefebvre, Mann, or whoever – the projects of turning Foucault into the 'same' – and instead we should recognise (and perhaps marvel at) the 'otherness' of his perspective on geography and postmodernism as something really quite 'alien' to all manner of current ways for proceeding as geographers.

2 Foucault's history into geography: theoretical manoeuvres

I am going to suggest that it is possible to identify what might be termed (albeit perhaps a little erroneously) a 'theoretical' input to Foucault's geography, and I will examine in some detail the theoretical manoeuvres that are involved here. This is not to imply that Foucault self-consciously or systematically inspects his conceptualisation of space, place, and geography, and neither is it to imply that we can distil any consistent but unstated principles that might be informing the conduct of his substantive historical inquiries. But what I am claiming is that a close reading of Foucault's arguments about history suggests a vision of how social life 'works' – a vision in which certain vocabularies (often spatialised ones) are employed to capture a certain ontology (an equally spatialised one) – which very definitely insists upon researchers of the past taking seriously the importance of space, place, and geography to the stories that they are endeavouring to tell. In part this is because Foucault recognises the simple but telling 'fact' that the phenomena, events, processes, and structures of history (however we may define them) are always fragmented by geography, by the complicating reality of things always turning out more or less differently in different places, and it occurs to me that Foucault's attempt to cope with this problem – a key if somewhat implicit dimension to his critique of 'total history' and attendant espousal of 'general history' – necessarily sees him embrace a spatialised perspective on what history actually *is*: a perspective that he sometimes depicts as a sensitivity to 'spaces of dispersion'. There is also a second argument to make about the importance of geography to the historical storyteller, but I will defer this to section 3 of the paper where I consider the role of geography in relation to Foucault's more substantive interests.

2.1 From 'total history' to 'general history'

Nietzsche once complained about intellectual exercises that insist on placing first – rather than last or even not at all – concepts that are paraded as general and hence 'highest', and his objection was that these concepts usually turn out to be not the highest but the 'emptiest', the 'thinnest', and little more than the 'last smoke of an evaporating reality' (in Dews, 1987, page 139). This attack on a priori modes of reasoning is not a wholly uncommon one, of course, although it is one with which a discipline such as geography has become highly uncomfortable thanks to its recent obsession with securing philosophical and methodological base camps – a positivistic belief in spatial laws, a Marxist belief in social relations impressing upon spatial form, a humanistic belief in people's subjectivities shaping places, a structurationist belief in the time–space constitution of engagements between agency and structure – from which empirical expeditions can be mounted.[2] And yet criticisms of the a priori surely need to be rehearsed by geographers, particularly given that the reality of geographical diversity is itself such a signal challenge to this particular mode of reasoning (as we will see shortly).

A useful route into this arena is indeed signposted by Foucault (1926–84), the celebrated French thinker who has probably done more than anybody to fuse philosophical, social–theoretical, and historical lines of inquiry, and of particular interest here is the critique that he develops of what he terms *total history*:

CHRIS PHILO

The project of total history is one that seeks to reconstitute the
overall form of a civilisation, the principle – material or spiritual
– of a society, the significance common to all the phenomena of
a period, the law that accounts for their cohesion – what is called
metaphorically the 'face' of a period.

(Foucault, 1972, page 9)

He then goes on to outline the sorts of ontological and methodological
strategies associated with the practice of total history, which have

supposed that between all the events of a well-defined spatio-
temporal area, between all the phenomena of which traces
have been found, it must be possible to establish a system of homogeneous relations: a network of causality that makes it possible
to derive from each of them, relations of analogy that show how
they symbolise one another, or how they all express one and
the same central core; it is also supposed that one and the same
form of historicity operates upon economic structures, social institutions and customs, the inertia of mental attitudes, technological
practice, political behaviours, and subjects them all to the same
type of transformation; lastly, it is supposed that history itself may
be articulated into great units – stages or phases – which contain
within themselves their own principle of cohesion.

(pages 9–10)

This is an extremely dense passage, throwing off veiled references to the
reasoning pursued by scholars of various theoretical persuasions – 'Whigs',
historians of 'Great People', 'psychohistorians', Hegelians, Marxists,
Annalistes[3] and others – all of whom impose a priori and often rather grand
historical visions upon the concrete phenomena and events of 'well-defined
spatio-temporal areas'. For Foucault, though, all such impositions are
suspect because they introduce a measure of order that arguably remains
alien to the details and the differences of history at particular times and
in particular places, and it is not difficult to see that he scorns this ordering
tendency because it inevitably smoothes over the specific confusions, contradictions, and conflicts which have been the very 'stuff' of the lives led by
'real' historical people, powerful and powerless alike. Foucault thus asserts
that total history operates through positing a 'central core' to the social
world – a centre which might encapsulate the words and deeds of 'heroes',
the traditions of culture, the machinations of capitalism, or whatever –
from which a 'homogeneous system of relations' supposedly spreads out
to govern all things, and it is obvious that such a positing of a centre
stands squarely in opposition to his own belief that 'nothing is fundamental: this is what is interesting in the analysis of society' (Foucault, 1982,

210

page 18). It should be clear, moreover, that Foucault's arguments here are both in line with and anticipatory of that wider current of thought (or 'attitude') now commonly referred to as *post-modernism*, in which the certainties of existing (modernist) intellectual projects – the certainty that there *is* a fundamental order to the world and that this order will be laid bare by orderly, rational research procedures – are thrown deeply into question (Cloke *et al.* 1991, chapter 6).[4]

Rather more might be said about this critique of total history and about its connections with postmodernist thought, but I will content myself here with a brief elaboration on how Foucault envisages total history dealing with both time and space. In the first instance, he is undoubtedly uneasy about the practice of slicing up the flow of real historical occurrences into ponderous *temporal* 'great units', and he obviously wishes to quarrel with the rigid periodisations that historians of various persuasions are prone to employ; and in the second instance – and despite not being so explicit on this count – he criticises historians for an insensitivity to the geography of the social world that manifests itself in stressing the homogeneity of events, phenomena, and their hypothesised determinations within *spatial* 'great units' (continents and perhaps countries) and in thereby ignoring the reality of smaller-scale areal differences and distributions. These twin criticisms are plainly underlain by a yearning to knock down the prized construct of *continuity*: the continuity that historians envisage as binding disparate events, phenomena, and hypothesised determinations together in overarching historical–geographical totalities, and the continuity that the same historians often envisage as binding one temporal–spatial 'great unit' both to its predecessors and to its successors (Foucault, 1972, pages 8–9). And for Foucault it is obvious that continuity and totality are seen as mutually reinforcing conceptual constructions.

Foucault's critique of total history is in itself not all that different from other assaults on the castles of coherence, but what *is* highly suggestive – at least to this reader – is the strategy whereby he proposes to negotiate the pitfalls of total history, and in this connection it is appropriate to consider his alternative conception of *general history*:

> The problem that now presents itself – and which defines the task of a general history – is to determine what form of relation may be legitimately described between these different series; what vertical system they are capable of forming; what interplay of correlation and dominance exists between them; what may be the effect of shifts, different temporalities and various rehandlings; in what distinct totalities certain elements may figure simultaneously; in short, not only what series, but also what 'series of series' – or, in other words, what 'tables' it is possible to draw up. A total description draws all phenomena around a single centre – a

principle, a meaning, a spirit, a world-view, an overall shape; a general history, on the contrary, would deploy the space of a dispersion.

(1972, page 10)

This is another difficult account, and it is not immediately obvious whether or not the project delimited amounts to more than a restating of the classic *Annalistes* approach to history,[5] but the distinctiveness of what Foucault is attempting can still be captured if it is remembered that behind his words lies the desire to write histories in which the paraphenalia of grand historical visions – the a priori teleologies and ontologies, the 'centralising' and the 'homogenising' – are overturned. Hutcheon (1988, pages 98–99) hence writes of both his 'assault on all the centralising forces of unity and continuity' and his parallel effort to expose the 'pretended' nature of such forces by insisting that – and the terms employed here should instantly arrest the geographical reader – the 'particular, the local and the specific' be instated in place of the 'general, the universal and the eternal'.

But what should be noted at once is that Foucault does not conceive of general history as a straightforward mirror image of total history, since such a mirror image would arguably demand a thoroughgoing empiricist endeavour purged of *any* theoretical, conceptual, or interpretative moment. It is true that in one essay he refers to his favoured approach to history – which here he describes as 'genealogy'[6] – as being 'gray, meticulous and patiently documentary' (Foucault, 1986b, page 76), but he is also perfectly aware that no historian can ever be a pure empiricist 'Dryasdust'[7] because the very practices of researching and writing inevitably bring a semblance of order – however unacknowledged or unwanted – to the empirical stockpile. And, as Derrida (1978a, pages 292–3) suggests through his philosophical reflections, it is surely impossible for any social scientific enterprise to choose once and for all between 'dreaming of deciphering a truth or an origin' (the grand theoretical and modernist dream of finding a centre to intellectual inquiries) and 'simply affirming play' or chaos (the hypothetical objective of pure empiricism; the ultimate destination of postmodernism as attitude). The route that Foucault takes across this confused terrain thus depends upon the acceptance that a layer of theoretical materials must be laid over the specific events and phenomena under study, but also in ensuring that the concepts deployed have not so much an a priori character – having decided in advance what is 'going on' in any particular situation in any given time and place – as the character of 'hovering' responsively above the empirical details revealed.[8] Rather than coming first, then, the grander theoretical statements of a given substantive study should gradually materialise as the study progresses, and this is what Foucault is driving at when remarking that

Genealogy . . . requires patience and a knowledge of details, and it depends on a vast accumulation of source material. Its 'cyclopean monuments' are constructed from 'discreet and apparently insignificant truths and according to a rigorous method'; they cannot be the product of 'large and well-meaning errors'.

(Foucault, 1986b, pages 76–7)

2.2 'Time is lost in space'?

What I want to argue is that Foucault's general history embraces a number of strategies designed to negotiate the snares of 'totalisation', and in so doing to capture more faithfully than can total history the fragmented (the particular, local, specific) ontology of social life in past times; and it seems to me that crucial to these strategies – if not reflected upon all that deliberately by Foucault himself – is the taking seriously of space, place, and geography as sources of fragmentation. And, to be more precise, I think that much can be learned from considering what he has to say about 'spaces of dispersion' (to use his own vocabulary), but before embarking upon such a consideration I will introduce something of the issues involved here by discussing several passages from one of Foucault's lesser-known texts: namely, his exegesis of the literature of Roussel – the turn-of-the-century French experimental playwright, poet, and novelist – in *Death and the Labyrinth* (1986a). Foucault (1986a, page 185), in an interview with Ruas, refers to this text as 'my secret affair' and as one that maybe 'doesn't have a place in the sequence of my books', but I have nonetheless found it helpful to reflect upon the sorts of claims that I have begun above to distil from the *Archaeology* in the light of what Foucault says here about Roussel's writing, particularly that contained in Roussel's 1904 work *La Vue*, described by Ashbery (in Foucault, 1986a, page xxi) as follows:

> *La Vue* (1904) is made up of three long poems: *La Vue, Le Concert* and *La Source*. In the first the narrator describes in incredible detail a tiny picture set in a penholder: the view is that of a beach resembling that of Biarritz, where Roussel spent his summers. The second poem is a description of an engraving of a band concert on the letterhead of a sheet of hotel stationery. In the third the narrator is seated at lunch in a restaurant . . . [but virtually all of the poem's fifty pages] describe a spa pictured on the label of a bottle of mineral water on the narrator's table.

Foucault is clearly fascinated by Roussel's painstaking descriptions of these tiny scenes – what Ashbery (in Foucault, 1986a, page xxii) refers to as Roussel's 'exasperatingly complete descriptions of uninteresting objects' – and he devotes much effort to capturing the way in which these descrip-

213

tions lay out before us a world of things, 'a space of luminous, patient, simple things' (page 115), in which little more can be done than to document the brute existence of these things relative to one another across the plane (the level or line) of the picture.

Foucault strives to capture this inability to say very much at all, and (perversely enough) is forced to say a great deal in the process:

> [The] small vignette on the letterhead is like the circular lens embedded in the souvenir pen, or the label on a bottle of Evian water, a prodigious labyrinth seen from above. Instead of concealing it, it naively places before one's eyes a network of paths and boxwood hedges, long stone walls, the masts, the water, those minuscule precise people going in all directions with the same fixed step. Language needs only turn to these silent figures to attempt through infinite accumulation to recreate that flawless visibility.
>
> (page 105)

There is a definite existential cast to Foucault's account here: existential in the sense of supposing that beyond the world of existence, beyond the apparent truth that things of all shape, size, consistency, quality simply *are* in the world, there is no deeper – or, at least, no ultimately knowable – 'essence' ('the essential') to be fathomed.[9] Foucault thus finds in Roussel a demand that 'the eye' describing a scene must preserve its contents in a 'state of being which ... [gives] each thing its ontological weight' (page 137), must 'let them "be seen" by virtue of their being' (page 106), and must in so doing allow 'a plethora of beings serenely [to] impose themselves' (page 108) free from the observer always seeking to 'penetrate' them. At the same time, what Roussel's descriptions also insist upon is not the establishment of a list of priorities whereby certain things in the scene are presented as somehow being more significant than other things – this is what is meant by the keeping of all things at the same register or level – and one aspect to this 'non-hierarchialising' of things is to respect the coequal existence of things large and small:

> There is a fundamental lack of proportion: seen in the same way are the porthole of the yacht and the bracelet of a woman chatting on deck, the wings of a kite and the two points formed by the tips of a stroller's beard raised slightly by the wind ... In this fragmented space without proportion, small objects thus take on the appearance of flashing beacons. It's not a question of signalling their position in this instance, but simply their existence.
>
> (1986a, pages 106–109)

And maybe there is a point of some salience here, both in terms of Ley's (1989) insistence on reinstalling 'human proportion' in social thought and in the context of calling for 'post-hierarchical' thinking and politics when moving beyond ascribing higher and lower values to the things (and particularly the many social 'others') in the world (for example, see Boyne, 1990).

Weaving in and out of Foucault's exegesis here a sense of his geography begins to emerge, as he relates at some length the extent to which Roussel's descriptions are in effect 'geographical descriptions' precisely because one of the few organising devices that Roussel permits himself – beyond that of trying to accumulate enough details to 'eliminate ... the distance [of language] from things' (Foucault, 1986a, page 136) – is found in his systematic spatial movement around the things pictured in his scenes:

> There is no privileged point around which the landscape will be organised and with distance vanish little by little; rather, there's a whole series of small spatial cells of similar dimensions placed right next to each other without consideration of reciprocal proportion ... Their position is never defined in relation to the whole but according to a system of directions of proximity passing from one to the other as if following the links in a chain: 'to the left', 'in front of them to the left', 'above, higher', 'further', 'further, continuing on the left', 'at the end of the beach', 'still close enough to them', 'a little more on the left on the other side of the arcade'. Thus spreads the sand of *La Vue*, in discontinuous grains, uniformly magnified, evenly illuminated, placed one next to the other in the same noonday sun.
>
> (1986a, page 107)

The route around things – which at the same moment is the actual arrangement (configuration, distribution) of things in space relative to one another, their nearness to one another or farness away, their 'to-the-rightness' or 'to-the-leftness', their 'aboveness' or 'belowness' – is here accorded considerable significance. Indeed, it serves to structure the description given, but its definite geography is also supposed to be one of the few properties of the set of things under scrutiny that can be accepted *without* shifting from empirical claims about their existence to more transcendental claims about their essence (their inner truths, their fundamentals). Foucault is perfectly aware that the elevation of this geography in the process of representation poses a threat to the more usual literary sensitivity to history – he acknowledges that in Roussel's scenic depictions 'time is lost in space' (page 110) and that there is an 'attempt to eliminate time by the circular nature of space' (page 78) – and in this respect Foucault's manoeuvre, through Roussel, does square with Soja's claims about Foucault's contribution to reasserting a role for space in social

CHRIS PHILO

thought. And yet, what must quickly be realised is that the (re)assertion of space for Foucault and Roussel is in part designed to circumnavigate essentialist modes of thought in which essences (deeper levels, layers) are revealed or work themselves out progressively though time. As Foucault puts it, 'the old structure of legendary metamorphosis' which dictates how things change themselves through time and thereby reveal their true essences is 'reversed' in Roussel, leaving only 'a joining of beings which carries no lesson: the simple collision of things' (1986a, page 84).

But this is not to declare that 'metamorphosis' is dead, because still present in Roussel's literature is a notion of 'metamorphosis' – or perhaps a term such as 'juxtaposition' better captures what is involved here – that is conceived of in terms of spatial relations rather than of temporal sequences. In the following passage Foucault (1986a, page 80) outlines contrasting notions of metamorphosis, although somewhat misleading (in the context of the present argument) is the fact that he explains both the temporal and the spatial versions of 'metamorphosis' by using *spatial* metaphors:

> Thus are constructed and criss-crossed the mechanical figures of the two great mythic spaces so often explored by Western imagination: space that is rigid and forbidden, surrounding the quest, the return and the treasure (that's the geography of the Argonauts and of the labyrinth); and the other space – communicating, polymorphous, continuous and irreversible – of the metamorphosis, that is to say, of the visible transformation of instantly crossed distances, of strange affinities, of symbolic replacements.

In the first case metamorphosis is change through time – in which things get changed during the 'quest' and subsequent 'return' with the 'treasure' (the true essences?) – whereas in the second case metamorphosis is found in the juxtaposition of things bumping up against one anther in space, and it is in this second and overtly spatial sense that Foucault sees Roussel leading his readers into an understanding of the transformations constitutive of the world. Not for Roussel the temporal sense of 'mice transformed into coachmen, nor pumpkins becoming coaches' (Foucault, 1986a, page 81), then, but rather 'the juxtaposition ... of two orders of being not close in the hierarchy which must cross a whole intermediary gamut in order to be joined' (page 81) or of 'the meeting of beings occur[ring] in the broad daylight of a discontinuous nature' (page 82). In other words, stress is placed upon how seemingly very different things (and things that might conventionally be conceived of hierarchically, such that one is reckoned to be 'superior' to the other: see above) may coexist, may touch, may fuse into one another, and the result is an insistence upon 'simultaneity' which both raises the possibility of unlike ('discontinuous') things jumbling together in a

216

fashion uncomfortable to 'the old principle of the continuity of beings' (page 81) and of a challenge to 'the hierarchic'.

The final point to appreciate here is that Foucault detects in Roussel's geographical descriptions a strange paradox between 'this infinitely chatty landscape' (page 115), this proliferation of words struggling to parallel the endless proliferation of things in even the tiniest of worldly spaces, and the unbearable 'silence' of these things in their stubborn refusal to give up their innermost truths: '[Roussel's] language turns towards things, and the meticulous detail it constantly brings forward is reabsorbed little by little in the silence of objects. It becomes prolix only to move in the direction of their silence' (page 105). And again:

> this world of absolute language is, in a certain way, profoundly silent. The impression given is that everything has been said, but in the depth of this language something remains silent. The faces, the movements, the gestures, even the thoughts, secret habits, the yearnings of the heart are presented like mute signs on the backdrop of the night.
>
> (page 113)

Even after Roussel has written his fifty pages on the picture adorning the bottle of mineral water – even after he has described all of the things that he can see there, including the geography, and has concocted stories about the personal lives and thoughts of the people frozen in the scene depicted – there is still a strange sense that we have indeed only remained on 'the surface of things' (the title of the chapter in *Death and the Labyrinth* where much of the above is discussed) and that anything beyond this surface is silence. Within the 'magical circle' (the tiny picture in the penholder, on the letterhead or on the bottle) the myriad things encountered and described 'appear in their insistent, autonomous existence, as if they were endowed with an ontological obstinacy which breaks with the most elementary rules of ... relation.[10] Their presence, like a boulder, is self-sufficient, free of any relation' (page 106).

This is a return to the earlier claims abut the existentialism of this scenic description, of course, but what I want to emphasise here is that informing this acceptance of silence is both an *expectation* that probably there is no deeper essence, truth, or whatever to be spoken beyond the determined 'isness' of things in the world (an 'isness' that can be infinitely described) and a *yearning* for this not actually to be the case. For much of the time Foucault and Roussel apparently accept that there is nothing outside of the proliferation of words about 'the surface of things' – that this proliferation pretty much captures in all of its comprehensiveness the total and only 'reality' of the things described – and the suggestion in this regard is that 'the discourse which describes them in detail is finally the one that explains them' (page 111). In this case, the silence of things on more 'essential' matters is perfectly comprehensible, for there is simply nothing else to say once Roussel has finished his description. But, and perhaps this is the key theme of *Death and the Labyrinth*, for both Foucault and Roussel there will always remain a longing, a hope, a desire, a will to hear words being spoken in the 'social space of silence' (Olsson, 1987): a yearning to discover that, after all, there is still something more to be said, and that this something more really does provide us with the 'key' permitting us to unlock the 'secrets' of things, existence, and (beyond this) essence, creation. A text published after his death and entitled *How I Wrote Certain of My Books* (1935) saw Roussel spell out the rules that had governed the composition of several of his better-known works, not including *La Vue*: and these rules (to do with using rhymes and series of mixed-up words that sounded similar to common phrases, book titles, or lines of poetry) had been completely unnoticed and 'silent' in the works concerned, and Foucault is left speculating whether it may be that there *really are* similarly unnoticed and 'silent' rules 'out there' – and just waiting to be specified – shaping not just the logic of Roussel's other texts, not just all literary compositions in general, but the whole patterning of human reality on this planet. Or is it perversely the case that Roussel's belated revelation of 'his secret' does nothing more than bring home to us the ultimate 'secrecy' or unknowability of the world; the sad awareness that there can only be 'secrecy' and silence, that Roussel's final holding up of a 'mirror' on his own oeuvre merely serves to confirm our deepest fears that the more we ask questions about what lies below 'the surface of things' the more 'the mirror deepens in secrecy'? (See Foucault, 1986a, page 2.)

2.3 'Spaces of dispersion' and 'systems of dispersion'

We begin to catch sight of Foucault's geography from his discussion of Roussel, so I would argue, and I think it particularly helpful to keep in mind his account of Roussel's route around the 'small spatial cells' of his 'chatty landscapes' when turning to a systematic presentation of what he means in the *Archaeology* by *spaces of dispersion*. I will return to the chief narrative of the paper, then, and suggest that one way of securing a handle on how Foucault envisages the 'working' of general history is to consider a notion introduced in one of the above quotations: namely, the notion of deploying the 'space of a dispersion', which presumably connects up to a secon notion – that of 'systems of dispersion' – as introduced by Foucault (1972) when discussing the analysis of discourse.[11] What Foucault seems to be proposing here is a form of *spatial ontology* which proceeds by imagining a hypothetical space or plane across which all of the events and phenomena relevant to a substantive study are dispersed; and this means that – to give an example close to my own current research – in a study of England's nineteenth-century 'mad-business' the researcher would envisage such things as asylums, upland environments, dirty towns, ardent reformers, the 1807 Select Committee, John Conolly, the *Asylum Journal*, a parliamentary debate, country walks, and Bentham's 'Panopticon' all being scattered over the space available. The wilful muddling up of items in this list – the playful juxtaposing of different categories of thing; the mixing of tangible and intangible, of natural and human, of collective and individual, of ongoing and time-bound – must be emphasised, as it is a strategy that right from the outset is an attempt to challenge the a priori ordering tendencies that so readily totalise historical inquiry.[12] In other words, the initial move in this conceptual 'limboland' is not to homogenise the components of relevance nor to chase them into one central compound, but to preserve and even to accent their details and the differences between them. It might be added that this abstract manoeuvre is even more radical than the 'levels' of thinking which Foucault appears to adopt elsewhere, and which sees him conceiving of history as 'moving on a number of different levels: the epistemological, the medical, the political, the pedagogical, the psychological, the economic and so forth' (Lemert and Gillan, 1982, page 43). It is revealing that his apparent spec-ification of levels still does not accept some neat division of social life into economic, political, social, and cultural strata,[13] and it is also revealing that each 'archaeological level' is granted some autonomy from others, but a 'levels model' still risks the positing of one category of events and phenomena as somehow being more fundamental than other categories – as somehow being more central – in a manner that prejudges the results of empirical inquiry. In short, a visualisation of myriad things being dispersed across a plane rather than being stacked up one on top of another comprises an excellent starting point for the general history that Foucault wishes to pursue.

And yet the envisaging of a space of dispersion is not tantamount to saying that all there is in the world is a chaos that the researcher can do no more than celebrate, because Foucault clearly supposes that there *is* some order in the dispersion waiting to be discovered, but that this order resides resolutely *in the things themselves* and not in any order theoretically imposed from without. This is not to imply that the researcher will need no theoretical imagination in teasing out the 'order of things', and neither is it to deny the value of tackling the case at hand with the hope of discov-ering the sorts of findings unearthed during the analysis of previous cases, but it is to maintain that no transcendental logic of how things are consti-tuted will be found at the heart of the substantive study. Indeed, beyond the broader claims that this study *will* reveal a region of differences and that there *will* be some discernible order or 'system' within these differ-ences,[14] Foucault reckons that it will not be possible to do more than specify what Dreyfus and Rabinow (1982, page 55) term 'local, changing rules'.

As these authors remark when examining Foucault's inquiry into 'statements' and 'discursive formations', although I believe that the principles expounded here are extendable to the wider project of general history:

> While the structuralist claims to find cross-cultural, ahistorical, abstract laws defining the total space of possible permutations of meaningless elements, the [Foucauldian] archaeologist only claims to be able to find the local, changing rules which at a given period in a particular discursive formation define what counts as an identical meaningful statement ... [T]he rules governing the system of statements are nothing but the ways the statements are actually related.
>
> (page 55)

This means that Foucault envisages not chaos but the connectedness of an order that is transient, and the upshot is that he signposts an avenue for inquiries which does not so much revel in dispersion as subject this dispersion to careful analysis free from any totalising retreat towards a priori constructs not rooted in the empirical materials at hand.[15] Crucial to Foucault's general history is hence the recovery of the 'local, changing rules' that in particular times and places govern, and in a sense simply *are*, the observable relationships between the many things under study.

But what can be said in addition is that these rules of a transient order are to do with space, and maybe to do with little more: they are the patterns *in* the spatial dispersion – the geography of how the things under study are scattered across the hypothetical plane, the distances between them (whether they stand together, nearby, or far apart) being indicative of the extent to which they differ from one another – and as such Foucault evidently supposes there to be an intelligible 'geometry' that enables the researcher to grasp a measure of order in how the things under study are connected. And yet this is not a fixed geometry, Euclidian or otherwise: it is a momentary geometry locked irrevocably into the fleeting character of the things themselves, and as such it is a geometry whose rootedness in the local time and place would be of little interest to the spatial scientist. It must immediately be acknowledged that, as presented here, Foucault's sensitivity to spaces of dispersion appears simply as a conceptual–metaphorical device designed to help us negotiate the traps of total history, but I would have no hesitation in hooking up this device to the treatment that Foucault provides in his substantive histories of tangible sites distributed across space. Indeed, I would argue that when Foucault gazes out on the social world of the past, he sees not the order of (say) a mode of production determining the lines of class struggle nor the order of (say) a worldview energising everything from how the economy functions to how the most beautiful mural is painted: rather, he sees the spaces of dispersion through which the things under study are scattered across a landscape and are related one to another simply through their geography, the only order that is here discernible, by being near to one another or far away, by being positioned in certain locations or associated with certain types of environment, by being arranged in a certain way or possessed of a certain appearance thanks to their plans and architectures. In part, then, the 'local, changing rules' to be recovered should contain detailed descriptions of the substantive spatial relations that the things of concern display with one another 'on the ground', and these descriptions should run alongside and inform any specification of more abstract rules of association and difference. Rather more needs to be said in this respect about Foucault's treatment of what he terms 'external spaces', of course, and having reached this point – and having begun to identify the substantive dimensions to Foucault's geography – it now becomes appropriate to turn more directly to such matters.

3 Foucault's geography in history: substantive manoeuvres

It now becomes appropriate to ask about the more 'substantive' manoeuvres present in Foucault's unravelling of how space and place have been inextricably bound up with the phenomena, events, people, ideas, and institutions that have been the very 'stuff' of his historical inquiries into madness and asylums, illness and clinics, criminality and prisons, sexuality and confession boxes (in short, his attempts to write the histories of 'social otherness'). In the following pages I want to suggest two sets of comments that might be made about Foucault's capturing of geography in history, both of which have a critical edge signposting an element of *mismatch* between the 'theory' of his geography and its actual 'practice', but what I will also go on to argue is that it is possible to look beyond these critical comments to a more favourable interpretation of what I term his evocation of 'substantive geographies'.

3.1 The geometric turn in histories of power

Even a casual glance at Foucault's substantive historical inquiries reveals a finely-honed alertness to *space*, or, to be more precise, to the way in which *spatial relations* – the distribution and arrangement of people, activities, and buildings – are always deeply implicated in the historical processes under study. In *Madness and Civilization* (1967) he draws various conclusions about what he terms the 'geography of haunted places',[16] for instance, whereas in *The Birth of The Clinic* (1976) he deals with the three different forms of 'spatialisation' involved in nosologies of disease, in the practices of pathological investigation, and in the provision of medical facilities or 'cure centres'.[17] Alternatively, in *Discipline and Punish* (1977) he explores the notion that 'discipline proceeds from the distribution of individuals in space' (page 141), and also describes in detail the physical and psychical control over individuals achieved through the manipulation of spatial relations in Bentham's notorious 'Panopticon' (see, too, the geographical papers by Dear, 1981; Driver, 1985a, 1985b; Philo, 1989a). In all of these works Foucault demonstrates through empirical detail the role played by spatial relations in the complex workings of *discourse, knowledge* and (crucially) *power*,[18] and it is thereby revealing that in one well-known interview he speculates that 'the history of *powers*' would at one and the same time amount to a history 'written of *spaces*' ('both these terms in the plural': Foucault, 1980a, page 149).

One way of characterising Foucault's projects here is to suggest that his sensitivity to spatial relations amounts to the introduction of a *geometric turn* into histories of 'social otherness', and it is also possible to find various commentaries portraying his studies as primarily concerned with excavating the geometries of power that have structured the historical experiences of the mad, the sad, and the bad. Indeed, writers such as Baudrillard (1987a) have talked at length about the fine capillaries of power discussed by Foucault, those microspaces through which power is supposedly both constituted and diffused:

> This time we are in a full universe, a space radiating with power but also cracked, like a shattered windshield still holding together . . . power is distributional; like a vector it operates through relays and transmissions . . . If we look closely, power according to Foucault strangely resembles 'this conception of social space which is as new as the recent conception of physical and mathematical spaces', as Deleuze says now he has suddenly been blinded by the benefits of science . . . The reference of power, which has a long history, is discussed again today by Foucault at the level of dispersed, interstitial power as a grid of bodies and of the ramiform pattern of controls.
>
> (pages 37, 42, 34 and 38)

In these quotes, Baudrillard revealingly connects up Foucault's work to a new form of *naturalism* that is arguably beginning to permeate social thought, the 'new physics' opened up by Heisenberg's uncertainty principle and the 'new chemistry' opened up by Monad's genetic code, and in this respect it is appropriate to mention both Major-Poetzl's (1983) explicit interpretation of Foucault in the light of changing natural scientific notions of spatial relations and Baudrillard's own equation of Foucault's work with Deleuze's 'molecular topology of desire'. For Baudrillard (1987a), it is thus possible to identify a curious coming together of diverse intellectual traditions from social science on the one hand and natural science on the other: two traditions 'whose flows and connections will soon converge – if they have not already done so – with genetic simulations, microcellular drifts and the random facilitations of code manipulators' (page 35).

We are hence left with an image of a social world spatially constituted through nodes and channels of power – fixed nodes where power is produced and crisscrossing channels along which power is diffused and collected – and it is for this reason that Foucault might aptly be termed the 'geometer of power', and it is for this reason too that Foucault could be said to slide into a form of social explanation not so very distant from those spatial scientists who aim to explain social life (everything from settlement patterns to cognitive processes) simply by charting the 'scientific' laws governing its constituent geometries.

This line of argument could be given in more detail by considering specific of Foucault's substantive inquiries, but let me confine my attention here to a handful of arguments that can be made about the geometric turn present within Foucault's first major text, *Madness and Civilization*. A revealing quote in this respect can be found where Major-Poetzl (1983) discusses the interpretation that Serres offers of this particular text:

> [Serres] interprets Foucault's categories of inclusion and exclusion in terms of spatial relationships, and . . . views Foucault's concept of unreason as a 'geometry of negativities'. The pre-Classical [pre]1600] period, Serres suggests, can be imagined as an original chaotic space in which madness had many points of contact with the world. The Classical [c1600–c1800] space, by contrast, was dualistic, with the space of unreason (hospitals [the *hôpitaux généraux* or prison – workhouses] and later asylums) functioning as a negative image of the space of reason (society, in particular the family).
>
> (page 120)

This is not the place to examine the geography of *Madness and Civilization* in any detail – in another paper I do offer such an examination (Philo, 1992b) – but what the above quote indicates is that the basic narrative line running through Foucault's text concerns the historical emergence in Western Europe of an impulse both social *and spatial* towards segregating people labelled as mad (as 'lunatic', 'insane', 'mentally ill') from the 'normal' round of work, rest, and play, often with the consequence that these people have ended up living out their days in houses of confinement both non-specialist (workhouses, prisons) and specialist (asylums, mental hospitals, mental health facilities).[19] Serres describes such houses of confinement as constituting the 'space of unreason', as the negative mirror image of the 'space of reason' allegedly to be found in the happy homes and streets of respectable families and communities, and he effectively finds in *Madness and Civilization* the story of a social geometry – the so-called 'geometry of negativities' – in which a certain human population (itself far from internally homogenous) becomes stigmatised and sociospatially excluded from society's 'normal' sites of interaction. A simple geometry of 'inclusion' and 'exclusion', of 'inside' and 'outside', it thereby projected on to the history of Western madness and nowhere is this more apparent than in Foucault's account of an historical continuity in the exurban location of social deviance, from the leper colony sited outside Medieval towns to the *hôpitaux généraux* and specialist asylums sited outside Early Modern and nineteenth-century urban areas. Read in this way, *Madness and Civilization* appears to fall into many of the traps of total history – its positing of clear historical stages in the story of a specific phenomenon (madness), its identification of key themes and continuities – and it also appears as if a sensitivity to space is very much written into the text, but not so much as an ongoing challenge to the totalising story being told as a geometric complement to this story. Moreover, and this may be a point of wider salience for this paper, Derrida (1978b; see also Boyne, 1990) criticises Foucault for seeking to hear and in part to write the 'language of unreason' – for trying to recover its 'truths' and to establish how these are compromised – from his own unavoidable position *within* the 'language of reason'. All that Foucault can achieve is to open up a discourse *about* madness which draws upon the precise conventions of rationality that in practice act to 'imprison' madness, so Derrida claims, and the suggestion might be that the geometric turn in Foucault's history of madness – the desire to impose simple spatial categories and, indeed, dualistic oppositions on the historical materials – is itself bound up with the continuing hegemony of reason (and perhaps of modernism as well) even in a text such as *Madness and Civilization*. These are important arguments, and they could be extended into a broader critique of an incipient equation of geometry, reason, and modernism present in all (or many) of Foucault's substantive historical inquiries, but I think that they do not paint the complete picture of how Foucault deals with space in his research on past madnesses, deviancies, and so on: and I will explain why I think this to be the case presently.

3.2 'Making the space in question precise'?

A second feature and perhaps problem of Foucault's geography arises in the way that he treats *place* in his substantive historical inquiries, and it occurs to me that – despite and maybe because of his geometric turn in focussing upon spatial relations – in much of his historical writing he does not show the concern for the associations of particular phenomena with particular material places, environments, and landscapes that might be expected given his theoretical stress on the importance of empirical details and differences. This weakness in his historical inquiries has not gone unnoticed, and in a 1976 interview with the French journal of radical geography *Hérodote* it was said:

> one finds in your work a rigorous concern with periodisation that contrasts with the vagueness and relative indeterminacy of your spatial demarcations. Your terms of reference are alternatively Christendom, the Western world, Northern Europe and France, without the spaces of reference ever really being justified or even precisely specified . . . [But, and as the interviewers then asked, how did this indeterminacy square with Foucault's complex time-based 'methodology of discontinuity'?] It is possible, essential even, to conceive such a methodology of discontinuity for space and the scales of spatial magnitude. You accord a *de facto* privilege to the factor of time, at the cost of nebulous or nomadic spatial demarcations whose uncertainty is in contrast with your care in marking off sections of time, periods and ages.
>
> (Foucault, 1980b, page 67)

In response to this Foucault began by commenting upon the difficulties associated with piecing together an analysis from documents in geographically dispersed archives, but he then readily agreed with his interviewers that 'There is indeed a task to be done of making the space in question precise, saying where a certain process stops, what are the limits beyond which something happens – though this would have to be a collective interdisciplinary undertaking' (page 68). But, even given this acknowledgement, the point must stand that – for all his paying of attention to details and to differences – Foucault has rarely followed his own advice of 'making the space in question precise', of specifying in some detail the particular places with their particular contextual characteristics where his histories work themselves out, perhaps to the overall detriment of these inquiries.

A slightly different window on the same set of issues is opened by Lemert and Gillan (1982), meanwhile, who effectively challenge Foucault for failing to pay sufficient attention to the material contexts of his histories. Indeed, they begin by underlining the sensitivity that Foucault *does* possess for time: a sensitivity that leads him to identify the dates of specific events and also to recognise the complexities of 'historical time' residing in the different temporal limits that can be attached to particular periods, in the differing temporalities of different 'archaeological levels' of reality (see page 219 above), and in the common lack of chronological correspondence between developments at different levels. This is Foucault's time-based 'methodology of discontinuity', and Lemert and Gillan argue that this methodology is clearly inspired by the tendency of *Annalistes* such as Braudel to view history not as a simple sequence of earlier events begetting later events, but as being located 'at the conjuncture of material, economic and social forces' (Lement and Gillan, 1982, page 11). And what these commentators also argue is that for Foucault, as for the *Annalistes*, 'history does not run through time' but emerges from the relations of a time that is 'spatialised'. Having charted this connection between Foucault and the *Annalistes* they then declare that, although Foucault may follow Braudel *et al.* in spatialising time, his own historical studies actually suffer from a *failure* to take on board the 'dialectic of space and time' central to Braudel's 'geohistory' (pages 97–98). In part this means that he fails to appreciate how Braudel takes seriously not only the complexities of datings, periodisations, and temporalities (which in an abstract sense leads to the spatialisation of time), but also takes a more substantive cognisance of the different rates at which different phenomena spread across the tangible spaces of plains, valleys, mountains, and suchlike. The result is that, 'in order to explain the rise and fall of markets, the location of cities and the growth of civilizations, it is necessary to embrace time as a passage of events across the extension of space' (page 97).[20] And this is not all, as Lemert and Gillan (pages 97–98) go on to explain:

> The dialectic of space and time is absent from Foucault's historical writings. Spatialised time makes it possible to introduce the concepts discursive formation, knowledge (*savoir*) and power. But discursive formations, knowledge and power do not adhere to locations, to systems of communication or to cultural networks tied to economic routes and marketplaces. Braudel's *The Mediterranean*, Bloch's *Feudal Society*, Le Roy Ladurie's *The Peasants of Languedoc*, George Lefebvre's *The Great Fear of 1789* have indicated that neither power, nor language, nor knowledge can be separated from geography. There is more, they suggest, to the problematic of power and knowledge than an epistemic strategy. Power and knowledge as exercised in rural and urban societies work in a space criss-crossed by trade routes, valleys and highlands, mountains and rivers.

The suggestion is therefore that – whereas for Foucault 'power and knowledge operate in the space of the body, not of geography' – for Lemert and Gillan (page 98) it is essential to recognise that the 'body in history is part and parcel of those anonymous structures such as land and sea routes, the plan and location of cities, climate and terrain'. This recognition evidently encompasses the above complaint that Foucault fails to 'make the space in question precise', but it also does more than this: it stresses that the operations of discourse, knowledge, and power are crucially bound up with the soils of nameable places; the material contexts of particular environments and landscapes, regions and countries. This is a signal claim, in that it indeed pinpoints the substantive importance of the geography *in* history – it goes beyond saying that we must take geography seriously because differences between places inevitably fragment the logic of historical 'metanarratives' – and in the process it opens up the possibility of rethinking the need for research into *geographical history*: into the way in which historical processes involving discourse, knowledge, and power (as well as all manner of other processes which might be conceptualised in non-Foucauldian terms) are always at work in and shaped by that real-world space 'criss-crossed by trade routes, valleys and highlands, mountains and rivers'.[21] These are again valuable arguments that could be extended in various ways, then, but once again I feel that their critical tone does not paint the whole picture of how Foucault conducts his substantive historical inquiries.

3.3 Foucault's 'substantive geographies'

Writing in 1985 about the potential of a realist human geography, Gregory (1985, pages 70–73: see also Cloke *et al.*, 1991, pages 161–164) drew upon the geometric notations of Haggett's *Locational Analysis* (1965) – 'the nodes, movements and networks, hierarchies and surfaces' – to suggest the outline form of the spatial structures associated with different 'levels' of social reality, but he went on to insist that 'what appeared there as *formal geometries* appear here as *substantive geographies*' (my emphases). What Gregory argues is that social life clearly does possess a geometry, but that we must always go beyond simply mapping the dots, circles, lines, hexagons, and trend surfaces: indeed, we must not suppose that the formal geometric languages we might use to represent these shapes and relations in the abstract will tell us anything of consequence about their reality 'on the ground', and from the outset we must regard these geometries as full of substance – as entirely substance-ridden – and must thereby always be thinking in terms not of the straight-line distance across an 'isotropic plane' between point A and point B but of, for example, the distance along difficult and hilly tracks between the cottage of Cornelius Ashworth and the local cloth market in Halifax (see also Gregory, 1982). As Gregory indicates (1985, page 73), '[c]ontent is [to be] poured into the inquiry at every level in a constant and creative process of discovery', and such continual attention to content will ensure that the geometries 'come alive' and become energised by the real everyday struggles that people face when seeking to negotiate their way around (often acutely unhelpful) geographies of production, reproduction, and consumption. In a sense, Gregory's ambition in this respect is no less than to rescue geography from geometry *without* jettisoning the hard-won insight that spatial structure makes an important difference to the conduct of social life, and it seems to me that this manoeuvre is similar to one that can be made in rescuing Foucault's geometric turn from a lapse into a formalism, a naturalism, and a modernism antithetical to the protocols of his general history.

I hence think that the spatial relations discussed throughout Foucault's histories of social otherness can best be understood not as formal geometries, but as substantive geographies where (say) the geometry of town–country relations in the history of madness or the geometry of prison plans in the history of criminality is immediately envisaged as being full of people, problems, ideologies, happenings, resistances, or whatever. For instance, a close reading of *Madness and Civilization* and related texts quickly reveals that we must look beyond the broad-brush account of a 'geometry of negativities' to recover the specificities of how particular mad, bad, and sad human populations have been identified, categorised, maybe stigmatised, maybe idolised, maybe included in everyday life, maybe excluded in houses of confinement in very different ways at different times and in different places. It is evident that in Western Europe since about 1600 there *has* been something of an impulse to segregate mentally disordered people, but positing an a priori geometric model of unreason always and everywhere excluded to the 'outside' of social life should only be a very preliminary step before engaging directly with the substance of past situations where certain people may have been taken to certain institutions for certain economic, social, political, and/or cultural reasons (and of course there have been many situations where segregation has in practice *not* been the favoured solution).[22]

And a similarly close reading of *Discipline and Punish* and related texts indicates that positing an a priori geometric model of 'panopticism', a model which supposes all manner of disciplinary institutions to be laid out to satisfy the 'inspection principle' of Bentham's 'Panopticon', can only act as a first base when writing the historical geography of prisons, asylums, workhouses, colonies for delinquents, and suchlike, many of which have departed dramatically from the 'Panopticon' model thanks to following other logics under other circumstances (Driver, 1985b, 1990; Philo, 1989a). Foucault therefore arrives at a treatment of space that is not completely beholden to geometry, and he is certainly not labouring with any notion of transcendental spatial laws as might a 'Haggettesque' spatial scientist, and what might now be added is that a careful scrutiny of his historical inquiries also unearths a treatment of place that is on occasion more attentive to details of precise location and context than my earlier criticisms imply. Thus, we find that two very specific places are featured in *Madness and Civilization*: namely, Paris, where the first *hôpital général* was founded in 1656 and where Philippe Pinel struck off the chains of the lunatics in Bicetre in 1796; and York, where William Tuke established a specialist asylum – the 'Retreat' – whose outwardly humanitarian regime was to prove such a model for subsequent phases of asylum-building throughout the world. And these places turn out to be more than just incidental to

Foucault's history: the urban geography of Early Modern Paris – its demo-graphic, economic, social, and political characteristics, as conjoined with its messy, often dirty, and diseased urban fabric – was central to the shaping of institutional responses to both madness and other deviancies (feeding into the *hôpitaux généraux* and subsequent specialist asylum provision for the mad); and the rural geography of the districts around late eighteenth-century York – the 'fertile and smiling countryside', as one visitor to Tuke's 'Retreat' put it (de la Rive, in Foucault, 1967, page 242) – was central to the formulation of a therapeutic regime that deliberately replaced the disturbing sights and sounds of town life with the surroundings of nature and the wholesomeness of agricultural labour. Similar claims might be made about the specific places that feature in other historical inquiries of Foucault's, such as the region around the Mettray reformatory colony in France, but there is no need to pursue this point further here.

There *are* undoubtedly some problems with Foucault's treatment of space and place in his histories of social otherness, in that there is a danger of his geometric turn effectively elevating an abstract sense of space above a concrete sense of place, but my own view is that the practice of his geography still manages to put enough content into the picture to prevent it becoming solely an exercise in formal geometry. What I would also add is that imagining Foucault as less the 'geometer of power' and more the patient 'archaeologist of substantive geographies' is something that appar-ently resonates with his own views, notably when he highlights the value of proceeding with a clear attunement to real, worldly spaces ('external spaces') full of substance-ridden things (people, animals, forests, rivers, slopes, buildings, roads, railways: the list is endless) all jumbled up together and related one to another through spatial relations:

> The space in which we live, which draws us out of ourselves, in which the erosion [passing] of our lives, our time and our history occurs, the space that claws and gnaws at us, is also, in itself, a heterogeneous space ... We do not live inside a void ..., we live inside a set of relations that delineates sites which are irreducible to one another and absolutely not superimposable on one another.
>
> (Foucault, 1986c, page 23)

As implied earlier, then, we arrive here at a translation from Foucault's conceptual–metaphorical feel for spaces of dispersion to a more tangible feel for substantive geographies. It is signal to underline the fact that a geometric turn is still clearly visible in the attention paid to 'relations that delineate sites' – and we can still envisage gazing down on the geometry of these sites, and maybe even mapping them as nodes, networks, and surfaces – but what must also be accented is how Foucault immediately recognises the rich content of this 'heterogeneous space' that 'claws and

gnaws at us' rather than blandly confronting us as a series of dots and
lines on a piece of paper. Furthermore, it may be significant that in the
above quote Foucault stresses the 'irreducibility' and 'nonsuperimposability'
of sites, and this claim could be interpreted as a genuine concern for
places – for the details of what 'takes place' in specific places, environments, and landscapes – which are always reckoned to differ one from
another as well as constituting crucial contextual influences upon the workings of history and social life.

4 'The geography of things'

> We could perhaps develop a model of drifting plates, to speak in
> seismic terms, in the theory of catastrophes. The seismic is our
> form of the slipping and sliding of the referential ... Nothing
> remains but shifting movements that provoke very powerful rare
> events. We no longer take events as revolutions or effects of the
> superstructure, but as underground effects of skidding, fractal zones
> in which things happen. Between the plates, continents do not
> quite fit together, they slip under and over each other. There is
> no more system of reference to tell us what happened to *the geog-
> raphy of things*. We can only take a geoseismic view.
> (Baudrillard, 1987b, pages 125–126,
> emphasis added)

In this quote Baudrillard visualises the social world in terms of what he
calls the 'geography of things': a view that arises from a deep scepticism
about the ability of theoretical endeavour adequately to represent the
'goings on' of the thing-realm (the realm of all objects beyond the hallways of theory, including other people), and a view that thereby supposes
this thing-realm to obey its own rules akin to deep 'geoseismic' logics and
forces whose workings will always remain unknowable to the subjectivity
of the researcher. I have begun elsewhere to examine these and other
aspects of 'Baudrillard's geography' (Philo, 1990), but for the purposes of
bringing this paper to a close I simply want to suggest that here – in this
account of the social world as a messy and (to the researcher's eye) disordered geography of 'plates', 'continents', or 'fractal zones' slipping, sliding,
and skidding into, under, and over one another – we encounter a view
that parallels and complements Foucault's theoretical alertness to spaces
of dispersion *and* his substantive attention to substantive geographies. And
it is a view that I would argue has much in common with the way in
which Roussel traces the geography of the 'small spatial cells' in his tiny
pictures, telling a range of stories as he does about the people and places
depicted in and connected by these cells full of incredible detail and differ-

ence. It is important to realise that both Baudrillard and Foucault arrive at this geographical way of looking at the social world as a result of their doubts about the great certainties of order, coherence, truth, and reason assumed by those intellectual exercises that give priority to the 'highest' and the 'grandest' concepts: and as such they abandon what might be termed *depth* accounts of social life, where more fundamental levels of social reality (whether these be conceived of as economic, psychological, or whatever) are called upon to explain less fundamental ones, and move instead to what might be termed a *surface* account, where the things of the world – the phenomena, events, people, ideas, and institutions – are all imagined to lie on the same level (whether they be advanced capitalism or the toy rabbit) in a manner that strives to do away with hierarchical thinking. And it should not be difficult to see that this criticism of depth accounts and the attendant move to surface accounts stands Baudrillard and Foucault squarely in line with the emerging suspicion of totalising theories identifiable as the attitude of postmodernism (see also the arguments in Cloke *et al.*, 1991; Gregory, 1989a, 1989b; Ley, 1989), and it is for this reason that I describe as 'postmodern' the theoretical and substantive geographies that are so intimately bound up with the twin attempts of these two writers to cope with the fragmentation and chaos remaining after the modernist certainties have been thrown into question.

In this paper I have concentrated on Foucault's geography, then, and have sought to show how his retheorising of history necessitates a sensitivity to space, place, and geography – to indicate how general history, his alternative to total history, depends upon thinking in terms of spaces of dispersion – and I have suggested that this sensitivity is not simply a conceptual–metaphorical device, but one which features in a more substantive 'register' throughout his historical inquiries into social otherness. This latter suggestion has then led me into a direct if preliminary consideration of the geography written into Foucault's historical works, and my conclusion here is that – although in practice a problem *can* arise because of a geometric turn that risks overplaying space and underplaying place – there is still clear evidence of him recovering substantive geographies where 'content is [indeed] poured into the inquiry at every level in a constant and creative process of discovery' (Gregory, 1985, page 73). It also seems to me, finally, that what Foucault provides us with here is a blueprint for a truly 'postmodern' geography: a postmodern geography in which details and difference, fragmentation and chaos, substance and heterogeneity, humility and respectfulness, feature at every turn, and an account of social life which necessarily brings with it a sustained concern for the geography of things rather than a recall for the formal geometries of spatial science. In his essay on '*Chinatown* Part Three' Gregory (1990) dwells upon what he terms a 'geometric imaginary' that permeates Soja's *Postmodern Geographies* (1989), and a key argument for Gregory is to warn about allowing such a geometric turn to subvert the gains of the postmodern 'reassertion of space in social theory' (and there is perhaps a similarity here with his warning against giving an overly geometric reading of spatial structure in realist human geography: see pages 224–30 above). A warning of this sort is also appropriate in relation to the geometric leanings of Foucault's geography, but it might be argued that, whereas Baudrillard's geometric representation of the US desert is taken by Gregory as a lens to reflect upon the geometry and latent modernism of Soja's supposedly postmodern geography, a more appropriate lens to reflect upon Foucault's geography is maybe given by the dense and seemingly chaotic (though meaningful and interpretable) tangle of words and pictures produced by Quoniam (1988) – the 'geographer, painter' – in response to Arizona's rugged desertscape. This alternative lens hints at a substantive geography, not a formal geometry, and at a way of thinking that could be described as antimodernist, postmodernist, or simply 'other' and beyond the categories.

Acknowledgements

I would like to thank Derek Gregory, Dagmar Reichert and Jennifer Robinson for their advice and criticism, and Nigel Thrift. This paper benefitted from the critical but fair comments of referees when it was published in *Society and Space*.

Notes

1 This phrase is taken from one of the referees of this paper, who notes – in reflection of my emphasis upon Foucault – that 'maybe we dredge the thoughts of a single luminary too much and neglect the many voices, the "other" voices who also have worthwhile ideas'. I agree entirely with this sentiment, although I would reply that the purpose in this paper is precisely to let (as far as is possible) the 'other' voice that is Foucault's speak in order that it not be neglected by geographers or (as I feel is more likely) simply translated into a language that is not its own.

2 It is telling that in one keynote commentary (Dear, 1988) a worry is expressed about the fragmentation of human geography (the discipline) as prompted by the 'postmodern challenge', which the author feels has led to the majority of geographers studying highly specialist topics and adopting an 'anything goes' attitude towards matters of philosophy and methodology. The appropriate response, so it is claimed, is 'to reconstruct human geography by realigning it with the mainstream of social theory' (page 271), and – although the aim here is avowedly not to hook up the discipline to a 'search for grand theory' (page 272) – the suggestion remains that progress will only be made if geographers attend first and foremost to questions posed in a realm of a priori theory.

3 It should be acknowledged that Foucault's target is not so much the *Annales* 'school' of French historians, even though these researchers are often referred to as practitioners of an *histoire totale*: rather, although some of his criticisms do apply to this 'school', it is also the case that his own project of general history owes much to *Annalistes* such as Braudel. Baker (1984, page 10) is hence correct

to assert that 'Foucault is integrating and extending the ideas of ... *Annalistes* about the character of historical transformations, building upon total history what he terms a general history', but such an observation must not obscure the fact that Foucault's objective is to forge a general history which actually stands in marked opposition to – and which in no sense seeks to build upon – what *he* means by total history.

4 Foucault's rejection of total history clearly stands alongside the anti-totalisation tracts prepared by a number of 'poststructuralist', or, as some might say, 'postmodern' thinkers (including Baudrillard, Deleuze, Derrida, Lyotard, and various Anglo–American cultural and literary theorists). These efforts are neatly outlined by one author who is actually restating a case for Marxism (Callinicos, 1982, page 112):

> From the standpoint of a philosophy of difference, which insists on the priority of multiplicity, which denies the possibility of a simple essence at the origin of things – the 'true substantive, substance itself is multiplicity', writes Deleuze – Hegel, because his system envelops difference in the Absolute Idea, is the enemy who must be defeated, must be destroyed.

5 See Lemert and Gillan (1982, page 11), who suggest that for Foucault as for the *Annalistes*, 'causality in history is not from human event to human event, but at the conjuncture of material, economic and social forces'. But the similarities go deeper in that Foucault's vocabulary for describing his general history – his references to elements, series, 'series of series', and temporalities – echoes loudly the vocabularies of *Annalistes* such as Braudel.

6 Foucault's so-called 'genealogical' inquiries – as inspired by Nietzsche – are usually taken simply to include his later work on issues to do with power, the clash of multiple forces, and the disciplining of society (in particular, see Foucault, 1977; 1979), but in the text where he first introduces his approach to 'genealogy' (Foucault, 1986a) he is actually making a series of claims very similar to those that appear in the introduction to his main 'archaeological' text (Foucault, 1972). This is perhaps unsurprising, given that the two works were prepared almost simultaneously, but it does indicate that for Foucault the conceptual distance between genealogy and archaeology is not as great as is sometimes thought.

7 To use the evocative phrase of Carlyle, as employed by Trevelyan (1948, pages viii–ix) when suggesting that even the most 'Dryasdust' historian is in truth a 'poet' grasping for an imaginative transcendence of the dusty empirical sources.

8 This claim is inspired not just by Foucault, but also by the arguments of Geertz (1973, pages 27–28) about how anthropologists should relate theoretical materials to empirical details:

> A repertoire of very general, made-in-the-academy concepts and systems of concepts ... is woven into the body of thick-description ethnography in the hope of rendering mere occurrences scientifically eloquent. The aim is to draw large conclusions from small, but very densely textured facts; to support broad assertions about the role of culture in the construction of collective life by engaging them exactly with complex specifics.

9 I am sure that Foucault would follow Sartre (1948, page 28) in supposing 'existence to precede essence', and that his attitude towards the things of the world is akin to that of Roquentin in *Nausea* (Sartre, 1964, page 185) peering at the

'black, knotty mass of a chestnut tree root, [and] realising that, faced with that big rugged paw, neither ignorance nor knowledge had any importance; the world of explanations and reasons is not that of existence'.

10 Though maybe not with the 'primitive' rules of spatial relations (the nearness and farness, the 'to-the-rightness' and 'to-the-leftness', the 'aboveness' and 'below-ness').

11 In this connection Foucault (1972, page 37) argues as follows:

> hence the idea of describing these dispersions themselves; of discovering whether, between these elements [the *contents* of 'things said' or of 'things written'], which are certainly not organised as a progressively deductive structure, nor as the *oeuvre* of a collective subject, one cannot discern a regularity, an order in their successive appearance, *correlations in their simultaneity, assignable positions in a common space* [my emphasis].

In Philo (1989b) I discuss at greater length the character of Foucault's thinking about 'discourse' in and around *The Archaeology of Knowledge*, and in so doing I make a number of observations that connect up to my arguments in this paper (as well as speculating on how an alertness to discourse might illuminate the study of locational decisionmaking in human geography).

12 A connection might be drawn here between this deliberately muddled ontology and the delight that Foucault (1970, page xv) takes in Borges's 'certain Chinese encyclopaedia', in which an apparently 'strange categorisation of animals shatters the 'familiar landmarks of thought' and thereby threatens to 'break up all the ordered surfaces and all the planes with which we are accustomed to tame the wild profusion of existing things'.

13 So many intellectual exercises – *Annaliste*, Marxist, structuralist, realist, and others – now proceed with a 'levels model' of the social world that few writers ever pause to consider what the implications would be of *abandoning* such a way of thinking (though, see Darnton, 1985).

14 In one sense this claim – that there are always differences, that these differences are always patterned in an intelligible fashion – is written into Derrida's conception of *différance*. As he writes (1982, pages 21–22), *différance* is the principle underlying the 'production of differences' – and particularly the production of those differences between words, concepts, and things which make possible acts of 'signification' and of 'conceptuality' – but adds that 'it is not a present being, however excellent, unique, principal or transcendent. It governs nothing, reigns over nothing and nowhere exercises any authority. It is not announced by any capital letter'. In other words, the 'motif' of *différance* proclaims that there *will* be some discernible order of 'system' within the differences of the world, but that there will be little to say about this order prior to its recovery from particular situations.

15 Dreyfus and Rabinow (1982, page 56) declare that Foucault is still conducting an 'analysis', as he deals with 'elements' and 'rules' (or with details, differences, and the connections between categories of statement [or of thing] and their context-dependent transformations' – seems very alien when set alongside most other varieties of scientific inquiry.

16 These 'haunted places' are those institutional spaces beyond the city walls to which lepers were consigned in Medieval times, and to which all manner of 'misfits' – the 'mad person' included – have been consigned in more recent times.

17 This text begins with the remark that 'this book is about space' (Foucault, 1976, page ix) and its first chapter is entitled 'Spaces and classes'. It is intriguing to

speculate what a medical geography based on a Foucauldian account of the 'spatialisation' of medical knowledge and practice would look like.

18 I would suggest that here discourse, knowledge, and power are equivalent to Geertz's 'very general, made-in-the-academy concepts' (see note 8), and that Foucault only makes these concepts work for him through their engagement with the realm of 'densely textured facts'.

19 The 'big story' that Foucault tells here – one which has been seriously criticised on both theoretical and empirical grounds, and one which he has since in part disowned – concerns the way in which throughout the history of Western societies there has been a progressive 'silencing' of madness, from a time (from antiquity to the Middle Ages) when people identified in one way or another as mentally 'different' were *included* within the social body – were tolerated, welcomed, even venerated; were allowed into a 'dialogue' with everybody else – to a time (from the Early Modern period onwards) when these individuals became *excluded* from the social body. Foucault argues that the latter situation has arisen because of a moral and medical fear of the 'otherness' of madness, a fear itself bound up with more material difficulties associated with a capitalist order hostile to seemingly 'nonproductive' and 'dependant' human populations. And he also argues that this fear has been manifested through the shutting up of mentally disturbed people in specialist asylums, mental hospitals, and mental health-care facilities. Moreover, for Foucault the twentieth-century psycho-analytical innovations of Freud and his followers have amounted to nothing but an intensification of the process whereby the voices of madness have been silenced: he claims that, for all the apparent listening which analysts do to the accounts given by patients of their own lives and experiences, in the final 'analysis' the a priori terminologies, concepts, and models of psychoanalytic theory end up imposing an artificial explanatory framework that pays little heed to the specific things said and felt by specific sufferers in specific circumstances.

20 Note the additional remark that for both Braudel and Lefebvre 'historical time is an extended time, a spatialised temporality' (Lemert and Gillan, 1982, page 97).

21 In Philo (1992a) I argue that we need to recover the project of 'geographical history' from the snares of environmental determinism, where the workings of the social world are reckoned to be explainable by reference to the natural characteristics of regional environments, and to formulate instead a type of historical inquiry sensitive to the very real difference that space and place in all of their complexity make to all historical 'stories' [whether 'large' ones about (say) the transition from feudalism to capitalism or 'small' ones about (say) the running down of the Dartmoor tin industry].

22 In Philo (1992b) I trace these more detailed and nuanced aspects of the narrative told in *Madness and Civilization*, and then investigate at length the gradual move to incarcerate mad people in England and Wales from Medieval times through to the second half of the nineteenth century.

References

Baker A R H, 1984, 'Reflections on the relations of historical geography and the *Annales* school of history', in *Explorations in Historical Geography* eds A R H Baker, D Gregory (Cambridge University Press, Cambridge): 1–27.

Baudrillard J, 1987a, 'Forget Baudrillard', in *Forget Foucault* J Baudrillard (Columbia University Press, New York): 7–64.

Baudrillard J, 1987b, 'Forget Baudrillard: an interview with S Lotringer', in *Forget Foucault* J Baudrillard (Columbia University Press, New York): 65–137.

Bonnett A, 1990, 'Key words' *Praxis* **19** 10–12.

Boyne R, 1990 *Foucault and Derrida: The Other Side of Reason* (Unwin Hyman, London)

Callinicos A, 1982 *Is There a Future for Marxism?* (Macmillan, London).

Cloke P, Philo C, Sadler D, 1991 *Approaching Human Geography: An Introduction to Contemporary Theoretical Debates* (Paul Chapman, London).

Darnton R, 1985 *The Great Cat Massacre and Other Episodes in French Cultural History* (Penguin Books, Harmondsworth, Middx).

Dear M, 1981, 'Social and spatial reproduction of the mentally ill', in *Urbanisation and Urban Planning in Capitalist Society* eds M Dear, A J Scott (Methuen, Andover, Hants): 481–497.

Dear M, 1988, 'The postmodern challenge: reconstructing human geography' *Transactions of the Institute of British Geographers: New Series* **13** 262–274.

Derrida J, 1978a, 'Structure, sign and play in the discourse of the human sciences', in *Writing and Difference* J Derrida (Routledge and Kegan Paul, London): 278–293.

Derrida J, 1978b, 'Cognito and the history of madness', in *Writing and Difference* J Derrida (Routledge and Kegan Paul, London): 31–63.

Derrida J, 1982, '*Différance*', in *Margins of Philosophy* J Derrida (Harvester Press, Hemel Hempstead, Herts): 1–27.

Dews P, 1987 *Logics of Disintegration: Post-structuralist Thought and the Claims of Critical Theory* (Verso, London).

Dreyfus H L, Rabinow P, 1982 *Michel Foucault: Beyond Structuralism and Hermeneutics* (Harvester Press, Hemel Hempstead, Herts).

Driver F, 1985a, 'Power, space, and the body: a critical assessment of Foucault's *Discipline and Punish*' *Environment and Planning D: Society and Space* **3** 425–446.

Driver F, 1985b, 'Geography and power: the work of Michel Foucault', unpublished typescript, Department of Geography, Royal Holloway and Bedford New College, Egham, Surrey.

Driver F, 1990, 'Discipline without frontiers? Representations of the Mettray Reformatory Colony in Britain, 1840–1880' *Journal of Historical Sociology* **3** 272–293.

Foucault M, 1967 *Madness and Civilization: A History of Insanity in the Age of Reason* (Tavistock Publications, Andover, Hants).

Foucault M, 1970 *The Order of Things: An Archaeology of the Human Sciences* (Tavistock Publications, Andover, Hants).

Foucault M, 1972 *The Archaeology of Knowledge* (Tavistock Publications, Andover, Hants).

Foucault M, 1976 *The Birth of the Clinic: An Archaeology of Medical Perception* (Tavistock Publications, Andover, Hants).

Foucault M, 1977 *Discipline and Punish: The Birth of the Prison* (Allen Lane, London).

Foucault M, 1979 *The History of Sexuality: Volume 1, An Introduction* (Allen Lane, London).

Foucault M, 1980a, 'The eye of power: conversation with J-P Barou and M Perrot', in *Power/Knowledge: Selected Interviews and Other Writings, 1972–1977, by Michel Foucault* ed. C Gordon (Harvester Press, Hemel Hempstead, Herts): 146–165.

Foucault M, 1980b, 'Questions on geography: interview with the editors of *Hérodote*', in *Power/Knowledge: Selected Interviews and Other Writings, 1972–1977 by Michel Foucault* ed. C Gordon (Harvester Press, Hemel Hempstead, Herts): 63–77.

Foucault M, 1982, 'Interview with Michel Foucault on space, knowledge and power' *Skyline* (March) 17–20.

Foucault M, 1986a *Death and the Labyrinth: The World of Raymond Roussel* (Athlone Press, London).

Foucault M, 1986b, 'Nietzsche, genealogy, history', in *The Foucault Reader* ed. P Rabinow (Penguin Books, Harmondsworth, Middx: 76–100.

Foucault M, 1986c, 'Of other spaces' *Diacritics* (Spring) 22–27.

Geertz C, 1973 *The Interpretation of Cultures: Selected Essays* (Basic Books, New York).

Gregory D, 1982 *Regional Transformation and Industrial Revolution: A Geography of the Yorkshire Woollen Industry* (Macmillan, London).

Gregory D, 1985, 'People, places and practices: the future of human geography', in *Geographical Futures* ed. R King (Geographical Association, 343 Fulwood Road, Sheffield S10 3BP): 56–76.

Gregory D, 1989a, 'Areal differentiation and postmodern human geography', in *Horizons in Human Geography* eds D Gregory, R Walford (Macmillan, London): 67–96.

Gregory D, 1989b, 'The crisis of modernity? Human geography and critical social theory', in *New Models in Geography, Volume Two* eds R Peet, N Thrift (Unwin Hyman, London): 348–385.

Gregory D, 1990, '*Chinatown* Part Three? Soja and the missing spaces of social theory' *Strategies* **3** 40–104.

Gregson N, 1989, 'On the (ir)relevance of structuration theory to empirical research', in *Social Theory of Modern Societies: Anthony Giddens and His Critics* eds D Held, J B Thompson (Cambridge University Press, Cambridge): 235–248.

Haggett P, 1965 *Locational Analysis in Human Geography* (Edward Arnold, Sevenoaks, Kent).

Hutcheon L, 1988 *A Poetics of Postmodernism: History, Theory and Fiction* (Routledge, Chapman and Hall, Andover, Hants).

Lemert C C, Gillan G, 1982 *Michel Foucault: Social Theory as Transgression* (Columbia University Press, New York).

Ley D, 1989, 'Fragmentation, coherence and limits to theory in human geography', in *Remaking Human Geography* eds A Kobayashi, S Mackenzie (Unwin Hyman, London): 227–244.

Major-Poetzl P, 1983 *Michel Foucault's Archaeology of Western Culture: Towards a New Science of History* (Harvester Press, Hemel Hempstead, Herts).

Ogborn M, 1990, '"A lynx-eyed and iron-handed system": the state regulation of prostitution in nineteenth-century Britain', unpublished typescript, Department of Geography, University of Salford, Salford.

Olsson G, 1987, 'The social space of silence' *Environment and Planning D: Society and Space* **5** 249–261.

Philo C, 1989a, '"Enough to drive one mad': the organisation of space in nine-teenth-century lunatic asylums', in *The Power of Geography: How Territory Shapes Social Life* eds J Wolch, M Dear (Unwin Hyman, London): 258–290.

Philo C, 1989b, 'Thoughts, words and "creative locational acts",' in *The Behavioural Environment: Essays in Reflection, Application and Re-evaluation* (Routledge, Chapman and Hall, Andover, Hants): 205–234.

Philo C, 1990, 'A letter to Derek Gregory on *Chinatown* and post-modern human geography', unpublished typescript, copy available from the author.

Philo C, 1992a 'History geography and the "still greater mystery" of historical geog-raphy', in *Rethinking Human Geography: Society, Space and the Social Sciences* (Macmillan, London).

Philo C, 1992b *The Space Reserved for Insanity: Studies in the Historical Geography of the English and Welsh Mad-business* PhD thesis, Department of Geography, University of Cambridge, Cambridge.

Quoniam S, 1988, 'A painter, geographer of Arizona' *Environment and Planning D: Society and Space* **6** 3–14.

Robinson J, 1990, '"A perfect system of control?" State power and "native locations" in South Africa' *Environment and Planning D: Society and Space* **8** 135–162.

Sack R D, 1986 *Human Territoriality: Its Theory and History* (Cambridge University Press, Cambridge).

Sartre J-P, 1948 *Existentialism and Humanism* (Methuen, Andover, Hants).

Sartre J-P, 1964 *Nausea* (New Directions, New York).

Soja E W, 1989 *Postmodern Geographies: The Reassertion of Space in Critical Social Theory* (Verso, London).

Trevelyan G M, 1948 *English Social History: A Survey of Six Centuries, Chaucer to Queen Victoria* (The Reprint Society, London).

Wolch J, Dear M, 1989 *The Power of Geography: How Territory Shapes Social Life* (Unwin Hyman, London).

11

PIERRE BOURDIEU

Joe Painter

Introduction

Pierre Bourdieu was born in 1930 in south-west France in the small town of Denguin, some 15 km north-west of Pau, capital of the ancient province of Béarn, in what was then the *Département des Basses-Pyrénées* (now *Pyrénées-Atlantiques*). In the early 1950s he began his academic career in philosophy, completing his *agrégation* in philosophy at the *École Normale Supérieure* in Paris, one of the elite '*grandes écoles*' that were much later to form the subject of his detailed study of the French higher-education system (Bourdieu, 1996). After a year teaching in a provincial *lycée*, in 1956 he was conscripted for military service with the French Army in Algeria during the bitter and brutal war against French colonialism (1954–62). Bourdieu's experience in Algeria set him on a path away from philosophy in the narrow sense towards anthropology and sociology and in 1958 he published his first book, *Sociologie de l'Algérie* (Bourdieu, 1962). He remained in Algeria after completing his military service to undertake teaching at the University of Algiers and further research, before returning to France. After teaching posts at the University of Paris-Sorbonne and the University of Lille, in 1964 he became Director of Studies at the *École des Hautes Études*. In 1981 he was appointed to the chair in sociology at the prestigious *Collège de France*. Much of his research activity has been conducted through the Centre for European Sociology, which he founded in 1968, and published in its journal *Actes de la Recherche en Sciences Sociales*.

Bourdieu's published output is extensive and wide-ranging. His writing is dense and complex and his works combine sophisticated social theory with enormously detailed empirical evidence, often drawn from very large social surveys. Until very recently, in Anglo-American social science his writings have been most influential in the fields of anthropology, educational research, and lately, cultural studies. In other disciplines, including geography, his work has, in the main, either been ignored or been referred to in passing without being used in any depth. This uneven reception of Bourdieu's ideas is due to the specific subject matter of his major writings.

239

The interest of anthropologists in his writings stems from his work in Algeria in the late 1950s. His earliest works on the sociology of Algeria (Bourdieu, 1962; Bourdieu et al., 1963; Bourdieu and Sayad, 1964) have been described by one critic as 'prosaic' (Jenkins, 1992: 24) and, although they remain a reference point for Bourdieu himself, they reveal little of the conceptual depth that characterizes his later work. By contrast, his ethnographic research on the Kabyle people of the Maghreb (a branch of the Berber ethnic group) prefigures much of his subsequent theoretical work, and accounts for much of the anthropological interest in his ideas. His work on the Kabyle people appeared over a period of years in a variety of journals and books. Perhaps the most important and widely cited is his structuralist account of the traditional Kabyle house written in 1963, but not published until 1970 (Bourdieu, 1973). After his return from Algeria, in 1960 Bourdieu undertook ethnographic work on marriage among the people of his native region of Béarn; this work too was important in influencing the development of his thinking (Bourdieu, 1972).

Educationalists have been drawn to the writings of Bourdieu because much of his empirical work focuses on the role of education in generating and reproducing social divisions, especially those of social class, and because education (in its broadest sense) plays a central role in several of his key theoretical concepts including his ideas of cultural capital, symbolic violence, habitus and field (see below). In *The Inheritors* (Bourdieu, 1979) and *Reproduction in Education, Society and Culture* (Bourdieu, 1977) Bourdieu uses surveys, case studies and statistical data to examine the reproduction of cultural privilege through the French educational system, and the role of 'symbolic violence' in securing that reproduction. In *Homo Academicus* (Bourdieu, 1988) and *The State Nobility* (Bourdieu, 1996) he examines the role of France's top educational institutions in producing and reproducing the French cultural and political elite, and the cultural mechanisms through which they do so.

Most recently, Bourdieu's work has generated interest in the area of cultural studies. Again this stems in part from his emphasis on the field of cultural production (including studies of photography, art, literature and sport) and in part from the broader role played by the concept of 'culture' in his theoretical approach. These concerns come together most clearly in *Distinction* (Bourdieu, 1984) in which Bourdieu analyses in great detail the relationships between social groups and social status on the one hand and taste in clothes, food, furniture, pastimes, music and so on, to show that value judgements about 'good' and 'bad' taste are deeply entwined with social divisions of class, wealth and power.

Given these particular empirical focuses to his work, it is not surprising that, in the English-speaking world, Bourdieu's work has generated most interest in these three fields. However, to see Bourdieu as 'merely' a sociologist of education and culture, albeit a very sophisticated one, is really to miss the point. As I have hinted, he focuses on these topics not as random case studies or because he just happens to be a specialist in these areas (in the sense that an historian might specialize in a particular period); rather education and culture are at the very centre of Bourdieu's conceptual approach to understanding social life in general. This means that his ideas can, in principle, be applied much more widely than has been the case to date in anglophone research. This chapter will examine the implications of this for understandings of space and spatial relations. First, though, a brief outline of the main concepts in Bourdieu's theoretical approach is required.

Key concepts

Transcending subjectivism and objectivism

Underlying Bourdieu's entire theoretical approach is what he perceives to be the vital need to transcend one of the longest-standing conceptual dichotomies in Western thought, namely subjectivism and objectivism. By 'subjectivism' Bourdieu refers to all those approaches to human life and action that locate the prime causes of social behaviour in individual free will, conscious decision-making and lived experience. By 'objectivism' he means those approaches that set out 'to establish objective regularities (structures, laws, systems of relationship, etc.) independent of individual consciousness and wills' and to explain social life in terms of such phenomena. In *The Logic of Practice* (Bourdieu, 1990), which provides the best overview of his approach, Bourdieu critically examines objectivism and subjectivism in turn. The objectivist tradition, most clearly exemplified for Bourdieu by structuralism, is flawed, he argues, because while it aims to treat the social world objectively, it fails to turn the same objective gaze on itself. As a result, objectivism ends up as a form of idealism, with the supposedly objective structures or regularities of the world dependent on the subjectivity of the objective observer. A similar paradox undermines the subjectivist tradition, epitomised for Bourdieu by rational-actor theory. Rational-actor theory appears to locate the causes of human actions at the level of individual decision-making. However, since it presupposes that actions are rationally motivated, they turn out not to be the product of subjective decisions at all, but rather the expression of binding, and thus objective, rationality. In other words, Bourdieu's immanent critiques of objectivism and subjectivism result in each position collapsing into the other.

Practice and habitus

In place of an unsustainable dichotomy between objectivism and subjectivism, Bourdieu proposes a 'theory of practice'. 'Practice' refers to the ongoing mix of human activities that make up the richness of everyday social life. According to Bourdieu, social practices neither represent the working out of objective social laws operating, as it were, behind the scenes, nor stem from the independent subjective decision-making of free human beings. Instead he argues that practices arise from the operation of 'habitus'. The concept of habitus is absolutely central to Bourdieu's work, but it is also difficult to grasp. It is the mediating link between objective social structures and individual action and refers to the embodiment in individual actors of systems of social norms, understandings and patterns of behaviour, which, while not wholly determining action (as in the objectivist model) do ensure that individuals are more disposed to act in some ways than others. Moreover habitus is both the product and the generator of the division of society into groups and classes. Habitus is thus shared by people of similar social status, but varies across different social groups. Bourdieu's rather abstract definition is as follows:

> The conditionings associated with a particular class of conditions of existence produce *habitus*, systems of durable, transposable dispositions, structured structures predisposed to function as structuring structures, that is, as principles which generate and organize practices and representations that can be objectively adapted to their outcomes without presupposing a conscious aiming at ends or an express mastery of operations necessary in order to attain them. Objectively 'regulated' and 'regular' without being in any way the product of obedience to rules, they can be collectively orchestrated without being the product of the organizing action of a conductor.
>
> (Bourdieu, 1990: 53)

John Thompson's (1991) explanation of this formulation is particularly helpful:

> The habitus is a set of *dispositions* that incline agents to act and react in certain ways. The dispositions generate practices, perceptions and attitudes which are 'regular' without being consciously co-ordinated or governed by any 'rule'. The dispositions which constitute the habitus are inculcated, structured, durable, generative and transposable – features that each deserve a brief explanation. Dispositions are acquired through a gradual process of inculcation in which early childhood experiences are particularly important. Through a myriad of mundane processes of training and learning, such as those involved in the inculcation of table manners ('sit up straight', 'don't eat with your mouth full', etc.), the individual acquires a set of dispositions which literally

242

mould the body and become second nature. The dispositions produced thereby are also *structured* in the sense that they unavoidably reflect the social conditions within which they were acquired. An individual from a working-class background, for instance, will have acquired dispositions which are different in certain respects from those acquired by individuals who were brought up in a middle-class milieu. In other words, the similarities and differences that characterize the social conditions of existence of individuals will be reflected in the habitus, which may be relatively homogeneous across individuals from similar backgrounds. Structured dispositions are also *durable*: they are ingrained in the body in such a way that they endure through the life history of the individual, operating in a way that is pre-conscious and hence not readily amenable to conscious reflection and modification. Finally, the dispositions are *generative* and *transposable* in the sense that they are capable of generating a multiplicity of practices and perceptions in fields other than those in which they were originally acquired. As a durably installed set of dispositions, the habitus tends to generate practices and perceptions, works and appreciations, which concur with the conditions of existence of which habitus is itself the product.

(Thompson, 1991: 12–13)

Habitus gives individuals a sense of how to act in specific situations, without continually having to make fully conscious decisions. It is this 'practical sense', often described as a 'feel for the game', that Bourdieu's theory of practice seeks to understand.

Cultural and symbolic capital

One of Bourdieu's most widely adopted ideas is that the concept of 'capital' should be seen not only in economic terms, but also as applicable to a range of other resources such as knowledge and status. He thus distinguishes between a range of forms of capital of which four are the most important. *Economic capital* refers to material wealth and is roughly equivalent to capital in the traditional sense of the term used by political economists, although Bourdieu's understanding is somewhat looser than the strict Marxist definition of capital. *Social capital* refers to the power and resources that accrue to individuals or groups by virtue of their social networks and contacts. This is seen most starkly in membership of elite clubs, but in principle applies to all kinds of social groupings. *Cultural capital* refers to knowledge and skills acquired in early socialization or through education. The possession of cultural capital is signified by formal educational qualifications. *Symbolic capital* refers to the representation of other forms of capital symbolically and is 'the form that the various species of capital assume when they are perceived and recognized as legitimate' (Bourdieu, 1989: 17). One of Bourdieu's key insights is that each form of capital can be converted to the other forms. For example, cultural capital can be converted into economic capital as when educational qualifications secure their holder a well-paid job. However, each form of capital also provides the resources for social struggles within its respective sphere independent of the other forms and without requiring conversion. Thus symbolic struggles, for example, have autonomy from economic struggles.

Field and strategy

The effectiveness of different forms of capital and the possibility of their autonomy arise partly from the organization of the social world into fields and subfields such as the economic field, the artistic field, the political field and so on. In the analogy of the game, the field might be thought of as the playing field or board on which the game is played. Fields are also sites of strategy and social struggle:

> For Bourdieu all societies are characterized by a struggle between groups and/or classes and class fractions to maximize their interests in order to ensure their reproduction. The social formation is seen as a hierarchically organized series of fields within which human agents are engaged in specific struggles to maximize their control over the social resources specific to that field, the intellectual field, the educational field, the economic field etc. and within which the position of a social agent is relational, that is to say a shifting position determined by the totality of the lines of force specific to that field.
>
> (Garnham and Williams, 1980: 215)

The field is thus a relational concept, a structured space of positions, which are determined by the uneven distribution of the various forms of capital:

> In analytic terms, a field may be defined as a network, or a configuration, of objective relations between positions. These positions are objectively defined, in their existence and in the determinations they impose upon their occupants, agents or institutions, by their present and potential situation in the structure of the distribution of species of power (or capital) whose possession commands access to the specific profits that are at stake in the field, as well as by their objective relation to other positions.
>
> (Bourdieu and Wacquant, 1992: 97)

244

In contrast to the more fluid, dynamic and embodied notion of practice or practical sense, Bourdieu frequently uses the concept of field in what appears to be a more static and deterministic fashion. Access to social positions within fields is decided by the possession of economic, social, cultural and symbolic capital, and each field has its own 'logic' (Bourdieu and Wacquant, 1992: 97). Despite Bourdieu's insistent rejection of objectivist approaches, there is a distinctly objectivist flavour to his account of the logic of fields which shapes not only the institutional components of the field, but also the human bodies within it:

> An institution, even an economy, is complete and fully viable only if it is durably objectified not only in things, that is, in the logic, transcending individual agents, of a particular field, but also in bodies, in durable dispositions to recognize and comply with the demands immanent in the field.
>
> (Bourdieu, 1990: 58)

The concept of the field is closely linked to the concept of capital outlined above. As Bourdieu puts it 'a capital does not exist and function except in relation to a field' (Bourdieu and Wacquant, 1992: 101). Individuals bring to the field both the embodied dispositions of the habitus and their stock of accumulated capitals. The power of different forms of capital varies according to the nature of the field. Economic capital is more powerful in the field of business, than in the field of religion, for example. Fields are the sites of social struggles 'aimed at preserving or transforming the configuration of [their] forces' (Bourdieu and Wacquant, 1992: 101). Bourdieu frequently likens these struggles to a game:

> We can picture each player as having in front of her a pile of tokens of different colors, each color corresponding to a given species of capital she holds, so that her *relative force in the game*, her *position* in the space of play, and also her *strategic orientation toward the game*, ... the moves she makes, ... depend both on the total number of tokens and on the composition of the piles of tokens she retains, that is, on the volume and structure of her capital.
>
> (Bourdieu and Wacquant, 1992: 99)

Although the idea of the game is only a heuristic device (albeit a recurring one in Bourdieu's work) there seems little to distinguish this formulation from more structuralist accounts or even, viewed another way, from the rational-actor model (vehemently rejected by Bourdieu) in which calculating actors make their plays based on a rational assessment of their objective positions in the game.

The analogy of the game links with another central concept in Bourdieu's framework, 'strategy'. For Bourdieu, a strategy is not the product of rational calculation, but can be oriented towards certain ends. This orientation arises from the 'feel for the game'. Just as some players have a more intuitive feel for the game, so some actors pursue more successful strategies. A strategy is thus a series of actions and practices that are directed towards certain goals, but (unlike conventional uses of the term 'strategy') not deliberately, self-consciously so. Goal-oriented practices arise not from acts of individual will, but from the operation of embodied dispositions, or habitus, within a particular field.

Power and symbolic violence

Building on the recognition that capital can take different forms, each providing its holder with resources, Bourdieu's theory of power emphasizes the variety of forms that power can take. In particular, Bourdieu is keen to stress the cultural and symbolic aspects of power (partly as a corrective to those theories, which have tended to see power as predominantly political or economic). All fields are 'fields of power' in which individuals and groups exist in relations of dominance and subordination by virtue of the uneven distribution of different forms of capital. When one group imposes a set of meanings, ideas and symbols on another (as happens continually in the education system, and in colonial situations, for example) this is referred to as an exercise in *symbolic violence*.

Bourdieu in geography

The use of Bourdieu's work by human geographers is marked by the same unevenness as his reception in the social sciences more generally. This has two implications. First, geographers have not adopted his ideas with the enthusiasm shown by their colleagues in education, anthropology and cultural studies. Second, insofar as geographers have been interested in his work, it has been applied (with some exceptions) in the fields of cultural and social geography, rather than, for example, political or economic geography. Bourdieu is one of those social theorists whom geographers cite frequently, but rarely engage with in any depth. This can perhaps be explained by the fact that while his ideas seem to hold out the promise of transcending many of the divisions in social-scientific (and human-geographical) thinking (such as structure and action, and subjectivism and objectivism), they are expressed in such dense and difficult prose that working with them in substantive research contexts seems a daunting prospect. The Social Sciences Citation Index provides a crude but suggestive indication of the impact of Bourdieu's work on geographical research

Table 11.1 Citations of Bourdieu's key works

Book	Number of articles listed in the SSCI that refer to the book	
	All Journals	*Geography Journals*
Outline of a Theory of Practice (first published in English in 1977)	1196	76
Distinction (first published in English in 1984)	1089	87
The Logic of Practice (first published in English in 1990)	189	8

Source: Social Science Citation Index search, May 1997

and writing. Table 11.1 shows the results of a citation search for three of Bourdieu's key works.

These figures suggest that while social scientists as a group have been marginally more interested in the more abstract *Outline*, geographers have shown a slight preference for the more substantive *Distinction*. However, such a search tells us little about the way in which Bourdieu's work is actually used by geographers. A more detailed survey of the articles themselves revealed that the vast majority of citations of these three works by geographers were brief references, rather than sustained commentaries or in-depth engagements. In particular, geographers refer frequently in passing to the suggestive concept of habitus, but rarely explore its implications for geographical theory and research in detail. This neglect is beginning to be remedied, and while it is unlikely (and probably undesirable) that a full-scale Bourdieusian school of geographical research will emerge, a number of geographers have started to use and adapt Bourdieu's ideas.

The geographies of practice and social action

Don Parkes and Nigel Thrift provide one of the earliest uses of Bourdieu's work in human geography. In *Times, Spaces and Places* Parkes and Thrift (1980: 91–3) develop an approach to human geography sensitive to time in all its dimensions (the book is subtitled *A Chronogeographic Perspective*). Drawing on his ethnography of the Kabyle people, they use Bourdieu's work to reveal the character of social time in what they term 'small scale societies'. Thrift (1983) and Allan Pred (1984) further develop Parkes and Thrift's emphasis on the importance of seeing social life as irreducibly embedded in space and time. Both these articles bear the imprint of what at the time was a strong interest among human geographers in structuration theory, although Thrift has subsequently argued that the section of his article that deals with structuration theory 'was simply a tag for what I had read thus far' (Thrift, 1995: 529) and that the main concern of his arguments lay in the theory of practice and the theory of the subject. Notwithstanding this recent clarification, Thrift's 1983 piece clearly enrols Bourdieu as a member of the 'structurationist school' (but see Thrift, 1996: 61 n.1). While it in no way misrepresents Bourdieu's ideas, it may have had the unintended effect of aligning Bourdieu in the minds of many anglophone geographers rather too directly with Anthony Giddens' structurationist project. As academic fashions changed, and structuration theory lost its popularity among geographers, the assumption that Bourdieu represented a kind of French structuration theorist may have contributed to his relative neglect by geographers.

Thrift returns briefly to Bourdieu in two more recent essays. The first, written with Steve Pile (Pile and Thrift, 1995), emphasizes Bourdieu's concern to develop a theory of practice, action and strategy as a 'ceaseless flow of conduct' that is embodied, intersubjective and situated in context (1995: 27–32). The second (Thrift, 1996) relates Bourdieu's theory of practice explicitly to Thrift's concern to develop 'non-representational' social theory; a point to which I will return below.

The theory of practice or practical action may well have been Thrift's primary concern, as it is Bourdieu's, but despite what Thrift describes as his 'call to arms' (Thrift, 1995: 528) very few geographers have pursued the challenge of developing a spatially sensitive theory of practice, and this is clearly one area in which a more thoroughgoing engagement with Bourdieu's ideas may well pay dividends in future. David Harvey, for example, adopts the notion of habitus explicitly in his painstaking elaboration of the complex relationships between the circulation and accumulation of capital through urbanization, and individual and class consciousness of the 'urban experience' (Harvey, 1989b). This concern is developed further in Harvey's *The Condition of Postmodernity* (Harvey, 1989a), although Derek Gregory takes Harvey to task for a one-sided reading of Bourdieu (Gregory, 1994: 406–10).

Gregory also briefly discusses Bourdieu's analysis of the Kabyle house (1994: 383–4). While Parkes and Thrift refer to Bourdieu's work on Kabyle temporality, Gregory's concern is with the spatiality of the house, which, he says demonstrates precisely Henri Lefebvre's claim that the representation of space in traditional societies is dominated by analogical space in which

> the physical form of the dwelling and the village itself typically represent and reproduce a divine body that is itself a projection, often in distorted or exaggerated form, of the human body [. . . the Kabyle house's] internal space is at once corporealized and gendered
>
> (Gregory, 1994: 383)

All of these geographical uses of Bourdieu's work on habitus and practice are interesting and suggestive, but none of them represent the kind of in-depth engagement that geographers have undertaken with the ideas of many other thinkers. There are, though, two examples of geographical writing where the concepts of habitus and practice are deployed more systematically. First, in a recent consideration of one of the most widely debated subjects in contemporary geography, Judith Gerber (1997) discusses the duality in Western thought between culture and nature. She argues that the three-fold division between the mental world, the social world and the physical world must be transcended and agrees with other commentators that this requires a new language. Part of the language, she suggests, is provided by Bourdieu's concepts of habitus and practice. Because habitus represents embodied dispositions it is at once physical and mental. Because habitus both generates and is generated through practice it is also social. Although Gerber's account is somewhat schematic, it does indicate the potential of Bourdieu's approach for informing geographical work. Second, I have suggested elsewhere (Painter, 1997) how the ideas of habitus, field and practice can be used to interpret urban politics, in ways that link the behaviour of individual political actors with wider institutional and political–economic processes.

The geographies of distinction and cultural and symbolic capital

By contrast with their very limited use of his theory of practice, geographers have shown slightly more sustained interest in the ideas of cultural and symbolic capital and the related notion of distinction (note though that the ideas of practice and capital are closely interrelated – they are considered separately here simply for convenience). Two recent studies published in geographical journals illustrate this trend. Both papers draw on *Distinction* to examine the gentrification of inner urban areas, but in rather different ways. Derek Wynne and Justin O'Connor (1998) focus on a new residential development in the city centre in Manchester, and use quantitative and qualitative empirical research to test the claim that 'post-modern culture is tied strongly to the emergence of a new middle class', the so-called 'new cultural intermediaries', and that these socio-cultural groups are the driving force in gentrification (Wynne and O'Connor, 1998: 844–5). They argue that this claim is grounded at least implicitly in the ideas about taste, social status and cultural capital that Bourdieu develops in *Distinction*. This is perhaps debatable since the empirical material on which *Distinction* is based relates to the 1960s and early 1970s somewhat before the supposed emergence of the new middle class and the widespread dissemination of post-modern cultural forms. Conversely, Bourdieu does develop the category of the 'new petite bourgeoisie' which 'comes into its own in all the occupations involving presentation and representation (sales, marketing, advertising, public relations, fashion, decoration and so forth) and in all the institutions providing symbolic goods and services' (Bourdieu, 1984: 359).

Wynne and O'Connor's methodological approach is similar in some respects to Bourdieu's own. Their quantitative survey asked for responses about 104 items related to cultural consumption, and data analysis involved searching for correspondences between patterns in individuals' expressed cultural preferences, lifestyle choices and judgements of taste. Their analysis suggests that there is considerable overlap in taste and cultural activities among the individuals in the sample, but that this overlap reveals 'middle-brow', rather than 'elite' cultural preferences, leading Wynne and O'Connor to challenge what they see as the widespread assumption that gentrification depends on the existence of a new middle class with very high levels of cultural capital. Although some of their critiques of Bourdieu seem misplaced (for instance, Bourdieu himself would certainly not expect the particular pattern of linkages between class and cultural taste in mid-1990s Manchester to reflect that of France in the late 1960s) the study is interesting as one of the few attempts in the geographical literature to adopt the style of empirical analysis used in *Distinction*.

Julie Podmore (1998) also draws on *Distinction* to examine gentrified lifestyles – this time in Montreal – but does so rather differently from Wynne and O'Connor. Podmore's criticism of the gentrification literature is that it has tended to see culture as an instrument of capital. This, she suggests, makes it difficult to explain the widespread adoption of the aesthetic and lifestyle of 'loft living' in many different cities, including those without a substantial stock of old industrial buildings or a strong culture-industry sector in the economy. The dispersion of the loft lifestyle across space depends, argues Podmore, on a shared disposition within certain social and cultural groups to favour the aesthetic taste, which it expresses – in other words it depends on habitus:

> If we examine which social groups make use of the loft as *habitus*, the centrality of the media, the material environment and physical location of the industrial inner city, the transregional quality of the SoHo loft becomes rather apparent. Loft dwellers, whether they are artists or corporate executives, generally have high levels of cultural capital; they are cultural elites, physically located in specific urban environments but more broadly connected to a global *habitus* of shared dispositions and social practices through the mass media and other communication technologies. Defining themselves in opposition to the postwar suburban middle classes, this social group makes use of the location of the inner city as a form of distinction as well as a social location.
>
> (Podmore, 1998: 286–7)

Using Montreal as an example, Podmore then goes on to elaborate in much more detail the nature of the symbolism and forms of distinction involved in loft living and the cultural mechanisms through which they are disseminated and which are constitutive of habitus.

Homo Geographicus?

The final category of adoptions of Bourdieu's ideas within the literature of contemporary anglophone geography relates to his work on the culture and sociology of intellectual and academic life itself. *Homo Academicus* was published in English in 1988. It represents an attempt by Bourdieu to put into practice his own edict that the sociological gaze should be turned reflexively onto the sociologist and the practices of sociological research and teaching. In the book he presents a critical social and cultural analysis of the world he himself inhabits: the French university system. According to Jenkins (Jenkins, 1992: 120), 'by the discipline's defining criterion of exoticism it may not be *real* anthropology, but it may yet turn out to be Bourdieu's *best* anthropology'. There is clearly much scope for this kind of reflexivity with geography (though some may object that the discipline already has its fair share of such work). Philip Crang, for example, suggests that the vogue for 'polyphonic' textual strategies in geographical writing can be partly understood with reference to Bourdieu's concepts of intellectual and academic capital. Thus while apparently displacing the academic author from their erstwhile position of power, in practice

> in the work on textual construction the expert or organic intel-
> lectual becomes the writer, crafting her texts, weaving his narrative.
> And in the 'new' cultural geography more generally, there is the
> inescapable cultural capital conferred by knowing about Aboriginal
> art ('cultured'), or the geography of the Manchester dance music
> scene ('verging on the trendy').
>
> (Crang, 1992: 546)

Keith Bassett (1996) uses Bourdieu's ideas to reflect on the position of intellectuals and universities. While recognizing that the account presented in *Homo Academicus* is very specific to France, Bassett argues that it can have wider application. Two particular features of Bourdieu's approach attract Bassett's attention. First, there is Bourdieu's insistence that the academic field is a field like any other, marked by relations of power and force and struggles for cultural, symbolic, academic and intellectual capital. Thus Bassett suggests that

> the ongoing restructuring of the English [*sic*] university system
> could be analysed in terms of a changing field of play, centring

around new forms of competition for research rankings and new mechanisms for accumulating cultural capital. Such an analysis could also focus on the changing nature of the academic habitus as the new forms of competition become internalised as new dispositions and practices by new generations of academics.

(Bassett, 1996: 522)

Second, there is Bourdieu's claim that the academic field is also not quite like any other because it has the potential to provide an autonomous space in which the (always reflexive and historically situated) pursuit of rational dialogue can be protected:

Bourdieu projects the idea of the university as the 'scientific city', a field where the institutional conditions for rational dialogue and undistorted communication can be developed and protected. The university should thus be the site where 'the most unavowable intentions have to sublimate themselves into scientific expression', and where 'the worst, the meanest, and the most mediocre participant is compelled to behave in accordance with the norms of scientificity in currency at the time' (Bourdieu and Waquant [sic], 1992, page 178). From such bases intellectuals can even begin to build 'an international of artists and scientists', as an independent political and moral force capable of intervening and influencing those who rule (Waquant [sic], 1993, page 38).

(Bassett, 1996: 521)

Bassett's account is still schematic – he does not, for example, develop the analysis of the changing structures, practices and habitus of British academia suggested in his article. In addition, while broadly sympathetic to Bourdieu's approach he does identify what he sees as a number of problems with it, including a lack of detail about the precise relationship between autonomy and engagement, a lack of attention to undergraduate education, and a supposed 'fundamental contradiction' (Bassett, 1996: 523) between Bourdieu's conception of the academic field as structured around struggles and the unequal distribution of cultural capital on the one hand and his insistence on the other that it is possible to 'escape' in Bassett's word from these forces to a space in which rational dialogue can be pursued. Nevertheless, Bassett's article is a rare attempt in the geographical literature to use Bourdieu's approach to understanding the academy.

Bourdieu's spatialities

As Bourdieu's major texts have appeared in English translations his work has attracted considerable attention from geographers and terms like habitus and cultural capital have entered the geographical vocabulary. In the main, though, references to these ideas by geographers have mostly been gestural or at best schematic. There are very few examples in the geographical literature of sustained engagements with Bourdieu's work whether exegetical or applied. This does not mean, however, that there is little in Bourdieu's conceptual framework or empirical research worthy of consideration by those interested in 'thinking space'. On the contrary, Bourdieu's work has the potential to provide a very rich source of ideas on space and spatiality. This concluding section suggests a number of ways in which a more sustained dialogue between Bourdieu's ideas and geography might be developed as well as pointing up some of the problems with the use of concepts of space in his work.

Bourdieu's own work already contains some (rather limited) substantive analyses of spatiality. Best known is his structuralist reading of the spatial organization of the Kabyle household (Bourdieu, 1973). In this wonderfully written account, Bourdieu reveals how the organization of Kabyle society is translated into the micro-geography of the house, as in the following example from the 1990 translation:

> The low, dark part of the house is also opposed to the upper part as the female to the male. Not only does the division of labour between the sexes (based on the same principle of division as the organization of space) give the woman responsibility for most of the objects belonging to the dark part of the house, the carrying of water, wood, manure, for instance; but the opposition between the upper part and the lower part reproduces, within the internal space of the house, the opposition between the inside and the outside, between female space – the house and its garden – and male space
>
> (Bourdieu, 1990: 273–4)

This kind of explicitly spatialized interpretation is not typical of Bourdieu's subsequent writings, but nor are issues of substantive spatiality wholly absent. For instance, as Bourdieu often acknowledges, his empirical research on French society is highly geographically-bounded and might be read as an extended ethnography of the space that is modern France, particularly in relation to its educational, academic and artistic life. There are also very occasional discussions of the significance of geographical variation *within* France. One example is his early anthropological work on matrimonial strategies among the Béarnais, which is highly sensitive to the specificities of locality. On the whole though, as Bourdieu has moved over time from the anthropological field to the sociological one he has tended to deal with France as an undifferentiated whole, with only very occasional discussions of sub-national geographical variation. One such occurs in *Distinction* (Bourdieu, 1984: 363–5), where he compares the distribution of cultural capital within the petite bourgeoisie in Paris with that in the same class fraction in provincial France:

> The dispositions of which the new petite bourgeoisie is the bearer find the conditions for their full development only in Paris. Cultural pretension – together with education, of which it reinforces the effects – is no doubt one of the factors conducive to appropriation of the advantages associated with proximity to the centre of cultural values, such as a more intense supply of cultural goods, the sense of belonging and the incentives given by contact with groups who are also culturally favoured. Consequently there is no other category in which the systematic differences between Parisians and provincials are more marked: differences in the intensity of the legitimate practices (museum visits etc.) and the range of competence (in music, for example); differences in the relationship to legitimate culture, with the sense of being an outsider to the world of painting or music always being more marked among provincials, other things being equal; differences, above all, in the ability to recognize – often without knowing them – smart opinions.
>
> (Bourdieu, 1984: 363)

In addition to these limited treatments of substantive geographies, Bourdieu also develops the concept of 'social space' at some length. According to Bourdieu, social groups are formed in and distributed across 'social space' (Bourdieu, 1985). However, it is clear that he intends this concept to be understood heuristically, as a space in thought ('the social world can be *represented* as a space' (Bourdieu, 1985: 723, emphasis added)), and not immediately translatable to what he calls 'geographical space', from which he says social space should be clearly distinguished:

> We can compare social space to a geographic space within which regions are divided up. But this space [i.e. social space – JP] is constructed in such a way that the closer the agents, groups or institutions which are situated within this space, the more common properties they have; and the more distant, the fewer. Spatial distances – on paper – coincide with social distances. Such is not the case in real space [i.e. physical space – JP]. It is true that one can observe almost everywhere a tendency toward spatial segregation, people who are close together in social space tending to find themselves, by choice or by necessity, close to one another in geographic space; nevertheless, people who are very distant from each other in social space can encounter one another and interact,

if only briefly and intermittently, in physical space, Interactions
. . . mask the structures that are realized in them. This is one of
those cases where the visible, that which is immediately given,
hides the invisible which determines it.

(Bourdieu, 1989: 16)

Three comments need to be made here. The first is that this should serve
as a caution against any assumption that just because Bourdieu (in common
with many other social theorists) uses a spatialized vocabulary his work
necessarily deals with substantive spatialities in the sense that geographers
understand them. The second is that Bourdieu's comparison of social space
with geographic space seems to be based on an assumption that geography masks the real nature of social relations. The implication here,
contrary to the careful contextualization evident elsewhere in his work, is
that the researcher must abstract from the contingencies of spatial proximity and distance to understand the true distribution of social power.
This may reflect in part Bourdieu's antagonism towards Geography as an
institutionalized academic discipline in France. The third is that Bourdieu's
understanding of geographic space is a rather more limited one than most
contemporary human geographers would accept. Geographic space tends
to be seen exclusively in terms of distributions, distances and arrangements. This limited view of space affects his 'social space' metaphor
too. This can be seen in a particularly telling way in the idea of the field.
We have already seen how Bourdieu's formulation of the field concept
can be criticized for its unacknowledged determinism and we can now
relate this criticism to his understanding of space:

The principle of the dynamics of a field lies in the form of its
structure and, in particular, in *the distance, the gaps, the asymmetries
between the forces* that confront one another . . . As a space of potential and active forces, the field is also a *field of struggles* aimed at
preserving or transforming the configuration of these forces. . . .
The strategies of agents *depend on their position in the field*, that is, in
the distribution of the specific capital, and on the perception that
they have of the field depending on the point of view they take
on the field as a view taken from a point *in* the field.

(Bourdieu and Wacquant, 1992: 101,
some emphases added)

If space is primarily conceived in terms of distance, distribution and separation, there is a tendency to see agents as individualized (more or less
separated) and as facing each other across the playing field. This kind of
representation has the unintended consequence of calling to mind precisely
the rational-actor model of social life, with its individual calculating agents,

that Bourdieu insists he wishes to reject. Moreover, critics have suggested that Bourdieu's concepts of cultural and symbolic capital can have the effect of economizing social life by turning cultural relations into another form of market competition. This perception too is bolstered by seeing the space of the field in terms of the spatial distribution of capital. The implication (made *explicit* in Bourdieu's game analogy) is that agents occupy positions in the field and pursue strategies determined by the quantitative distribution of capital across the space of the field. A more nuanced, dynamic and relational conception of geographic space might feed through into the metaphorical space of the field and thereby mitigate some of the more deterministic and economistic features of this element of Bourdieu's framework.

These features can also be related to geography/spatiality in a rather different way. Whether intentionally or otherwise, Bourdieu's empirical work seems often to be primarily focussed on cultures and groups in French society that are intensely hierarchical and tightly regulated, such as the higher education system. Despite Bourdieu's insistence that the theoretical framework grounded in these studies is of general applicability, it seems likely that some of the particular features of these notably status-conscious and rule-bound cultural worlds are worked into Bourdieu's more general theoretical arguments, so that they become marked by their geographical origins and not always as broadly applicable as is claimed. Conversely, de Certeau (1984) argues that Bourdieu's theory of practice depends on an exoticism – that it is the studies of 'the other' (i.e. of Kabylia and Béarn) that most characteristically ground the idea of habitus (de Certeau, 1984: 58). Moreover, according to de Certeau, Bourdieu's concepts of practice and strategy are limited and dominated by 'an economy of the proper place' (de Certeau, 1984: 55) in which both the maximization of capital (material and symbolic) and the development of the body (through the acquisition of habitus) are 'entirely peculiar to the closed space in which Bourdieu examines them and to the way in which he observes them' (de Certeau, 1984: 55).

As this suggests, and as Thrift points out (1996: 15), de Certeau's critique of Bourdieu is framed in explicitly spatial terms, and it is clear that attempts to 'think Bourdieu spatially' will require *critical* engagements with his work rather than any straightforward 'geographical translation' of the framework (whatever that might look like). Perhaps one of the most promising lines of future development is to use Bourdieu's ideas to pursue the elaboration of a genuinely contextual social theory (Thrift, 1983; 1996). The concepts of habitus, practice and field are central here. Habitus, as the embodiment of acquired (and biological (Gerber, 1997)) dispositions, provides a potentially fruitful way of understanding the spatial and temporal embeddedness of social life (Friedland and Boden, 1994). As we have seen, despite its suggestive qualities, references to the idea of habitus in geographical writing to date have been mainly gestural; few authors have really *used* it in depth. Indeed to some extent the term has become a shibboleth, functioning in some accounts as an unexplained explanatory factor – a kind of *deus ex machina* or independent variable which cannot itself be understood or accounted for. Here is de Certeau again: 'Bourdieu's texts are fascinating in the analyses and aggressive in their theory. . . . Scrupulously examining practices and their logic . . . the texts finally reduce them to a mystical reality, the *habitus*, which is to bring them under the law of reproduction' (1984: 59).

This mysteriousness – real or apparent – at the heart of Bourdieu's theory means that simple genuflections in the direction of habitus are quite insufficient, providing only the appearance of explanation. A more sympathetic reading than that offered by de Certeau might link Bourdieu's struggle to grasp the nature of habitus with the more general problems inherent in 'non-representational thinking' (Thrift, 1996: 6). For Thrift, practice defies representation and any adequate theory of practice must pursue non-representational modes of knowing. Practice, practical sense and practical consciousness, which are all about our 'going on in the world', are in some senses ineffable. Once they are captured (represented?) in language and text they die. It is perhaps this that explains the sensation that Bourdieu's work circles and approaches the moment of practice without ever quite reaching it.

Field, another potentially misleading geographical metaphor, can also be understood as substantively spatialized, inasmuch as power *is* distributed spatially as well as socially. These are not new claims, of course, but as I have suggested, few geographers have yet pursued their implications in any depth. Given the critical comments made above concerning the use that Bourdieu makes of the field concept and its relations with capital, there may be scope for geographers to elaborate a less determinist notion of the field drawing on a wider range of understandings of geographic space that that used by Bourdieu. The idea of the field could, I think, be recast without much difficulty around a more complex spatiality, involving multiple and overlapping spaces, network approaches as well as theories of space that emphasize discontinuity, fragmentation and contradiction. Geographers might also want to blur the crude distinction in Bourdieu's work between (metaphorical) social space and (physical) geographic space. If society and space are understood as co-constituting then fields are socio-spatial (and socio-temporal) phenomena, opening up the potential of a more thoroughly spatialized theory of practice.

Following from this, further in-depth investigations of uneven socio-spatial distribution of cultural and symbolic capital provide another future avenue for research. Here too we need to extend Bourdieu's rather narrow reading of geographical space. While he is correct to insist that there is no possibility of mapping social space directly onto geographic space (they are not homologous), he is arguably mistaken in dismissing geographic space as merely an obfuscatory veil. If the spatio-temporal embedding of practice is as important as the concept of habitus suggests, then a critical appropriation of Bourdieu's approach by geographers and other spatial theorists can both enrich Bourdieu's concept of capital in its various forms and offer the prospect of improved understanding of contemporary social life.

References

Bassett, K. (1996). Postmodernism and the Crisis of the Intellectual: Reflections on Reflexivity, Universities, and the Scientific Field. *Environment and Planning D: Society and Space, 14*, 507–527.

Bourdieu, P. (1962). *The Algerians* (Ross, A C M, Trans.). Boston, MA: Beacon Press.

Bourdieu, P. (1972). Les Stratégies Matrimoniales dans le Système des Stratégies de Reproduction. *Annales, 4–5*, 1105–27.

Bourdieu, P. (1973). The Berber house. In M. Douglas (ed.), *Rules and Meanings. The Anthropology of Everyday Knowledge. Selected Readings.* Harmondsworth: Penguin.

Bourdieu, P. (1977). *Reproduction in Education, Society and Culture* (Nice, R, Trans.). London: Sage.

Bourdieu, P. (1979). *The Inheritors: French Students and their Relation to Culture* (Nice, R, Trans.). Chicago, IL: Chicago University Press.

Bourdieu, P. (1984). *Distinction: a Social Critique of the Judgement of Taste* (Nice, R, Trans.). London: Routledge.

Bourdieu, P. (1985). The Social Space and the Genesis of Groups. *Theory and Society, 14*(6), 723–44.

Bourdieu, P. (1988). *Homo Academicus* (Collier, P, Trans.). Cambridge: Polity.

Bourdieu, P. (1989). Social Space and Symbolic Power. *Sociological Theory, 7*(1), 14–25.

Bourdieu, P. (1990). *The Logic of Practice* (Nice, R, Trans.). Cambridge: Polity.

Bourdieu, P. (1996). *The State Nobility: Elite Schools in the Field of Power* (Clough, L C, Trans.). Cambridge: Polity.

Bourdieu, P., Darbel, A., Rivet, J.-P., and Siebel, C. (1963). *Travail et Travailleurs en Algérie.* Paris and The Hague: Mouton.

Bourdieu, P., and Sayad, A. (1964). *Le Déracinement: la Crise de l'Agriculture Traditionelle en Algérie.* Paris: Les Éditions de Minuit.

Bourdieu, P., and Wacquant, L. (1992). *An Invitation to Reflexive Sociology.* Cambridge: Polity.

Crang, P. (1992). The Politics of Polyphony: Reconfigurations in Geographical Authority. *Environment and Planning D: Society and Space, 10*, 527–549.

de Certeau, M. (1984). *The Practice of Everyday Life.* Berkley, CA: University of California Press.

Friedland, R., and Boden, D. (1994). NowHere: an Introduction to Space, Time and Modernity. In R. Friedland and D. Boden (eds), *NowHere: Space, Time and Modernity.* Berkeley, CA: University of California Press.

Garnham, N., and Williams, R. (1980). Pierre Bourdieu and the Sociology of Culture: an Introduction. *Media, Culture and Society, 2*, 209–3.

Gerber, J. (1997). Beyond Dualism – the Social Construction of Nature *and* Social Construction of Human Beings. *Progress in Human Geography, 21*(1), 1–17.

Gregory, D. (1994). *Geographical Imaginations.* Oxford: Blackwell.

Harvey, D. (1989a). *The Condition of Postmodernity.* Oxford: Blackwell.

Harvey, D. (1989b). *The Urban Experience.* Oxford: Blackwell.

Jenkins, R. (1992). *Pierre Bourdieu*. London: Routledge.

Painter, J. (1997). Regulation, Regime and Practice in Urban Politics. In M. Lauria (ed.), *Reconstructing Urban Regime Theory: Regulating Urban Politics in a Global Economy* (pp. 122–43). Thousand Oaks, CA: Sage.

Parkes, D., and Thrift, N. (1980). *Times, Spaces and Places: a Chronogeographic Perspective*. Chichester: Wiley.

Pile, S., and Thrift, N. (1995). Mapping the Subject. In S. Pile and N. Thrift (eds), *Mapping the Subject: Geographies of Cultural Transformation*. London: Routledge.

Podmore, J. (1998). (Re)reading the 'Loft Living' Habitus in Montreal's Inner City. *International Journal of Urban and Regional Research*, 22(2), 283–301.

Pred, A. (1984). Place as Historically Contingent Process: Structuration Theory and the Time-geography of Becoming Places. *Annals of the Association of American Geographers*, 74, 279–297.

Thompson, J. B. (1991). Editor's introduction. In P. Bourdieu, *Language and Symbolic Power*. Cambridge: Polity.

Thrift, N. (1983). On the Determination of Action in Space and Time. *Environment and Planning D: Society and Space*, 1(1), 23–57.

Thrift, N. (1995). Classics in Human Geography Revisited: Author's Response. *Progress in Human Geography*, 19(4), 528–30.

Thrift, N. (1996). *Spatial Formations*. London: Sage.

Wacquant, L. (1993). From Ruling Class to Field of Power: an Interview with Pierre Bourdieu on La Noblesse d'Etat. *Theory, Culture and Society*, 10, 19–44.

Wynne, D., and O'Connor, J. (1998). Consumption and the Postmodern City. *Urban Studies*, 5–6, 841–64.

12

THE TROUBLED SPACES OF FRANTZ FANON

Steve Pile

Introduction: situating Fanon

It is probably inauspicious, but I will begin this chapter with a confession. I admit that I have misinterpreted Frantz Fanon: I thought that one of his anecdotes referred to an event in Martinique, rather than in France. Nevertheless, the consequences of this mistake are what this chapter is about: that is, what difference does it make to think theory in different places or through different understandings of space? More than this, however, the displacement of Fanon from one country to another asks questions about what it means to think politics spatially. Thus, the question this chapter addresses is whether it makes any difference to 'thinking' and 'politics' to move them, as it were, from one place to another.

In some ways, this is not a new question to ask of Fanon. His writings require that the reader notice the situatedness of his interventions. Indeed, he demands it. However, writers who have drawn on Fanon, almost invariably, end up reproducing a paradox. Henry Louis Gates puts it this way:

> Thus, while calling for the recognition of the *situatedness* of all discourses, the critic delivers a Fanon as a global theorist *in vacuo*; in the course of an appeal for the specificity of the Other, we discover that his [sic] global theorist of alterity is emptied of his own specificity; in the course of a critique of identitarian thought, Fanon is conflated with someone who proved, in important respects, an ideological antagonist. And so on.
>
> (Gates, 1991, page 459)

It has proved conspicuously difficult to deliver a situated account of Fanon's analyses of the colonial situation and of his incendiary revolutionary rhetoric. Gates is right to argue that the tendency in accounts of Fanon has been to 'globalise' his theories. Paradoxically, however, commentators commonly assert that there are many Fanons (usually there are three, often chosen from – amongst others – colonial psychiatrist, psychoanalyst, diplomat, revolutionary ideologue and Sartrean philosopher). The paradox plays out this way. Having suggested that there are many Fanons, the critic then feels able to pick the Fanon they want to and, from there, elevate the chosen one Fanon into a Global Theorist. In this, it is common to side either with the Fanon that wrote *Black Skin, White Masks* (1952) or with the man that wrote *Studies in a Dying Colonialism* (1959) and *The Wretched of the Earth* (1961). Two, Three Fanons? Which Fanon to choose? Perhaps a choice is not necessary. Perhaps, instead, I should begin to situate Fanon, not just within intellectual traditions of thought or within anti-colonial struggles, but also in his geography. For, as Said points out, 'just as none of us is outside or beyond geography, none of us is completely free from the struggle over geography' (1993, page 6). So, the issue becomes not just what did Fanon think, but where was he and what did it mean (for him) to struggle over geography?

On 20 June 1925, Frantz Fanon was born in Fort-de-France on the French West Indian island of Martinique.[1] By the time he was 17, Martinique was under Nazi control and he made his escape to Dominica. There, he volunteered for military service and trained to become a soldier in the French Army. He fought in North Africa and Europe and was awarded medals for gallantry. After the war, Fanon returned to Martinique, where he campaigned on behalf of his one-time school friend, Aimé Césaire, during Césaire's candidacy for the presidency of Martinique. Soon, however, Fanon decided to take advantage of scholarships offered to French war veterans to study in France. Although initially deciding to study dentistry, Fanon opted instead for psychiatry, which he studied in Lyon. At that time, Lyon was a hot-bed both of student radicalism and of racism. By early 1948, Fanon had already edited the first issue of a magazine, *Tam Tam*, for black students.

As Caute says, 'throughout his life, Fanon was plagued and embittered by his encounters with racism. As a young man he had believed that he could break through the colour barrier on the strength of his education and personal capacities' (1970, pages 8–9).

In many ways, Fanon's work enacts – or travels in the space between – his rage against racism and his hope that people will listen to reason. This hope was almost dispelled by his experiences in Algeria. Although Fanon had wanted to work in Senegal, after his graduation in 1952, he was offered a post in the Psychiatry Department at Blida-Joinville Hospital in Algiers. By the time of his arrival, in 1953, Algeria was already in political turmoil. By 1954, the Algerian war of independence had begun in earnest. Fanon became increasingly embroiled in the conflict, eventually resigning his post at the hospital in 1957 when it became untenable.

After a brief spell in Tunis, Fanon was eventually sent by the Algerian government to become their ambassador in Ghana.

However, Fanon's health was already failing. He had contracted leukaemia and, after a short period of treatment in the Soviet Union, was eventually forced – with great reluctance – to seek medical treatment in Maryland, USA, where he died on 6 December 1961 at the age of 36. His body was laid to rest on an Algerian battlefield. It is not just Fanon's body that marks a battlefield, his body is also a map of war, his soul the site of struggle. Exceptionally, Fanon's work dramatises *both* the struggle to free lands from external oppression *and* the necessity of freeing interior landscapes from the psychic realities of racism. While these aspects of Fanon's work are pursued throughout his work, they have had rather different intellectual after-lives.[2]

In the sixties, his work was applied in nationalist struggles around the world. At this time, it was Fanon's revolutionary ideas which gained most attention from scholars, especially those working in political theory and development studies. It was his peculiarly explosive mix of nationalism and socialism that inspired such people. Since the eighties, post-colonial theorists returned to Fanon because of their interest in his analysis of the psychodynamics of colonial administrations, both amongst the colonised and colonisers. What particularly attracted people was Fanon's refusal to allow the 'normal' categories of colonial life – such as 'black' or 'white', 'native' or 'foreigner' – to be authentic or stable. So, Fanon's work has been differently influential, but in this chapter I will take exemplary stories from both the earlier and later Fanons in order to think space politically. Across the early/late divide, Fanon's ideas were designed to show that colonial understandings of 'other' subjugated peoples were fundamentally flawed and could not provide the basis for liberation, either from external authority or from internal colonisation.

For Stuart Hall, Fanon's writings attempt

> to subvert the structures of 'othering' in language and repre-sentation, image, sound and discourse, and thus to turn the mechanisms of fixed racial signification against themselves, in order to begin to constitute new subjectivities, new positions of enunciation and identification, without which the most 'revolu-tionary' moments of national liberation quickly slide into their post-colonial reverse gear (Algeria being one of the most troubling and heart-wrenching instances).
>
> (1996, pages 19–20)

If Fanon refused to allow revolutionary subjectivities to ossify into fixed identities, then this also has the consequence of producing (what many commentators, almost despairingly, refer to as) an 'exceptional instability' in his analytical and political rhetoric (Gates, 1991, page 470). I am troubled by this all-too-quick 'slide' from the production of new radical subjectivities towards their freezing into the unmoving subject positions that characterise practices of domination and exclusion. Is there, in Fanon's thought, a sense of place and nation that might be used to prevent the troubling and heart-wrenching slide into barbarism of even the most democratic of liberation movements? Perhaps, in addressing this question, it will be necessary to transform commonplace understandings of place and nation (following Massey, 1994). And perhaps this will be achieved by *situating* Fanon's ideas, by *localising* their globalisation.

In this chapter, I will situate and localise Fanon's ideas by looking more carefully at two emblematic passages in his work. Each, in turn, has been taken by critics to outline universal features, first, of the lived experience of black people under colonialism and, second, of the role of national consciousness in anti-colonial struggle. What I will demonstrate is that these seemingly 'global' stories actually are born of 'local' circumstances and, further, that shifting their location changes radically the character of radical politics. It is not just that geography matters, it is also that the way space is conceptualised modulates the kind of politics that can be thought of as radical.

Dislocations: the lived experiences of colonialism

In chapter 5 of his first book, *Black Skin, White Masks* (1952), Fanon begins thinking about the daily experiences of black people in colonial situations and, more particularly, about the paradox that 'blackness' is only apparent in relation to 'the white man' (page 110). It is this chapter, and especially the so-called 'Look, a Negro!' passage (pages 111–115), that has attracted most attention amongst contemporary post-colonial critics (see Hall, 1996). I will show that thinking through the location of this passage provides further insights into the relationship between place and the politics of identity.

In *Black Skin, White Masks*, Fanon argues that black skin becomes the outstanding sight (and site) of difference and inferiority in a white-dominated world. This is to say that colonial power operates through a corporeal schema, which grades bodies according to the colour of the skin. It should be noted that blood, hair, bones, and so on, are also chained (through the body) to skin in these racist corporeal schemas. More tellingly, Fanon suggests that the colonial situation puts a mirror up to the face of the black man and the reflection tells him that he is inferior and other.[3] The effects of this are far-reaching. Because colonised peoples (can only) recognise themselves as other (to themselves) and because white people are apparently superior and ideal, they are compelled to enact a script which is not their own; to behave according to values and norms which are not theirs; to perform according to standards that they have not set; and, to both identify with and internalise these values, norms and standards as if they were their own. Thus, despite Fanon's constant demand to be (acknowledged as) black, he despairs that 'out of the blackest part of my soul, across the zebra striping of my mind, surges this desire to be suddenly *white*' (page 63). By identifying with – and desiring – the position and power of the white man, the black man ends up by seeing himself as 'not-white', 'not-Master' and 'nowhere'. In this way, the black man is both alienated from himself and absolutely depersonalised by 'the grotesque psychodrama of everyday colonial life in colonial societies' (Bhabha, 1986, page 71). It is important to bear in mind that, at this point in his argument, Fanon's theory is primarily concerned with black men's experiences, that is with the simultaneous construction of masculinity, sexuality and race.[4]

For Fanon, the colonial regime's imposition of skin hierarchies not only defines the visibility of the body, and also territorialises the body, but it is also woven by the white man 'out of a thousand details, anecdotes, stories' (1952, page 111). He shows that black male identity is forged out of a set of identifications that are inherently anxious – simultaneously fearful and desiring. These identifications smuggle senses of self – black and white – across a fictional, though foundational, black/white border. The black/white epidermal schema is not just imposed from the outside, but it is also inscribed in the movements of people, in their actions, thoughts and feelings. But it is the black who moves under the constant scrutiny of the fearful/fear-full master's 'blue' eyes:

> 'Look, a Negro!' It was an external stimulus that flicked over me as I passed by. I made a tight smile.
> 'Look, a Negro!' It was true. It amused me.
> 'Look, a Negro!' The circle was drawing a bit tighter. I made no secret of my amusement.
> 'Mama, see the Negro! I'm frightened' Frightened! Frightened! Now they were beginning to be afraid of me. I made up my mind to laugh myself to tears, but laughter became impossible.
>
> (pages 111–112)

Fanon cannot laugh because he knows that the little boy's fear is based on the white production of the black body. Fanon is apparently made visible by the skin of his body, but he is actually cloaked in stereotypes, popular fictions and myths. He is simultaneously visible and invisible, marked and erased, certain and uncertain – he certainly has a black body, but there is a deep uncertainty about what this might be (see Bhabha, 1990, page 44). His body had been placed by the white boy into a racialised and racist corporeal matrix.[5]

Fanon had stopped being amused, not because he had lost his place in the world, but because he had found it: he was the person spoken about, the black body, the body which carries so many associations for whites. These connotations, which have nothing to do with him, shroud him. Haunted by presence of these white ghost stories, Fanon is sick. He is 'completely dislocated' and 'absolutely depersonalised' by this experience. He has been separated from his body and both his body and his soul have been incarcerated within the prison-house of white desires, fantasies and fears: 'I discovered my blackness, my ethnic characteristics, – and I was battered down by tom-toms, intellectual deficiency, fetishism, racial defects, slave-ships, and above all, above all else: "Have a banana"' (page 112, modified translation).

Fanon is split, both phobia and fetish (see Bhabha, 1986, page 78). And he is aware of being severed from his body. This amputation makes Fanon endure 'a haemorrhage that splattered my whole body with black blood' (page 112). His eviscerated body is never allowed to be equal to the white man's: across many lines, he suffers the deadly cuts of the racist grid of meaning, identity and power. 'My body was given back to me sprawled out, distorted, recolored, clad in mourning in that white winter day. The Negro is an animal, the Negro is bad, the Negro is mean, the Negro is ugly' (page 113).

Violence, fear, desire, hatred swirl around poles of absolute difference marked by skin: where 'white' is as good as Tarzan and Jane, while 'black' is cowardly and savage, licentious and profane. As Fanon moves around the world, he feels that he is 'being dissected by white eyes, the only real eyes' (page 116). By these eyes, Fanon is *fixed* into place, into his fiction-alised body.[6] Thus, the embodied colonial mirror severs Fanon from his own image, from his own body, and spreads him out in front of himself as an other. There is nothing *essential* about this. There is no *Reason* for it. As Fanon astutely observes, 'The Negro is not. Any more than the white man' (page 231). Nevertheless, these fictitious and fluid black/white cate-gorisations of difference become the hard co-ordinates of oppression and repression. Desire and fear, identification and abjection go hand in hand.

'Look, he is handsome, that Negro ...'
'Kiss the handsome Negro's ass, madame!'[7]
Shame flooded her face. At last I was set free from my rumi-nation. At the same time I accomplished two things: I identified my enemies and I created a scandal. A grand slam. Now one would be able to laugh.

(page 114, modified translation)

Fanon seeks to resolve the situation by scandalising his enemies. The woman says how handsome, but he shames her by putting his 'abject otherness' in her face. He confronts her with her desire and laughs at it. Far from offering a resolution to the problem of racism, however, this encounter is full of trouble. The little boy remains afraid of the black man, the woman is shamed and, even while Fanon laughs triumphantly, there is nothing funny about this situation. Nevertheless, Fanon has broached the vicissitudes of fear and desire which underpin everyday experiences of racism – for both whites and blacks. These vicissitudes are dramatised by the characters in the story: the little boy enacts white astonishment and fear in the face of blackness, while the white women openly expresses an unconscious desire for the black man, and Fanon himself is not just a body but also a soul yearning for equality and recognition.

It is common in post-colonial theory to take this anecdote to be symptomatic of the daily life of black people in (all) colonial situations. The outcome of this is to see a constitutive ambivalence at the heart of a colonial psychodrama, where the whites both desire and fear the blacks, while the blacks want to be white and loathe their blackness. The consequence for thinking about politics is that it must not only address itself to the decolonisation of the territories occupied by the colonisers, but also to the decolonisation of the interior landscapes of body and soul. The white mask must be thrown off. The question I would like to ask is whether it makes a difference *where* this story takes place. Does the interpretation of the situation change if the story, as it were, travels? And do the political implications change? It is possible to imaginatively reconstruct this story in different locations. Given Fanon's biography, it could take place in Martinique, in France or in Algeria. I will take each of these possibilities in turn.

We could imagine that this encounter takes place in Martinique (as many theorists tend to). But we would have to imagine this with a little difficulty. The little boy first has to be astonished by the presence of a black man, then he has to find it fearful. It would be surprising that the boy would react this way if he had grown up in Martinique, for it can be assumed that he would be accustomed to seeing and meeting black people. Indeed, white supremacy in this situation would lead us to expect that the little boy would be being nursed by a black woman (which, I might add, produces its own characteristically ambivalent psychodynamics . . .). Nevertheless, it may be that Fanon has appeared in an area reserved for whites, this would explain the surprise and fear. Thus, if this is a street scene, blacks would outnumber whites in most public spaces, except in those places where whites excluded blacks. It is possible, then, that Fanon has transgressed white boundaries of social and spatial exclusion. Even so, neither of these situations would necessarily frighten the boy. The boy's fear alerts us to the specificity of the circumstances under which he might be afraid, even in a colonial situation where there are only a small minority of white people. The establishment of white (p)reserves and the possibility of 'reverse invasion' by blacks nevertheless demonstrates not only that white control over these spaces is tenuous and partial, but also that the colonisers might be rooted out and expelled (even if this is not so easy or quick to achieve in practice). The anti-colonial struggle might, then, be mobilised by a politics of territorial control: black against white; the white man removed, the mask cast down and ground to dust. The white man would be no more. No more than the black man.

On the other hand, it is much more likely that Fanon's story is set in France – and does not, after all, dramatise the definitive ambivalences of everyday life *in* the colonies. We can note, for example, that Fanon is travelling by train, in uniform, that it is freezing cold and snowing. Most likely, then, this situation describes the experience of a black man who, having been led to believe in French values of equality, fraternity and liberty, is suddenly confronted by brute racism – not just in the conscious mind of a full-grown adult, but in the unconscious fantasy life of a small boy: 'Mama, the Negro will eat me' (page 114, translation modified).

In post-war France, it might be expected that the boy would be familiar with – and even delighted by – men in uniform. The sight of a black man might, however, be a complete surprise (especially outside the major cities). In his shock, as Fanon suggests, the boy could have instantly conjured up all too familiar images of cannibals, as seen in all those Tarzan movies, or perhaps in Imperial exhibitions, or even in his school books. But he might also have become even more afraid by the uncanny appearance of the uniform; that is, the boy could be even more terrified because the cannibal is *well* disguised as a French soldier. And it is not just the boy who fantasises about the black man. Confronted by a black man, who embodies for her a whole series of hyper-masculine sexual stereotypes, the woman seems to have been lured into speaking her innermost desire.

Unlike Martinique, in France it is Fanon who is outnumbered. The subject positions – and, therefore, the politics – that are available to him are circumscribed differently. In Martinique, Fanon might cast off the white mask, but in France it is *de rigueur*. In adopting the white mask, Fanon seeks to become like the white man – or, at least, invisible to the white man – in a white landscape. Thus, *Black Skin, White Masks* may say more about the experiences of black people in the Imperial heartland, than about psychodynamics in the colonies. In France, anti-colonial politics would have to take a different shape to that in (even French) colonies. Radical politics, here, would have to demolish interior landscapes of racism as a way of transforming the spaces of social exclusion and discrimination.

There is one last place I would like to site this tale: in Algeria. This time the little boy's surprise and the woman's desire is provoked, not by the surprise at seeing the colonised 'out of place', but by the presence of a black man in an Arab country. This situation is not so unlikely as it might at first appear. It will be remembered that, during the war, Fanon served in North Africa. At this time, he suffered racial abuse – not only from whites, but also from Arabs. If this story is set in Algeria, then Fanon is schematised as too black by both whites and Arabs. Further, his presence in Algeria is also complicated by his arrival as part of the coloniser's forces. Fanon is part of the colonising army: in Algeria, he is one of the oppressors. Even if the story was set at a time when Fanon was a psychiatrist, he was still working for the colonial administration. Moreover, Fanon would not even have been in Algeria, if it were not for the migration patterns installed by French colonialism. Now, Fanon is caught in the middle: he is both colonised and coloniser. If Algeria is to be truly freed from the ambivalences and oppressions of everyday colonial life, then surely it would have to be freed from both whites and blacks. Moving this story to Algeria shows that Fanon is implicated in the colonial project; that is, in reproducing colonialism. As Memmi (1971) suggested, Fanon's position is impossible.

I do not wish to give the impression that these political responses are the only ones imaginable in these places or that any underlying 'truth' of Fanon's thought is completely compromised by changing situations. My purpose is much more local than this. I only intend to demonstrate that changing the location of Fanon's writings matters to the politics of that location. If it is not my purpose to reveal the truth of the situation or settle the politics in each place, I would like to make this point.

The location of the story matters in so far as the political implications are necessarily embedded within the possibilities that any place offers. But, even this begins to suggest that these locations have nothing to do with one another. Yet, Fanon's life shows that places are connected up (or not), through different histories, through different geographies. Situating Fanon does not simply involve moving him from one country to another or assuming that such migration makes no difference. It shows that Martinique, France and Algeria are tied to each other in specific ways – and these ties are inherently spatial. Through tracing these spatial relationships, it is possible to uncover the ways power relations act at a distance, to demonstrate the partiality of supposedly universal or global ties, and to identify the political possibilities that these different locations do (and do not) offer. If the stories so far have implied locations – personal, political, geographic – that seem discrete and separate, then Fanon's life dramatises other aspects of lived, political spaces; spaces always in relation to others. These spaces remain unresolved in Fanon's thinking and they, in their turn, need to be rethought; rethought through tracing the troubled spaces that constitute relations between people and peoples.

In the last scenario, I situated Fanon in Algeria – and I left him in an impossible situation: both coloniser and colonised. Only if territories and locations are seen as bounded, border-guarded and internally homogenous, could the suggestion be made that Fanon was entirely 'out of place'. If the territory is, however, striated by histories of migration, by geographies of contact, and by internal differences, then Fanon's experiences belie the assumption that being from somewhere provides a natural or authentic grounding for an emancipatory politics of identity and nation (see Posnock, 1997; also Fuss, 1995; for a conflicting view, see Memmi, 1971). Instead, it is necessary to take into account the struggle over geography; or, alternatively, the struggle to remake geography.

Land and Freedom: anti-colonial struggles and national consciousness

At the outset of Chapter 3 of his last book, *The Wretched of the Earth* (1961), Fanon boldly asserts, 'History teaches us clearly that the battle against colonialism does not run straight away along the lines of nationalism' (page 119).

This might not seem like such a curious remark, but it signals a problem at the heart of anti-colonial struggles which aspire to free *both* the people *and* the territory from the oppressors. The problem that Fanon identifies is that once the nationalist liberation struggle has succeeded in removing the colonial administration, then two pitfalls await. First, those nationalists who take over the positions of power in the post-colonial administration themselves become a powerful élite, whose interests may very well differ from those of 'the people'. Second, the post-colonial nationalist élite soon finds it expedient to negotiate with the former colonial power and very quickly the newly-freed nation is tied into new kinds of colonialism. Fanon offers a solution to this problem: that all nationalist struggles have also to be socialist revolutions (see White, 1996). And, in the main, commentators have read *The Wretched of the Earth* as a call for social and political transformation. However, I would like to look more closely at the ways in which Fanon conceives of the place and space of 'the nation' in liberation struggles. Crucially, Fanon deploys apparently contradictory assessments of place of the nation in emancipation, but these might be rethought and realigned by thinking of the nation through different understandings of space.

In Fanon's view, the struggle to free the land from colonial oppression must build on nationalism. Further, in his later writings, he also argues that an authentic liberation struggle must be violent: the land can only be purified of the oppressor's touch by being bathed in blood. Nationalism, however, is not enough of a guarantee that the people are freed from oppression. Fanon worries that

> National consciousness, instead of being the all-embracing crys-
> tallization of the innermost hopes of the whole people, instead of

being the immediate and most obvious result of the mobilization of the people, will be in any case only an empty shell, a crude and fragile travesty of what it might have been.

(1961, page 119)

While nationalism may be a necessary component of any anti-colonial struggle, Fanon searches for ways in which black élites can be prevented from utilising nationalist ideals for their own purposes. The question, for Fanon, is what would a liberatory nationalism be like. If national élites can make frenzied appeals to national unity in their own interests, then Fanon uses national consciousness as an antidote to the co-option of nationalist ideals. First, he argues that national élites should be in the service of the people, rather than the people simply acting as the power base for the nationalist party. The objective of independence in anti-colonial struggles, therefore, is to free all of the people from domination from whatever source – even from oppression by their own leaders. Second, Fanon insists that the country's economy should be nationalised both because this puts the means of production in the hands of the people and because it prevents recolonisation by capitalism (too often, aided and abetted by national élites). Finally, national consciousness is produced by the mobilisation of the whole of the people, rather than being the shouted slogans of nationalist party ideologues.

Whatever we might think of the prospects for these ideas, they are forged in a particular understanding of the nation-state and political economy. Significantly, Fanon is presuming that there needs to be some kind of consensus *within* the nation's borders as to who are its citizens and where the borders of the country lie. This assumption is bolstered, in his reasoning, by the sense that national consciousness forges an internally integrated nation in the course of a violent war of liberation: the people become the nation and the nation becomes the people. Thus, in the struggle for national freedom, Fanon believes that individual experiences become understood within a wider frame of reference.

The living expression of the nation is the moving consciousness of the whole of the people; it is the coherent, enlightened action of men and women. The collective building up of a destiny is the assumption of responsibility on the historical scale. Otherwise there is anarchy, repression and the resurgence of tribal parties and federalism.

(1961, page 165)

It might be doubted, however, that the people can be relied upon to move towards shared, coherent, enlightened ideals, even – or, especially – in the midst of a vicious struggle for national liberation. Thus, the struggle to

270

change geography need not assume that the nation is a homogeneous space occupied by an undifferentiated people. Further, it was the colonial powers that scratched national borders onto the surface of the earth, according to their own geopolitical imperatives. Without needing to take account of local and regional circumstances, without regard to racial, ethnic, religious and cultural particularities, these border lines arbitrarily cut into, cut out and chopped up different peoples. In his privileging of the nation, Fanon himself has been criticised (somewhat unfairly) for ignoring differences amongst, for example, Algerian peoples. Nevertheless, it is now possible to see that forging a national consciousness during the anti-colonial struggle, rather than being a necessary feature of liberation, might actually be a Trojan Horse enabling the internal colonisation of the country by specific regional or class élites.

Fanon certainly sticks to his nationalist guns: national consciousness must articulate and crystallise the hopes and desires of the people. However, there is no one People. And this is just as true of the former colonies as of the former colonising nations. Fanon's thinking about the spaces of the nation is not so monochromatic as to ignore either internal power relations or wider circumstances within which nations are embedded. If he believed that nationalism was necessary for the anti-colonial struggle, then he also believed that it had to be abandoned in favour of deeper social and political objectives. It was only these objectives that would guarantee that the revolution would be in the hands of the people, rather than the people being in the hands of the revolution. In this sense, Fanon is suspicious of appeals to unity that presume that there are essential commonalities between people. Indeed, he scorned attempts to ground and to unify struggles through seemingly natural spatial scales: whether through regionalism, nationalism, or Africanism.

Geographic scales are troubling spaces. Fanon decries all forms of regional and sectional parochialism, yet relies on national parochialism to forge an anti-colonial struggle and to identify the subjects of revolution. The land and the people are ambiguously placed in his revolutionary theory: they are both what needs to be freed and also the agents through which liberation is to be defined and achieved. Paradoxically, Fanon privileges the nation in his revolutionary theory, yet the nation is simultaneously the scale of oppression. In Fanon's abstract dialectics, freedom is doomed to vacillate between the land and the people, since land and people rarely coincide. However, it might be possible to re-imagine these relationships if a different spatial understanding of 'nation' and 'people' are bought to bear.

After a discussion of national culture in *The Wretched of the Earth*, the publishers reprint an address which Fanon made to the Second Congress of Black Artists and Writers in 1959. Before continuing, I should note that Fanon's writing is always an intervention in a political and social situation. And, in this light, many of his writings on nationalism and national consciousness should be read as a call for unity in the face of a colonial oppressor that is attempting to divide the people and rule the territory. In this particular statement, however, Fanon is seeking a different kind of politics. And here it is possible to read a different sense of the relationship between people, nation and liberation.

In the course of political struggles, Fanon recognises that a guerrilla warfare involving flexible strategies and tactics will have to be adopted. However, he also consistently argues that new revolutionary subjectivities have to be formed. Not only will national traditions have to be abandoned or modernised, but new links will have to be made between people who might not see themselves as belonging to the struggle (for whatever reason). From this perspective, the nation might only be a first – and ultimately expendable – link in a chain that binds the people together. However, for Fanon, the nation might also act as a link into wider connections between people, elsewhere.

> Individual experience, because it is national and because it is a link in the chain of national existence, ceases to be individual, limited and shrunken and is enabled to open out into the truth of the nation and of the world.
>
> (1961, page 161)

While national consciousness might enable people to connect their freedom with the freedom of the land, liberation can also open up connections between people and the wider world. If decolonisation of interior landscapes involves new forms of individuality, then these revolutionary subjectivities are born not only in (spatial) comradeship with other people within the national territory but also in relation to those beyond national boundaries. In this opening out, Fanon wishes neither to erase nor to fossilise national differences, but to enable ways in which different peoples might meet each other as different people (see pages 197–199). Nevertheless, in his thinking, it appears that the price of maintaining differences between nations is the erasure of differences within the nation. Nevertheless there might be other relationships between the individual, other people, the nation and the world. Fanon, himself, hints as much:

> The responsibility of the native man [sic] of culture is not a responsibility vis-à-vis his national culture, but a global responsibility with regard to the totality of the nation, whose culture, merely, after all, represents one aspect of that nation.
>
> (page 187)

Perhaps, in this, it is possible to discern a sense that differences within and between nations are maintained, even during the struggle, and also that radical politics carries a responsibility to others. Instead of thinking about the nation as an isotropic, bounded surface within which all people are the same, it might be possible to see the nation as a porous and mutable space, where people take responsibility for their connections to others, taking responsibility for acknowledging and even enjoying their differences. And this rethinking alters what it means to suggest that Fanon is a 'global theorist'. If Fanon is seen as implacably opposed to fascism, colonialism and capitalism wherever they might be, then liberation struggles must be extended to a subversion of the spaces (places and scales) that relations of power produce and maintain (in order to reproduce power relations). If colonialism produces a national territory which it seeks to control, then destabilising both the production of borders and the internal stereotypical sameness of the peoples in that territory would be necessary components of the struggle over geography. And, it might be, that there are many meanings of 'over' in this struggle over geography: such as, for, across, outwards from, on the other side of, covering, beyond, and so on. Each of these meanings evokes different ways that politics may be imagined geographically, different ways that geography might be remade.

Geography, from this perspective, is not a dull, obstinate, unchanging fact of everyday life. It is produced out of the relationships between people. Locations, territories, colonies, nations, the world are not natural scales, but produced through power relationships between individuals, groups and peoples. Political struggles are not fought on the surface of geography, but through its very fabric/ation. In yearning for new ways of relating to one another, in searching for new forms of subjectivity, Fanon's revolutionary theory also necessitates that space is produced differently. And it is on this point that I would like to conclude this chapter.

Conclusion: situating politics

At the outset of this chapter, I asked if it made a difference to think about the situatedness of someone's writings or to think about their thought using different conceptions of space. I addressed this question, first, by situating Fanon's 'Look, a Negro' story in different countries and then by looking at his understandings of nation and national consciousness and their role in anti-colonial wars of liberation. I have also hinted that, in most commentaries on Fanon, it is usual to separate out the earlier and the later Fanons, but I would like to bring these Fanons' places and spaces together. Before this, however, a word of warning.

As Gordon, Sharpley-Whiting and White point out, Fanon has been attacked for being 'misogynous, homophobic, anti-black, anti-Caribbean, anti-Arab, and petit bourgeois' (1996, page 6). It would be facile to argue that anyone who had attracted this much criticism must have something going for them. It would be better to say that there can be no return to Fanon that fails both to recognise that substantial problems are generated by his stubbornly dualistic and hierarchical imagination and also to acknowledge that times – and places – have changed. It would be better, then, to re-interpret Fanon while bearing in mind that his rhetoric was intended to intervene in and transform an existing situation. These situations may no longer pertain. On the other hand, there is substantial evidence that some things have not moved very far: for example, the beating of Rodney King in Los Angeles contains much that is similar to the 'Look, a Negro' situation (see Gooding-Williams, 1993), while Taiwo's (1996) analysis of contemporary Nigeria suggests that Fanon's prophecies both of the replacement of white élites by national élites and of the neo-colonialist return of white capitalism have, for the most part, come horribly true (see also Watts, 1997).

Therefore, as Gates has suggested, it is necessary to situate Fanon's politics. In this way, it might be possible to understand the so-called radical instabilities of his writings. Far from seeing these instabilities as a problem, they must be seen as the troubled interventions of a person struggling to deal with the incommensurable spaces within which he found himself – spaces which, it has to be said, were not of his own making. Fanon's political rhetoric attempts to make sense of, and to change, the unreasonable world he was situated in. More than this, he sought to uncover the sources of power relations in society that not only produced conditions of political and economic inequity and injustice, but also involved the grotesque psychodrama of everyday racism – whether for whites or blacks. Only by providing both sides of the story – the personal and the political – did Fanon think that it might be possible to set loose new kinds of people, who might be able to find better ways of being in the world.

If political thinking leads us to conclude that neither place nor space have anything to do with it, then Fanon is a bad choice of Global Theorist, where this would involve a universalising, disembodied or unsituated knowledge. For Fanon struggled to find freedom in a 'global consciousness' that respects differences between and amongst peoples and places, partly by refusing to allow either to become *the* (authentic) agent of change or *the* (exclusive) site of resistance. Instead, Fanon sought to question the production of difference(s), in place, between people. Consequently, I would argue, it is necessary to deploy new understandings of both place (like a country) and space (like national consciousness). Not only does moving his 'Look, a Negro!' encounter from one place to another transform the situation, it also alters the kinds of politics that can be imagined. Not only does thinking about the nation spatially involve reconceptualising the whereness of the nation, it also changes the kinds of political identities that are built up in nationalist struggles. Revolutions are not just about struggles for freedom from oppression within a particular territory, nor just about wars in pursuit of control over a territory delimited by national borders, they also comprise rights and responsibilities which are established *with respect to others*.

Against my argument, Fanon does privilege the nation as the proper scale of revolutionary struggle. But I do not have to read Fanon against Fanon to suggest that emancipatory politics cannot be guaranteed, nor authenticated, by appeals to a politics of turf: the local isn't always right, just as the global isn't the only game in town. Instead, it is necessary to think through the ways in which people are associated with one another or, as importantly, disassociated from each other. From Martinique, through France, to Algeria, Fanon attempted to understand the political possibilities implicit in these troubled spaces. He invoked a politics of identity and place consistently suspicious of claims to authenticity, whether located in identity or in place. It is here that it is possible to reimagine a politics of location which is not simply about the politics of 'where you are', but about the ways in which 'where you are' is bound up in other spaces, other places, other people.

With neither an authentic ground to liberatory struggles, nor a privileged site of resistance, the possibility of forming political communities might appear impossible. This problem would be resolved, however, by recognising that political communities will involve both 'alliances and collaborations across divisive boundaries' and, despite internal rifts, a sense of 'spatial comradeship' (see Mohanty, 1991, page 4). This spatial comradeship would not, then, operate in a political location that is presumed to be isolated from all others, but act on the grounds that politics is social, politically and personally situated in relation to others. Thus, it is on the basis of struggling to change geography that political communities might be established. Partly through a distrust of social relations so familiarly grounded in locations, borders, margins, frontiers, territories, and so on. Partly by recognising that political communities cannot be self-grounding or self-legitimating, because these can all too readily install and ossify power relations within them.

Fanon, I am arguing, was edging towards a sense of spatial comradeship in which the politics of location co-ordinates not only your place in the world, but also a wider set of connections with others, who may be in very different places, or indeed in the same space. In this light, Fanon's work suggests that nationalist politics are not inevitably reactionary and bigoted; that is, that using a politics of the nation does not inevitably install nationalist élites in positions of power or become the basis for legitimating the worst acts of barbarism. Instead, the ambiguous political spaces of Fanon's thinking will remind us that the line between acts of liberation and acts of barbarism is thin. Such ambiguities imply that the justice of political outcomes cannot be settled in advance of political struggle. While it may be advantageous to suggest that there are universal principles of human justice, such political claims are properly located in the expediencies of the situation.[8] There can be no guarantee that action at a particular scale will be any more successful than any other, or that thinking about power or liberation at a particular scale will be any more progressive than any other. Instead, Fanon's life suggests that it is necessary to plot out the webs of spatial relationships in which people are entangled. Only then might it be possible to see ways of getting beyond troubled spaces.

Notes

1 For accounts of Fanon's life, see Caute, 1970, Geismar, 1971, and Gendzier, 1973. There is some doubt over Fanon's date of birth.
2 For example, recent collections dealing with aspects of Fanon's work include Gordon, Sharpley-Whiting and White, 1996, and Read, 1996. Also, Isaac Julien's film, titled *Frantz Fanon* (1996), is rich in historical detail, dramatic reconstructions and political analysis.
3 The analogy of 'the mirror' is not idle, Fanon is consciously drawing on and altering Lacan's early psychoanalytic work on the mirror stage of childhood development (page 160, footnote 28; see also Vergès, 1997).
4 This account has been described as misogynist and homophobic – on the former, see Doane, 1991, and Young, 1996, but for a contrasting perspective see Sharpley-Whiting, 1996; on the latter, see Mercer, 1995.
5 I find it curious that, almost without exception, it is assumed by subsequent writers that the child in the story is a girl. Sometimes the boy is stripped of his gender altogether. For example, in the film, *Frantz Fanon*, the encounter is dramatically reconstructed in such a way that the infant appears on-screen so quickly that it is impossible to tell what sex he/she is. Meanwhile, an on screen commentator refers to the boy as a 'child'. Fanon is, however, unambiguous. He describes the child as 'le petit garçon', 'le beau petit garçon' and 'le petit garçon blanc' (1952, French edition, page 117). These are translated by Charles Lam Markmann accurately and I remain uncertain as to why the boy is feminised or neutered in post-colonial commentaries.
6 On such fixing into the body – is this fixionalising?! – see also Mercer, 1989.
7 The original French version says 'Le beau nègre vous emmerde, madame!'. This might be better translated as 'The handsome Negro is covering you with shit, madame!'. It might be worth noting the formal tone of 'vous' in Fanon's reply, whereas blacks were invariably addressed by whites using an informal 'tu'. It seems words express and enact power relations between whites and blacks, even in the moment of anger and loathing.
8 See Harvey, 1996, on this.

References

Bhabha, H. (1986) 'The Other Question: Difference, Discrimination and the Discourse of Colonialism', in R. Ferguson, M. Gever, M-h. T. Trinh and C. West (eds) (1990) *Out There: Marginalization and Contemporary Cultures*, Cambridge, Massachusetts: MIT Press, 71–87.
Bhabha, H. (1990) 'Interrogating Identity: Frantz Fanon and the Postcolonial Prerogative', in *The Location of Culture*, 1994, London: Routledge, 40–65.

Caute, D. (1970) *Fanon*, London: Fontana.

Doane, M. A. (1991) 'Dark Continents: Epistemologies of Racial and Sexual Difference in Psychoanalysis and the Cinema', in *Femmes Fatales: Feminism, Film Theory, Psychoanalysis*, London: Routledge, 209–248.

Fanon, F. (1952) *Black Skin, White Masks*, 1986, London: Pluto Press.

Fanon, F. (1959) *Studies in a Dying Colonialism*, 1989, London: Earthscan.

Fanon, F. (1961) *The Wretched of the Earth*, 1967, Harmondsworth: Penguin.

Fuss, D. (1995) *Identification Papers: Reflections on Psychoanalysis, Sexuality and Culture*, London: Routledge.

Gates, H. L. (1991) 'Critical Fanonism', *Critical Inquiry*, 17(3), 457–470.

Geismar, P. (1971) *Frantz Fanon*, New York: Dial Press.

Gendzier, I. (1973) *Frantz Fanon: a critical study*, New York: Pantheon Books.

Gooding-Williams, R. (1993) ' "Look, a Negro!" ', in R. Gooding-Williams (ed.) *Reading Rodney King/Reading Urban Uprising*, London: Routledge, 157–177.

Gordon, L. R., Sharpley-Whiting, T. D. and White, R. T. (eds) (1996) *Fanon: a Critical Reader*, Oxford: Basil Blackwell.

Hall, S. (1996) 'The After-life of Frantz Fanon: Why Fanon? Why Now?', in Read 12–38.

Harvey, D. (1996) *Justice, Nature and the Geography of Difference*, Oxford: Basil Blackwell.

Massey, D. (1994) *Space, Place and Gender*, Cambridge: Polity Press.

Memmi A, 1971, 'La vie impossible de Frantz Fanon', *Esprit* (September), 248–73; translated and published in 1973 as 'The impossible life of Frantz Fanon', *Massachusetts Review*, 14(Winter), 9–39.

Mercer, K. (1989) 'Skin Head Sex Thing: Racial Difference and the Homoerotic Imaginary', in *Welcome to the Jungle: New Positions in Black Cultural Studies*, 1994, London: Routledge, 189–219.

Mercer, K. (1995) 'Busy in the Ruins of Wretched Phantasia', in R. Farr (ed.) *Mirage: Enigmas of Race, Difference and Desire*, London: Institute of Contemporary Arts/Institute of International Visual Culture, 12–55.

Mohanty, C. T. (1991) 'Cartographies of Struggle', in C. T. Mohanty, A. Russo and L. Torres (eds) *Third World Women and the Politics of Feminism*, Bloomington: Indiana University Press, 1–47.

Posnock, R. (1997) 'How It Feels to Be a Problem: Du Bois, Fanon, and the 'Impossible Life' of the Black Intellectual', *Critical Inquiry*, 23(Winter), 323–349.

Read, A. (ed.) (1996) *The Fact of Blackness: Frantz Fanon and Visual Representation*, London: Institute of Contemporary Arts and Institute of International Visual Arts.

Said, E. (1993) *Culture and Imperialism*, London: Vintage.

Sharpley-Whiting, T. D. (1996) 'Anti-black Femininity and Mixed-race Identity: Engaging Fanon to Reread Capécia', in Gordon *et al.*, 155–162.

Taiwo, O. (1996) 'On the misadventures of national consciousness: a retrospect on Frantz Fanon's gift of prophecy', in Gordon *et al.*, 255–270.

Vergès, F. (1997) 'Creole Skin, Black Mask: Fanon and disavowal', *Critical Inquiry*, 23(Spring), 578–595.

Watts, M. (1997) 'Black Gold, White Heat: State Violence, Local Resistance and the National Question in Nigeria', in S. Pile and M. Keith (eds) *Geographies of Resistance*, London: Routledge, 33–74.

White, R. (1996) 'Revolutionary Theory: Sociological Dimensions of Fanon's *Sociologie d'une Révolution*', in Gordon *et al.*, 100–109.

Young, L. (1996) 'Missing Persons: Fantasising Black Women in Black Skin, White Masks', in Read 86–101.

Part 3

REFIGURING SPACES
IN THE PRESENT

13

SOME NEW INSTRUCTIONS
FOR TRAVELLERS

The geography of
Bruno Latour and Michel Serres

Nick Bingham and Nigel Thrift

> reason today has more in common with a cable television
> network than with Platonic ideas. It thus becomes much less
> difficult than it was in the past to see our laws and our
> contracts, our demonstrations, and our theories, as stabilised
> objects that circulate widely, to be sure, but remain within
> well laid out metrological networks from which they are inca-
> pable of exiting – except through branchings, subscriptions
> and decodings.
>
> (Latour 1993: 119)

I. Introduction

Today actor-network theory is everywhere, and with it the names of Bruno
Latour and Michel Serres. This paper is an attempt to provide: an explanation of how actor-network theory came into being and what it is (Section
I); an account of its distinctive mode of thinking spaces and times as
mediated travellers itineraries (Section II); an example of the method
of actor-network theory at work – namely the construction of Latour
and Serres themselves (Section III) – and some very brief conclusions
(Section IV). Why do we believe that Latour and Serres should be included
in this book? Because, they have sought to repopulate space and time
with all the figures that have been stripped away by an idea of abstract
division, by concentrating instead on movement, on process, on the constant hum of the world as the different elements of it are brought into
relation with one another, often in new styles and unconsidered com-
binations. In other words they are attempting to rediscover the *richness* of
the world and

to find richness, one only has to turn toward the world itself, to
the wind, the foam, the snow-capped mountains in the background,
the earnest miniature city behind the harbour. 'Objective' time
and 'subjective' time are like taxes exacted from what peoples the
world, they are not all that those multitudes do and see and mean
and want. We are not forced to choose forever between losing
either the feel of time or the structural features of the world.
Processes are no more in time than in space. Process is a third
term . . .

(Latour 1997a: 172)

Where then does actor-network theory come from? It is possible to argue
that it has three main points of origin. The first is the sociology of science.
By the 1980s, the sociology of science had produced a strong programme,
which was intended to argue that science was a social construction *tout
court*. To begin with, actor-network theory was intended as 'a more direct
and less laborious way to write the strong programme' (Latour 1988a: 23)
by treating the natural and social sciences *symmetrically*.

Explaining a science means that we should be able to stabilise
with it more equal relations in such a way that we learn from it
about society and use our own disciplines to teach a few things
to the science we are dealing with. This more equal status should
be our touchstone even though, in the case of physics, such a
programme may appear ludicrous. The fecundity of an account
in this newly redefined strong programme will be assessed by our
ability to transform the definition of the social until it is *on a par*
with the very content of the science studied, and exchanges properties with it.

(Latour 1988a: 26)

Thus Latour tried to develop a *language*, an 'infra-physical' language, which
could 'translate' the natural and the social sciences, by showing that each
one comes, so to speak, equipped, that each one is composed of innumerable figures, some 'mechanical', some 'social', some 'fictional', which
actor-network theory aims to shuttle between, and act as a shuttle between.
But gradually Latour came to position actor-network theory in opposition
to constructivist approaches since, for him, actor-network theory questioned notions of society just as much as notions of Science. Actor-network
theory could therefore be described as a realist approach which emphasises the high degree of contingency of the world, a language for both
upsetting and collecting the collective.

The second source was French intellectual culture. French theory of
science is based upon *epistemology*. Figures like Bachelard and Canguilhem

are 'ultimately more concerned with the epistemological status of science than with its historical location' (Bowker and Latour 1987: 718), that is with issues like rationality, ruptures of thought, and so on. In this regard French epistemology 'may have a much richer repertoire' (Bowker and Latour 1987: 726) but its emphasis on the norm of rationality, and on a history of this norm in scientific discourse, can be disabling (Rabinow 1996). In certain senses, as Bowker and Latour (1987: 740–741) point out, this is quite an odd state of affairs. After all,

> France is the country where the link between scientists and the managerial, political, intellectual establishments is the strongest but it is, of all countries, the one which has least developed a field of social studies of science that links the establishment and science. This could in itself offer us a clue: epistemology in France is the only real way of talking politics, of delivering what holds all of us together – that is to say, the *concept*. Since politics is so rational and universal anyway, why not use the language of rationality and universality to talk about both science and politics? In particular, when epistemology talks about 'ruptures from common sense' it offers a nice political model for explaining why for everyone in France epistemology is placed at the top of the hierarchy, and would also explain why the idea of politicising science seems more than just absurd, totally vain.
>
> This doesn't mean that the French are the more able to understand the social shaping of science – quite the contrary. What shapes the facts, the raw data, is not for them society or culture, but something else: theory. The theory-ladenness of facts is the staple of French philosophy of science. This means that since they easily defeat empiricism they believe they can, without further ado, embrace theory. Society is thus short-circuited. More precisely, the French use the argument against empiricism to discourage in advance any field research in social studies of science, since it is always necessary to have a theory to inform your data. Nothing will be learned from the empirical study of the way science is produced since every scientific argument is theory-laden anyway. Thus it is better to do epistemology (for this at least deals with theory) than sociology.

Latour's work can then be seen as a kind of rebellion against this way of proceeding by producing a language which constantly short-circuits epistemology through a qualified empiricism. In certain senses, the rebellion is half-hearted. To begin with it is not as if Latour believes that all is well outside France. Though utterly committed to the importance of empirical studies, he is also extremely sceptical of, for example, Anglo-Saxon varieties of empiricism (which is why he has devoted so much time to providing a different account). And he cleaves to certain French values such as idiosyncrasy – 'the word "idiosyncratic" being positive on this side of the Channel and rather derogatory on the other (thus constituting yet another source of misunderstanding!)' (Bowker and Latour 1987: 730).

The third source of inspiration was (and is) Michel Serres; a writer who is regarded as idiosyncratic even in a culture that values idiosyncrasy! (See Latour 1988b, Serres and Latour 1995.) The problem with Serres' work is that it resists all characterisation as, in a sense, actor-network theory also attempts to do. It goes everywhere. In a nutshell, what Serres does that is so striking is

> to develop the argument there is no metalanguage: there is no superiority of religion over science, or science over literature. What he seeks to do is to find the structure that articulates a particular religious, scientific or literary system, and to know how it works. The beauty of his texts is that you never know who is right. It might be Lucretius or it might be the Bible – not because they are prescientific and thus empirically wrong, but because suddenly they are made accurate, as precise as results in biology or mathematics. This is an inversion we do not expect. It is assumed within the Anglo-Saxon world that if science has material, religious and social dimensions then somehow its truth is debased: Serres displays the poetry and beauty of the truths that subtend religion and science. In this rethinking of discourse, there is mingling of styles: anecdote, allegory and rigorous demonstration are found side by side.
>
> (Bowker and Latour 1987: 731)

What is it that Latour gets from Serres? Out of many influences, four come to mind. The first is Serres' anthropological bent. Two of Serres' leading influences were René Girard and Georges Dumezil, both anthropologists who were considered atypical, even eccentric and what Serres produces might be considered as a kind of anthropology, but of travel through discourse. Second, there is Serres' attitude to time and space. Thus, for example, Serres is 'absolutely indifferent to temporal distances' (Serres and Latour 1995: 44). For him, 'time does not flow according to a line. . . . nor according to a plan but rather according to an *extensive* complex mixture, as though it reflected stopping points, ruptures, deep wells, chimneys of thunderous acceleration (rendings, gaps) – all sown at random, at best in a viable disorder' (Serres and Latour 1995: 57). Time is like the weather, turbulent, folded and twisted, 'as various as the dance of flames in the brazier' (Serres and Latour 1995: 58). But people constantly confuse time with the measurement of time and so miss these qualities.

Then, third, there is Serres' suspicion of grand analytical categories. His is a method of rapprochement and rapport between categories:

> *Metaphor*, in fact, means 'transport'. That's Hermes's very method: he exports and imports; thus, he traverses. He invents and can be mistaken – because of analogies which are dangerous and even forbidden – but we know of no other route to invention. The messenger's impression of foreignness comes from this contribution: that transport is the best and worst thing, the clearest and the most obscure, the craziest and the most certain.
>
> (Serres and Latour 1995: 66)

This, then, is a method of rapid movement, and congruent 'comparativism', it is the method of the space between, of conjunction of bringing into proximity. All happens in the movement from place to place, as a vector. 'So I don't make my abstractions starting from some *thing* or some *operation* but through a relation, a rapport. A reading of my books may seem difficult, because it changes and moves all the time' (Serres and Latour 1995: 104). Then fourth, there is the emphasis on the object, and especially the deluge of objects manufactured since the industrial revolution and the new relations they have created. Serres wants to write a new 'contract' with these objects which have given us powers undreamed of by previous generations.

What then, does actor-network theory consist of? It's 'essence' is an 'infra-physical' language for mapping out the traces of networks through an anthropology of the figures that set them going and keep them at work. Each of these terms requires further characterisation. *All* 'networks' – the term is chosen carefully to produce an image of the constant back and forth motion of a circulating entity – require a certain degree of management to produce some kind of stable form.[1] These 'circulations' require certain activities to be delegated to 'recruits'; they require a certain degree of faithfulness to themselves (displacement with deformation); they require a certain notion of what is taken to be real (fact) and what is taken to be unreal (fiction); and so on. These are the minimum conditions of their existence. So Latour suggested a history of science which would consider it as

> The history of centres which are growing through the management of traces that have three main characteristics: they are as mobile, as immutable and faithful, and as combinable as possible. The circulation back and forth of these 'immutable mobiles' have *networks* – that is two-way paths leading from the centre to the now documented lens. These networks are constantly repaired against interruption by maintaining *metrological claims* that keep the

frames equivalent. To define these centres in the most general way, I have called them *centres of calculation*. The main point of their history is that no distinction has to be made between economics, science, technology or even the arts when we follow how each of these three characteristics is enforced.

(Latour 1988b: 21)

In turn, this basic intra-physical framework allows us to sense, through its analytic continuity, certain things which were previously concealed by the categorised ways of thinking we have developed. First, size. The abiding principle here is that 'size is not a property of characters, only of networks and their relations' (Latour 1988b: 30). For too long our explanation has been based upon a hierarchical assumption that a larger character must explain a smaller one. For example, 'sociologists always want to add the social context and they think that in a case study, something is amiss if there is no larger scale entity to explain the whole thing' (Latour 1988b: 30). But

> in practice, however, the characters presented in their accounts, which bear the name of 'social structure', 'longue', 'durée', 'large scale influences', 'over arching interests' and the like, are not bigger than the little ones they try to explain. A giant in a story is not a bigger character than a dwarf, it just does different things. The same two metre-square print may represent a battlefield or an apple; no one will say that the first is bigger and more encompassing than the second.
>
> (Latour 1988b: 30)

Big, in other words, 'does not mean "really" big or "overall" or "overreacting" but connected, blind, local, mediated, related' (Latour 1999: 18). Second, abstraction, in this account, is no longer the mental production of 'higher' categories but something different; it is the result of mediations between one frame of reference and another, many of which would consist of 'mundane' operations.

> The 'big picture' is not given in one frame of reference, but in going from one frame to all the others through a network. Operations like thinking, abstracting, building pictures, are not *above* other practical operations like setting up instruments, arraying devices, laying rods, but are *in between* them. The vocabulary often used by cognitive and social sciences to describe mental operations is misleading. Abstraction does not designate a higher level of figuration but a fast circulation from one repertoire to another. It is not a property of mind, it is a property of reference.
>
> (Latour 1988b: 35)

In turn, we can see that, third, Latour makes no strong distinction between actors in networks: 'they need not be human characters they can be anything' (Latour 1988b: 5). These 'actants' are figures which are able to make shifts in space and time; hence their capacity to act. In one example, for example, Latour gives a list of some of the most important actants in Einstein's work. Ravens, trains, clouds, men with rigid rods, lifts, marble tables, molluscs and of course clocks and rulers', and so on (Latour 1988b: 2–7). But this diversity is no problem. Rather it is what Latour wants to reintroduce: instead of a purified world of categories, he sees a hetero-geneous world of hybrids.

And this leads to the fourth principle of method. Actor-network theory is a comparative *anthropological* analysis of the modern world which follows the network and is 'faithful to the insights of ethnomethodology' (Latour 1999: 19).

> Once she has been sent into the field, even the most rationalist ethnographer is perfectly capable of bringing together in a single monograph the myths, ethnosciences, genealogies, political forms, techniques, religions, epic and rites of the people she is studying. Send her off to study the Arapesh or the Achuar, the Koreans or the Chinese, and you will get a single narrative that weaves together the way people regard the heavens and their ancestors, the way they build homes and the way they grow yams or manioc or rice, the way they construct their government and their cosmology. In works produced by anthropologists abroad, you will not find a single trait that is not simultaneously real, social and narrated.
>
> If the analyst is subtle, she will retrace networks that look exactly like the sociotechnical imbroglios that we outline when we are pursue microbes, missiles or fuel cells in our own western soci-eties. We too are afraid that the sky is falling. We too associate the tiny gesture of releasing an aerosol spray with taboos pertaining to the heavens. We too have to take laws, power and money into account in order to understand what our sciences are telling us about the chemistry of the upper atmosphere.
>
> (Latour 1993: 144)

But in order to do comparative anthropology of the modern world, our definition of 'the modern world' has to be altered. Latour wants enlight-enment *without* modernity. Like Serres, he sees the world as one in which many, many quasi-objects proliferate, and sensing these new 'monsters' is critical to understanding what is going on. There is much more 'space' than our old discontinuous ways of thinking have allowed us to see.

The mediators have the whole space to themselves. The enlightenment has a dwelling place at last. Natures are present but with
their representatives, scientists who speak in their name. Societies
are present but with the objects that have been serving as their
ballast from time immemorial . . . The imbroglios and networks
that had no place now have the whole place to themselves. They
are the ones that have to be represented.

(Latour 1993: 144)

What might such a depiction of the world mean for thinking space and
time? Let us turn to Latour and Serres' geography to find out.

II. Towards a philosophical geography: the spaces and times of Serres and Latour

For Michel Serres and Bruno Latour 'geography' can either be 'tyrannical'
(Latour 1997b: 3) or 'philosophical' (Serres in Critchley 1996: 3), either
'reductionist' or 'irreductionist' (Latour 1988b), and can either do violence
to the world or do justice to it (Serres and Latour 1995). Understanding
what is lost and gained by following the respective paths marked out by
these dichotomies is central to understanding their conception of space
(and time) (see Bingham 1996).

If we ask how often geography is 'tyrannical' for Serres and Latour, the
answer must be 'too often'. Too often, that is, the world is understood –
whether implicitly or explicitly – solely in terms of proximity–distance,
defined as a homogeneous space, a gridlike surface in which the path
from the local to the global is always already given and unproblematic (in
this sense, most contemporary narratives of 'globalisation' are simply the
degree zero of a long tradition of thinking (Thrift 1995)). This purified,
ordered vision is a long-term result, as Serres and Latour see it, of the
'global victory of a local phenomenon' (Gibson 1996: 14): an extension,
that is, of the 'space of measure and transport' (Serres 1982a: 52) that
forms the basis of Euclidean geometry and later cartographic geography,
a space in which all may be calibrated and quantified without complica-
tion or confusion.

Space and time here act as what Latour calls 'primitive terms': either
'Newtonian *sensoria*' or 'forms of perception' (1997a: 174), depending which
side of the 'Great Divide' between Nature and Society one favours as an
explanatory anchor (Latour 1993). In either case, such universal a prioris
sustain the same imaginary: as Andrew Gibson (commenting on Serres'
work) has put it, 'geometry vitrifies spaces and freezes duration. It ensures
the repetition of the identical and the rule of the same' (1996: 14). The
positive moment that emerges from their critique of the mode of thinking
that Gibson summarises, is what animates Serres and Latour's project of

inventing a 'philosophical' geography. First and foremost, their aim is to make real difference thinkable by restoring to consideration the multiplicity of the world – what Latour calls the 'Middle Kingdom' (1993) after Serres' (1982b) notion of the 'excluded third' – that the tyranny of the moderns has so effectively deleted.

What this requires is nothing less than a gestalt shift according to which space and time are no longer conceived of as existing 'independently as an unshakeable frame of reference *inside which* events and places would occur' (Latour 1987: 228, emphasis in original), but, conversely, the result of inter-action, '*consequences* of the ways in which bodies relate to one another' (Latour 1997a: 174, emphasis in original). It is not empty abstractions which are primary here, but the many and varied '*other entities* that are necessary for maintaining us in existence' (*ibid*.: 186, emphasis in original). In particular, what takes centre stage is the circulation of certain of these 'other entities'. As Latour himself argues,

> Gods, angels, spheres, doves, plants, steam engines, are not *in* space and do not age *in* time. On the contrary, spaces and times are traced by reversible or irreversible displacements of many types of mobiles. They are generated by the movements of mobiles, they do not frame these movements.
>
> (1988a: 25)

'Generated', because a given 'displacement' is never merely a 'smooth passage' (Latour 1997a: 175). Reaching one position from another always requires a great deal of work, the intervention of all sorts of 'bits and pieces' (Law 1994), bits and pieces which are rarely, if ever, well-behaved 'intermediaries' transporting faithfully. Much more likely, they are ill-mannered 'mediators' 'defining paths and fates on their own terms' (Latour 1997a: 175, also Latour 1993). The product of transformation and not the containers for transmission, spaces and times are outcomes of the combination and recombination of a full world.

The reconceptualisation that arises from these moves is best described as topological: the 'dangerous flock of chaotic morphologies' (Serres 1982a: 53) subdued by what Serres calls 'thanatocracy' and we have called a 'tyrannical geography', is revived here, with Euclidean space only one amongst (many) others. 'In a world made of mediations, of transformation by deformation', as Latour puts it (1997a: 178), spaces and times proliferate:

> my body [for example] lives in as many spaces as the society, the group or the collectivity have formed: the Euclidean house, the street and its network, the open and closed garden, the church or the enclosed spaces of the sacred, the school and its spatial

varieties containing fixed points, and the complex ensemble of flow-charts, those of language, of the factory, of the family, of the political party and so forth.

(Serres 1982a: 45)

Equally,

all times converge in this temporary knot: the drift of entropy or the irreversible thermal flow, wear and ageing, the exhaustion of initial redundancy, time which turns back on feedback rings or the quasi-stability of eddies, the conservative invariance of genetic nuclei, the permanence of a form, the erratic mutations of aleatory mutations, the implacable filtering out of all non-viable elements, the local flow upstream towards negentropic islands – refuse, recycling, memory, increase in complexities . . . What is an organism? A sheaf of times?

(*ibid:* 75)

My body – or anything else for that matter – is a weaving: an intersection of the tattered multiplicity into which it is plunged (Serres 1982a: 45). Or, to put it another way, as Latour has recently done (1997a), 'space' and 'time' are less important than the always unique acts of 'timing' and 'spacing' by which place-events are 'folded' or 'pleated' into existence.

Topological, then, because, in contrast to 'metric theory' (Serres and Latour 1995: 102), primacy in the sort of 'philosophical geography' that Serres and Latour are seeking to construct is granted not to substance-nouns or even process-verbs, but what they call prepositions or relations. It is relation-prepositions, according to Serres, that 'spawn objects, beings and acts, not vice versa' (*ibid*.: 103, 107), a situation, as he explains in one of his most recent works, that topology – as 'the science of proximities and ongoing or interrupted transformations' (*ibid*.: 105) – is uniquely well placed to articulate:

To do this it employs the closed (*within*), the open (*out of*), intervals (*between*), orientation and directionality (*toward, in front of, behind*), proximity and adherence (*near, on, against, following, touching*), immersion (*among*), dimension . . . and so on, all realities outside of measurement but within relations.

(Serres 1994: 71, translated and quoted in Boisvert 1996: 64, emphasis in original)

And if the world is topological, then so too must be its description. In this way, Serres and Latour begin to confuse the traditional distinction between what the world is actually like (the ontological question), and what

290

can actually be known about that world (the epistemological question). As Gerard de Vries writes of Latour, both these writers are engaged in trying to 'figure out the nature of a world in which knowledge plays a role' (1995: 3). Or, as Latour himself puts it in a discussion of actor-network theory:

> This solution becomes common sense once it is accepted that an account or an explication or a proof is always added to the world, it does not subtract anything from the world. Reflexivists as well as their pre-relativist enemies dream of subtracting knowledge from the things in themselves. ANT keeps adding things to the world and its selection principle is no longer whether there is a fit between account and reality – this dual illusion has been dissolved away – but whether or not one travels'.
>
> (1997b: 8)

The metaphor of travel employed here is important. For, if – as Serres and Latour suggest – being-in-the-world consists of linking incommensurable space-times, then this is how knowledge too must proceed. Not in order, as Serres has recently explained, to 'imitate' or 'justify' that world, but in order to 'understand' it ('and, desperately, perhaps, to know how – to be able – to direct its course' (Serres and Latour 1995: 114)). To follow this path, the path of 'compatibility' (*ibid.*), the path where 'to read and to journey are one and the same act' (Serres 1974: 14, translated and quoted in Harai and Bell 1982: xxi), leads to a very different conception of the theoretical terrain from that to which we are accustomed. The 'landscape' that emerges

> contains pits, faults, folds, plains, valleys, wells, and chimneys, solids like the earth and fluids like the sea. The metaphor is geophysical here; it could be mathematical. In any case, the model is complex. Here and there, locally, I identify fractures and discontinuities, elsewhere, on the contrary, relations and bridges.
>
> (Serres 1977: 200, translated and quoted in
> Harai and Bell 1982: xxii)

From an all-too-familiar mould in which 'myth', 'literature', and 'science' (*ibid.*) are hierarchised and held to be mutually exclusive methods of representing the world amongst which no dialogue is possible, Serres and Latour seek to exploit the channels of communication – what the former calls the 'Northwest Passage' (1997) – between apparently alien modes of knowing. This approach, encyclopaedic in its purest form – not in the sense of pursuing the contemporary and ultimately conservative tendency towards a 'philosophy of fragments' (Serres and Latour 1995: 120), but in the sense of having the courage to construct a 'fragile synthesis' (*ibid.*: 122) – is a best attempt to do justice to what we have seen they regard as a full and com*pli*cated (folded) world, of perhaps expressing 'the very tissue in which objects, things themselves, are immersed – the all-encompassing and diabolically complex network of inter-information' (Serres 1972, translated and quoted in Harai and Bell 1982: xxiii).

Encyclopaedic, then, and yet Serres and Latour's project heralds a very different kind of 'Enlightenment' to the foundational systems of the last two hundred years, which offered 'clear and distinct knowledge, scientific unity, [and] the triumph of reason' (Serres 1989: 32) by assuming 'a transparent space in where a single law reigned, that of light or the sun's power: nothing new under the sun' (Serres 1994: 109, translated and quoted in Boisvert 1996: 65). We might, they suggest, experience illumination in another way: as

> a fairly soft and filtered light that allows us better to see things in relief, through the effects of contrast produced by rays and shadows that melt together, that are mixed, nuanced . . . This is the way that we see ordinarily, really, daily – with our bodily eyes in concrete surroundings.
>
> (Serres and Latour 1995: 154)

This is enlightenment as 'scintillation' (Serres 1989: 32), bringing with it 'tentative knowledge' (*ibid.*).

In the pluralistic world sketched by Serres and Latour therefore, we must hope and seek not for a 'blinding revelation', but rather a 'flicker of recognition' (our terms). To this end, since the geography is, as we have seen, obviously philosophical, so too must the philosophy obviously be geographical. Our best bet, that is to say, is, as Raymond Boisvert, commenting on one of Serres' most recent books (*Atlas* (1994 in French, untranslated at the time of writing)) has put it, 'to construct maps' (1996: 65). Maps not as 'mirrors' of a pregiven world, but as 'modes of access, ways of orienting ourselves to the concrete world we inhabit' (*ibid.*: 65). More specifically, considering particularly the word's double etymology, we might talk of 'legends': legends as stories in their own right, and legends as aids in reading a map, of making sense of a world.

Certainly, the fact that characters such as 'the parasite' (Serres 1982b), 'the weaver' (Serres 1982a, Latour 1993), 'Daedalus' (Latour 1994), and 'angels' (Serres 1995) populate the work of Serres and Latour is no coincidence. In every case they are offered as companion-figures capable of helping us navigate a world which – as this section has hopefully made clear – unfolds by way of mediation, transformation, and circulation. As a way, that is, of getting to grips with a world always on the move. Referring to *the* legend by which his life's work has been organised – in terms at once of structure, content, and style – Serres, as is so often the case, puts this much better. Speaking of the god of communication of ancient Greece, who, as the writer Italo Calvino has concisely summarised

> with his winged feet, light and airborne, astute, agile, adaptable, free and easy, established the relationships of the gods amongst themselves and those between the gods and men, between universal laws and individual destinies, between the forces of nature and the forms of culture, between the objects of the world and all thinking subjects
>
> (1992: 52),

he argues that

> Hermes, by constantly renewing himself, becomes continuously our new god, for as long as we've been humans – not only the god of our ideas or our behaviour, or of our theoretical abstractions, but also the god of our works, of our technology, of our experiments, of our experimental sciences. Indeed, he is the god of our laboratories ... he is the god of our biology ... he is the god of computer science ... of commerce ... of the medias ... Hermes comprehends [this situation] – through his role, his figure, and his movements – but curiously, as a person and not as a concept, as a multiple and continuous transport, and not as a foundation or a starting point. We have to imagine a foundation with wings on its feet.
>
> (Serres and Latour 1995: 114)

III. A stylish journey: the multiple itineraries of Serres and Latour

Hermes, then, by embodying Latour's frequent (1987, 1996a, 1997c) exhortation to 'follow the actors', provides us with a chance of learning of the world. What might we learn by applying this same 'slogan' to Serres and Latour themselves? Hopefully, as this section will attempt to demonstrate, a certain understanding of the hermetic method itself. Or perhaps – and notwithstanding the title of the second chapter in *Conversations* ... (Serres and Latour 1995: 43–76) on which much of the following is basically an alternative take – that should be *anti*-method. For as Harai and Bell note in their excellent introduction to the English version of *Hermes* (Serres 1982a), the former term is

> problematic because it suggests the notion of repetition and predictability – a method that anyone can apply. Method also

293

implies mastery and closure, both of which are detrimental to invention. On the contrary, Serres' method invents: it is thus an anti-method.

(1982: xxxvi)

Successfully getting at what is truly interesting and unique about this (or indeed any other) way of proceeding requires a certain amount of invention in itself. Certainly, after Latour's (1988b) dismantling of the hagiography that has underpinned conventional accounts of Louis Pasteur, appealing in any shape or form to the 'great man' [*sic*] theory for our own explanation would (rightly) look rather foolish. Instead, we again prefer the metaphor of explication (Latour 1997c: 72 n.15), and seek to draw out and trace a small but pertinent number of the 'contacts, neighbour-hoods, encounters, and relations' (Serres 1997: 144) of which Serres and Latour ('topological and temporal' *ibid.*: 148, like us all) are constituted. In this, we are basically following Harai and Bell (1982), on the one hand, and Paul Harris (1997), on the other, in using the vocabulary of journeys introduced in the previous section to consider the trajectories of their creators (in these two cases Serres alone, in ours obviously Latour as well). Both for us are well labelled as 'itinerant theorist[s]' by Harris (*ibid.*: 37), again not so much because of the importance of thinking in terms of starting- or end-points, but because of the sense of the 'parcours' (Gibson 1996: 16) that the phrase communicates, and the heterogeneity that any such 'course through' brings together.

To help us further in thinking through these itineraries, we beg, borrow, and steal certain notions from Latour's aforementioned study of *The Pasteurization of France* (1988b), in particular three interwoven implications of the lines 'A man cannot do a great deal on his own. What he can do, however, is to move' (*ibid.*: 67), (lines whose truth – presumably – holds beyond their gendering in the text). Although in some ways a risky approach – 'Repeating a method – what laziness' (Serres 1997: 100) – we feel that the both the similarities and differences that the comparison allows to emerge means that it will be a helpful one in this context.

The first aspect of the movement of Serres and Latour, then, that the frame of the Pasteur study brings to the fore is the way in which both position and re-position themselves with respect to their surround-ings. Just as the 'Pasteurians place themselves in relation to those forces of hygiene that I have described, but do so in a very special way' (Latour 1988b: 60), so too the type and number of connections made by Serres and Latour *vis à vis* the disciplinary specialisations that organise the late twentieth-century intellectual landscape mark them out as exceptional by anyone's standards. However, just as Latour writes in the case of the hygienists that the word 'strategy' is 'too rational to account for the oper-ations in question' in that case (*ibid.*: 60), here too 'it is enough to speak of "displacement"' (*ibid*.: 60). In Serres' case, for example, we might speak of the displacements involved in his training:

> Educated as a philosopher, Serres says that he began by studying geometry as Plato recommended. Afterwards, he continued in more concrete domains: physics, biology, and the sciences of man [*sic*]. In the last area he became especially interested in anthropology, more specifically, in the history of religions . . . Thus Serres' itinerary is encyclopaedic, covering the three great modes of knowledge: philosophic, scientific, and mythic.'
>
> (Harai and Bell 1982: xv)

Or those amongst his own oeuvre:

> Serres has laboured in several fields, including molecular biology and science fiction, topology and painting, linguistics and anthropology. In his writings he has wandered from ancient Rome to the disastrous Challenger launch, from the flooded banks of the Nile to polar ice flows in the Northwest Passage; he has passed by Oedipus's fateful crossroads and disappeared down manholes into bubbles of chaos
>
> (Harris 1997: 37)

Or even within one work:

> *Atlas* discusses, for example, chaos theory, commercials, virtual reality, the Belgian comic book *Tintin*, mythology, the political creation of a public, the history of religions, classical mechanics, interactive computer networks, kimonos, distance education, and astronomy.
>
> (Boisvert 1996: 63–64)

If Latour's travels seem conservative by comparison, they are only relatively so: trained as philosopher and as an anthropologist, after field studies in Africa and California, he has moved effortlessly from laboratories (Latour and Woolgar 1979) to paintings (1988a), and thinks nothing of using ancient Greek philosophy to expose the sterile foundations of the contemporary so-called 'Science Wars' (1997d). It is also he – in *We Have Never Been Modern* (1993) – who has perhaps most clearly articulated that the catholicism of these displacements (those of Serres as well as his own), far from being an indulgence, is rather a very necessary response to the crumbling of 'the modern constitution' that has propped up our culture of purification and critique. As he has recently put it 'no progress will be made . . . if the whole settlement is not discussed in all its components:

ontology, epistemology, ethics, politics, and theology' (1997e: xii). Bound-aries must be crossed.

There are, however, many boundaries to cross, and, as we have already noted there is only so much one (or even two) can do on their own. Hence the need to think in terms of the second aspect of movement which Serres and Latour both demonstrate and help us analyse. For Pasteur, this was the 'certain type of displacement that enabled him to translate and divert into his movement circles of people and interest that were several times larger' (1988b: 67). According to Latour, Pasteur was a 'genius' at both 'getting allies while he moved' and 'getting himself attributed with the source of the movement' (*ibid.*: 71). For Serres, however, this sort of 'empire-building' by which everything is referred back to a centre, is simply not an option, so intimately is this sort of movement bound up with the func-tioning of the very model of Critique he is attempting to escape (Koch 1995: 11). As he describes one of his own 'rules':

> Always avoid all membership: flee not only all pressure groups but also defined disciplines of knowledge, whether a local and learned campus in the global and societal battle or a sectorial entrench-ment in scientific debate. Neither master, then, nor above all disciple.
>
> (1997c: 136)

Given this position – one which is further clarified throughout *Conversations* . . . – we would have to agree with the wonderfully expressed notion of Donald Wesling, that Serres' many conceptual inventions, to several of which we have already referred, 'must now seem secondary to his inven-tion of Bruno Latour' (1997: 198). For Wesling, Latour was 'predisposed' to be Serres' 'expositor, interviewer, and radical agent' (*ibid.*: 199): certainly by being more willing to undertake the movement-work of translation and diversion, the former has extended the latter in several senses, not least in terms of interested (or should that be interess-ed) parties. Located liter-ally in the Centre for the Study of Innovation of the Ecole De Mines, Paris, and hence firmly (if more metaphorically) with the social sciences for which Serres purports to have so little time, Latour has been able to function as something of a mediator for the ideas of the man who has obviously influenced him so profoundly (a description which is by no means intended to downplay his own creative contribution). Through his own works (both the written publications and the oral presentations that he always seems to be giving), his collaborations with colleagues both from inside and outside the CSI (for example Latour and Woolgar 1979, Callon and Latour 1992, Akrich and Latour 1992, Latour and Hennion 1995, Teil and Latour 1995, Strum and Latour 1987), and, of course, the book of interviews with him (1995), Latour, then, has managed to transport some of Serres' more radical insights into the most unlikely of settings. The very fact of this chapter's existence is in many ways testament to the efficacy of his inter-vention (coming between), an illustration both of the current extent of the network 'Serres–Latour' (even if it does not – yet – encompass 'the whole world' as did the network 'Pasteur' (Latour 1988b: 69)), and the way in which that same network is increasingly leading readers 'recruited' by Latour back to Serres, thereby providing the latter with a far more widespread audience that he originally possessed (certainly in the Anglo-American context).

All of which – by attributing a degree of Machievellianism to Latour in particular which is hugely unfair – is to reduce the multiple paths of Serres and Latour to a single broad highway that somehow exhibits at once both internal consistency and progress. But, if they teach anything, it is that no journey is ever this simple, and so, to complicate the route somewhat, we want to re-introduce at this juncture the last aspect of move-ment common to them through the form of Pasteur. This is captured by Latour's description of the latter 'step[ping] sideways' in order to confront some difficult problem that interests more people than the one he had just abandoned', hence constituting 'each time a new discipline in which he has 'some success'' (1988b: 68). Now, this account obviously has some resonances with the displacements that we have already noted which have taken Serres and Latour 'through' a number of usually distinct subject-terrains, and ways in which they have so often (just as Pasteur) thereby transformed 'applied' problems into 'fundamental' ones (cf. *ibid.*: 68). However, while for Pasteur such sideways movements were an integral part of a project of empire-building that we referred to earlier, for Serres and Latour the same sort of movement is at least as much concerned with destabilising their own positions in the network as with lengthening it.

Something of what this might mean is unravelled by Serres in *Conversations* . . . in reply to a question by Latour about what makes his commentaries 'different from others':

> The commentaries I used to criticise could be called imperialistic . . . because they used a single passkey to open all doors and windows; they used a passkey that was psychoanalytical or Marxist or semiotic, and so on . . . To me, however, singularities were important, local details for which a simplistic passkey was not suffi-cient. On the contrary, what was necessary was a tool adapted to the problem. No work without this tool.
>
> (Serres and Latour 1995: 91–92)

Latour replies that he can 'easily understand this need to retool, to recast the tools of analysis each time one tackles a new object' (*ibid.*: 92), and it

297

is by following Serres in doing this that he has striven (contra Pasteur) to avoid being held 'accountable' for the network which, as we have already seen, he has been so instrumental in extending. This has become most noticeable apropos the actor-network theory with which his name has become synonymous in many circles (despite his rare use of the term). While willing to offer a 'few clarifications' (Latour 1997b), Latour has also recently stressed the necessity of now moving on: actor-network is 'one of the many words we have to invent and use and drop after a while' (Latour in Crawford 1993: 262–263), '[p]owerful against structures as well as essences and moralising, [heterogeneous networks] become empty when asked to provide policy, pass judgement, or explain stable features (Latour 1996b: 304), we are 'after' actor-network theory (Latour 1997b). Moving sideways once more, perhaps Latour is beginning to follow Serres in his own self-defined passage (Serres and Latour 1995: 100) to another period, one more fully attuned to 'a certain [acritical] type of invention' (*ibid.*: 100, see also Crawford 1993, Koch 1995).

To conclude then, in summarising the itineraries and movements of this section, we may say of Serres and Latour – as Latour (via Dagognet) says of Pasteur – that they 'innovated by linking together' (Latour 1988b: 69). Is this enough, though, to elucidate what it is to be 'Serresian' or 'Latourian', their particular contribution? For do not we all innovate by linking together all of the time (and is this not one of Serres' and Latour's major points)? Perhaps we would be better off understanding what such a definition might mean by grasping it (once again after Latour on Pasteur (1988b: 94)) 'as a term of *style*' (his emphasis). Notoriously difficult to define, but of relevance to this discussion (see Serres and Latour 1995: 100, Deleuze 1973, Massumi 1997), a productive way of thinking the notion through might be as what Latour has recently called a 'regime of delegation' (1996b: 304), that is one of the 'limited number of ways' in which 'an indefinite number of entities' may 'grasp one another' (*ibid.*). As for the best way as to express that style, we will delegate that task (albeit via a mediator) to the man (or one of them) himself. Summarising what Serres has recently (1996) called the 'procedural', Marcel Henaff writes

> *Procedural*: this term has its origins in *procedo*, the act of walking, or rather moving forwards, step by step. This also means to advance among the particularity of sites and conditions. Can one define a way of thinking based on such a model? Is it not precisely what proper philosophy denounces as empiricism? Not even that, for at the end of its journey, empiricism intends to rejoin the universal it did not posit at the beginning. We are dealing here with something quite different – that is, taking seriously the particularities of the sites, the unpredictability of circumstances, the uneven

patterns of the landscape and the hazardous nature of becoming. In short, again: how to think the local? Which means: is there a science of the particular?

(1997: 72)

IV – Some brief conclusions

Not surprisingly, actor-network theory has its critics. The criticisms are of a number of kinds. One is that actor-network theory is really a new kind of totalising theory, wrapped up in modest trappings a 'theory' of the non modern which still manages to have a very definite theory of modernity (Rabinow 1999). Another is that the sizzle of the event is missing, the dizzy (and often embodied) force of conjuncture that is so evident in a writer that both Latour and Serres profess to admire, namely Gilles Deleuze (Thrift 1999). Yet another is that actor-network theory ignores the 'quite real effectivity of victimisation' (Wise 1997: 39); it is studiously neutral and, as a result, it bypasses questions of unequal power.

Whatever the force of these criticisms, actor-network theory has done one thing, and done it well. It has opened up spaces which have been closed down. By following circulations, it has produced a sense of a world of partial connection in which all kinds of constantly shifting spaces can co-exist, overlap and hybridise, move together, move apart. Latour and Serres do not argue that we need to replace one spatial temporal frame, say the Euclidean, with another, say the relativistic. It is rather that in their world no such frame can exist at all, except as a metrological construction which only goes so far and so fast. Their world is fluid.

In turn, this sense of a multiplicity of swooping and diving spaces allows us to recast place as well (Hetherington 1997; Thrift 1998). For places do not just trace out the traces of spaces, they have an active role which is inscribed in their activity. In their multitude of differences places are the means by which hybrids register each other as hybrids and, in allowing the performance of their difference, face change.

Note

1   Latour (1999: 15) now regrets the term:

> When the term 'network' was first introduced like Deleuze and Guattari's term Rhizome [it] clearly meant a series of transformations – translations, traductions – which could not be captured by any of the traditional terms of social theory. With the new popularization of the word network it now means transport without deformation, and instantaneous, unmediated access to every piece of information.
>
> That is exactly the opposite of what we meant.

References

Akrich M, Latour B, 1992, 'A Summary of a Convenient Vocabulary for the Semiotics of Human and Non-Human Assemblages', in Bijker W, Law J (ed.) *Shaping Technology/Building Society: Studies in Sociotechnical Change* (MIT, Cambridge): 259–264.

Bingham N, 1996, 'Object-ions: From Technological Determinism Towards Geographies of Relations', *Environment and Planning D: Society and Space* 14.6: 635–657.

Boisvert R, 1996, 'Remapping the Territory', *Man and World* 29: 63–70.

Bowker G, Latour B, 1987, 'A Booming Discipline Short of a Discipline: (Social) Studies of Science in France'. *Social Studies of Science*, 17: 715–748.

Callon M, Latour B, 1992, 'Don't Throw out the Baby with the Bath School!: a Reply to Collins and Yearley', in Pickering A (ed.), *Science as Practice and Culture* (Chicago University Press, Chicago): 343–368.

Calvino I, 1992, *Six Memos For The New Millennium* (Jonathan Cape, London).

Crawford T H, 1993, 'An Interview with Bruno Latour', *Configurations* 1.2: 252–263.

Critchley S, 1996, 'Angels in Disguise: Michel Serres' Attempt to Re-Enchant the World', *TLS*, 19/1: 3–4.

De Vries G, 1995, 'Should We Send Collins and Latour to Dayton, Ohio?', *EASST Review* http://www.chem.uva.nl/easst 14.4: 1–8.

Deleuze G, 1973, *Proust and Signs* (Allen Lane, London).

Gibson A, 1996, *Towards a Postmodern Theory of Narrative* (Edinburgh University Press, Edinburgh).

Harai J, Bell D, 1982, 'Introduction: Journal à Plusiers Voies', to Serres 1982a: ix-xl.

Harris P, 1997, 'The Itinerant Theorist: Nature and Knowledge/Ecology and Topology in Michel Serres', *SubStance* 26.2: 37–58.

Henaff M, 1997, 'Of Stones, Angels, and Humans: Michel Serres and the Global City', *SubStance* 26.7: 59–80.

Hetherington K, 1997, 'In Place of Geography: the Materiality of Place', in Hetherington K., Munro, R (eds) *Ideas of Difference*. (Blackwell, Oxford) 183–199.

Koch R, 1995, 'The Case of Latour' *Configurations* 3.3: 319–347.

Latour B, 1987, *Science in Action* (Harvard University Press, Cambridge).

Latour B, 1988a, 'Visualisation and Reproduction', in Fyfe G, Law J (eds) *Picturing Power: Visual Depiction and Social Relations* (Blackwell, Oxford): 15–38.

Latour B, 1988b, *The Pasteurization of France (with Irreductions)* (Harvard University Press, Cambridge).

Latour B, 1993, *We Have Never Been Modern* (Harvester Wheatsheaf, London).

Latour B, 1994, 'On Technical Mediation: Philosophy, Sociology, Genealogy', Common Knowledge 4: 29–64.

Latour B, 1996a, *Aramis: Or the Love of Technology* (Harvard University Press, Cambridge).

Latour B, 1996b, 'Social Theory and the Study of Computerised Work Sites', in Orlikowski W, Walsham G, Jones M, DeGross J (eds) *Information Technology and Changes in Organisational Work* (Chapman and Hall, London): 295–307.

Latour B, 1997a, 'Trains of Thought: Piaget, Formalism, and the Fifth Dimension', *Common Knowledge*, 6.3: 170–191.

Latour B, 1997b, 'On Actor-Network Theory: A Few Clarifications', http://www.keele.ac.uk/depts/stt/stt/ant/latour.htm

Latour B, 1997c, 'A Few Steps Towards an Anthropology of the Iconclastic Gesture', *Science in Context* 10.1: 63–83.

Latour B, 1997d, 'Socrates' and Callicles' Settlement – Or, the Invention of the Impossible Body Politic', *Configurations* 5.2: 189–240.

Latour B, 1997e, 'Stengers' Shibboleth', Foreword in Stengers I, 1997, *Power and Invention* (University of Minnesota Press, Minneapolis).
Latour B. (1998) 'A Relativist Account of Einstein's Relativity' *Social Studies of Science*, 18: 3–44.
Latour B, 1999, 'On Recalling ANT', in Law, J, Hassard, J (eds) *Actor Network Theory and After* (Blackwell, Oxford): 15–25.
Latour B, Hennion A, 1995, 'A Note on Benjamin', *Stanford Humanities Review*.
Latour B, Woolgar S, 1979, *Laboratory Life: The Social Construction of Scientific Facts* (Sage, London).
Law J, 1994, *Organising Modernity* (Blackwell, Oxford).
Massumi B, 1997, 'The Political Economy of Belonging, and the Logic of Relation', in Davidson C (ed.), *Anybody* (Anyone Corporation, New York): 174–188.
Rabinow P, 1996, *Essays on the Anthropology of Reason* (Princeton University Press, Princeton).
Rabinow, P, 1999, *French DNA: Trouble in Purgatory* (University of Chicago Press, Chicago).
Serres M, 1972, *Hermes II: L'Interférence* (Minuit, Paris).
Serres M, 1974, *Jouvences: Sur Jules Verne* (Minuit, Paris).
Serres M, 1977, *La Naissance de la Physique Dans la Texte de Lucrèce* (Minuit, Paris).
Serres M, 1982a, *Hermes: Literature, Science, Philosophy* (John Hopkins University Press, Baltimore).
Serres M, 1982b, *The Parasite* (John Hopkins University Press, New York).
Serres M, 1989, 'Literature and the Exact Sciences, *SubStance* 18.2: 3–34.
Serres M, 1994, *Atlas* (Juilliard, Paris).
Serres M, 1995, *Angels: A Modern Myth* (Flammarion, Paris).
Serres M, 1996, *Eloge de la Philosophie en Langue Française* (Fayard, Paris).
Serres M, 1997, *The Troubadour of Knowledge* (University of Michigan Press, Ann Arbor).
Serres, M, Latour, B (1995) *Conversations on Science, Culture and Time* University of Michigan Press, Ann Arbor.
Strum S, Latour B, 1987, 'The Meanings of the Social: From Baboons to Humans', *Social Science Information* 26: 783–802.
Teil M, Latour B, 1995, 'The Hume Machine: Can Association Networks Do More Than Formal Rules?', *SEHR* 4.2: 1–15.
Thrift N, 1995, 'A Hyperactive World', in Johnston R, Taylor P, Watts M (eds) *Geographies of Global Change* (Blackwell, Oxford): 18–35.
Thrift N J, 1998, 'Steps to an Ecology of Place' in Massey D, Allen J, Sarre P (eds) *Human Geography Today*. (Polity Press, Cambridge).
Thrift N J, 1999, 'Afterwords'. *Environment and Planning D. Society and Space* 2000 forthcoming.
Wesling D, 1997, 'Michel Serres, Bruno Latour, and the Edges of Historical Periods', *Clio* 26.2: 189–204.
Wise J M, 1997, *Explaining Technology and Social Space* (Sage, London).

14

EDWARD SAID'S IMAGINATIVE GEOGRAPHIES

Derek Gregory

> Just as none of us is outside or beyond geography, none of
> us is completely free from the struggle over geography. That
> struggle is complex and interesting because it is not only
> about soldiers and cannons but also about ideas, about forms,
> about images and imaginings.
>
> (Edward Said: *Culture and imperialism*: 7)

I Rethinking geography

I have borrowed my title from Edward Said, one of those rare critics for
whom a geographical imagination is indispensable. 'What I find myself
doing,' he once declared, is 'rethinking geography.' Now professors of
comparative literature do not usually speak like this, and when Said goes
on to suggest that '. . . we are perhaps now acceding to a new, invigorated
sense of looking at the struggle over geography in interesting and imaginative ways', then it is, I think, time for us to consider what he has in
mind.[1]

Geography is a recurrent motif in Said's writings, and commentators
from disciplines other than our own have recognized his deep interest in
space and spatiality. From anthropology, we are reminded that 'the creation
of geographies – the recognition and understanding of symbolic territories – is central to Said's work' and that, even when he writes in the
abstract, 'Said is moved to use geographical imagery'. From sociology,
he is seen as constructing a 'cartography of identities', disclosing the formation of a geographical imaginary that supplements the 'Euro-modernist
interest in time with an equivalent understanding of space and spatiality'.[2]
Yet if Said's work can be read as charting the changing constellations of
power, knowledge and geography – the phrase is his, not mine[3] – inscribed
within British, French and American imperialisms, the fact remains that
his project has received remarkably little attention from our own discipline.[4]
Said himself has repeatedly drawn attention to geography's complicity in

Orientalism and in the wider cultures of imperialism, but it is only very recently that a critical historiography capable of addressing these same issues has emerged within our discipline.[5] What gives this new body of work its critical edge, like Said's, is its refusal to confine these entanglements to distant and dusty archives. It may be comforting to believe with L.P. Hartley that 'the past is a foreign country: they do things differently there'; but it is also thoroughly deceptive. Many of the assumptions of the colonial past are still abroad in the neocolonial present. 'Geography militant', as Conrad once called it, was revealed with unspeakable clarity in the Gulf war of 1990–91, for example, Smith's 'first GIS war', and the colonial investment in geography as a kind of earth-writing is evident in the more mundane but none the less extraordinary arrogance with which the Royal Geographical Society celebrates its union with the Institute of British Geographers by scrawling its signature on its new membership card across part of the Arab world.[6]

But geography is about more than the will-to-power disguised as the will-to-map, and I want to accentuate its critical inflections. More specifically, I should like to begin a constructive exploration of Said's geographical imagination: its grounding, its constitution, its implications and its silences. Running through my discussion will be a dialectic between 'land' and 'territory'. These two words have been invested with multiple meanings, at once political and cultural, and Said uses them (or something very much like them) in ways that are perhaps not commonplace in geography. But his deployments are, I think, unusually creative: in effect, he charts a series of mappings, sometimes discordant and sometimes compounded, through which places and identities are deterritorialized and reterritorialized. He describes landscapes and cultures being drawn into abstract grids of colonial and imperial power, literally displaced and replaced, and illuminates the ways in which these constellations become sites of appropriation, domination and contestation. This is to paint with broad brush-strokes, but I hope to show that Said's inquiries into the historical predations of Orientalism, colonialism and imperialism and his writings on the contemporary plight of the Palestinian people are recto and verso of the same processes of inscription, through which power, knowledge and geography are drawn together in acutely physical ways. Like Homi Bhabha, I think Said's politico-intellectual trajectory can be characterized as a move between the West Bank and the Left Bank. I want to think about these two sites together, and retain the imbrications between them, in order to consider a simple question: where does Said's geography come from?

II Palestine and the politics of dispossession

The first set of answers is biographical or, as I suspect he would prefer me to say, experiential. Edward Said was born in 1935 in Talbiya, in west

Jerusalem, Palestine, into one of the oldest Christian communities in the world. His childhood was shaped by the disciplines of an unmistakably Anglican tradition – a student at St George's, an Anglican mission school in Jerusalem, he was baptized in the same parish. When Said was born, Palestine had been under British administration for 15 years. After the first world war and the collapse of the Ottoman empire, the League of nations had placed the newly independent Arab states under British or French mandate because, in the words of Article 22, they were deemed to be 'inhabited by peoples not yet able to stand by themselves under the strenuous conditions of the modern world'. In the autumn of 1947, following Britain's peremptory announcement that it would withdraw from its mandate within six months, Palestine dissolved into turmoil. In the course of the bloody war that ensued, as the Zionist Haganah and Irgun fought Arabs for territory, between 600 000 and 900 000 Palestinians fled their homes; among them were Said and his family.[7] Most of the refugees settled in Egypt, Jordan or Lebanon. Said continued his education at Victoria College in Cairo, another quintessentially British institution, and then, in 1951, moved to the USA to finish his secondary education. He subsequently studied English and history at Princeton and completed his doctorate in comparative literature at Harvard, where he wrote his thesis on another brilliant exile, Joseph Conrad.[8]

It is surely no wonder that Said should later devote so much of his working life to a critical appreciation of the western canon for, as my thumb-nail sketch implies, its history is, in part, his story. In effect, he compiles an inventory of what, following Gramsci, he calls the 'infinity of traces' left upon him, 'the Oriental subject', by 'the culture whose domination has been so powerful a factor in the life of all Orientals'.[9] But he also challenges his metropolitan audience to rethink their own cultural history, to cede their 'own-ership' and connect its privileges and assumptions to the busy commerce of colonialism and imperialism.

Of course, Said's 'voyage in' did not mean that he left his other cultural baggage behind; but the frictions of distance between Britain and the USA on one side and Palestine on the other make its recovery unusually problematic.[10] It would be impertinent for me to suggest how the flight from Palestine affected Said. 'Most of what I can recall about the early days,' he writes, 'are obscure boyhood memories of a protracted exposure to the sufferings of people with whom I had little direct connection.' Once he left Palestine he admits he was still further 'insulated by wealth and the security of Cairo'. He lived with his parents on the island of Zamalek, 'an essentially European enclave where families like my own lived: Levantine, colonial, minority, privileged'.[11] This means that he can only recover the connections between biography and history in his native land through a collective recitation, a series of disconnected performances enacted within the dispersed imagination of a displaced community – what he calls 'the intimate mementoes of a past irrevocably lost [that] circulate among us, like the genealogies of a wandering singer of tales' – and island chain of what Benedict Anderson would call an 'imagined community'.[12]

Such a project is moved by an agonizing dialectic of redemption and incompletion, and it is impossible to read Said's meditation on Jean Mohr's photographs of Palestinian lives, or his moving account of his own visit to 'Palestine-Israel' in the summer of 1992, without recognizing how deeply his (re)constructed sense of biography and history – his sociological imagination – is embedded in the shattered human geographies of Palestine.

1 After the Last Sky

The occasion for his collaboration with Mohr is particularly instructive. In 1983 Said was a consultant to the International Conference on the Question of Palestine. He persuaded its United Nations sponsors to commission Mohr (whose earlier work with John Berger he had much admired) to take a series of photographs of Palestinians to be hung in the entrance hall to the Geneva conference. The intention, I assume, was to remind the participants that 'the question of Palestine' was not some abstract conundrum, to be resolved by remote formularies, but an intensely practical question (in the original sense of that phrase), spun around webs of meanings created by particular people in a particular place. When Mohr returned, he and Said found that a condition had been attached to the exhibition: 'You can hang them up, we were told, but no writing can be displayed with them.' If geography is indeed a kind of writing – literally, 'earth-writing' – then this prohibition is hideously appropriate, as the Palestinian poet Mahmud Darwish explains in one of his early poems:

> We have a country of words. Speak speak so I can put my road
>   on the stone of a stone.
> We have a country of words. Speak speak so we may know the
>   end of this travel.

Darwish was one of the Resistance poets and, not surprisingly, the prohibition on 'earth-writing' was enforced in their homeland too. Many of them were arrested or forced into exile, but they continued to write poems that spoke directly of the anguish of dispossession. Said took the title of his collaboration with Mohr from another of Darwish's poems – 'Where should we go after the last frontier? Where should the birds fly after the last sky?' – and, in the text he eventually wrote to accompany Mohr's photographs, he too acknowledged the strategic–subversive copula of 'earth-writing'. No simple Palestinian geography is possible, or even permissible, Said seemed to say: 'We are "other" and opposite, a flaw in the geometry of resettlement and exodus.'[13]

305

In *After the Last Sky*, Said returns again and again to a Palestine riven by the tension between geography as territory and geography as land. This is mirrored in Mohr's triptych of images that move between the planar geometries of an Israeli settlement on the West Bank and the organic root-edness of a Palestinian village, and it reappears in Said's own, more general reflections:

> The stability of geography and the continuity of land – these have completely disappeared from my life and the life of all Palestinians. If we are not stopped at borders, or herded into new camps, or denied reentry and residence, or barred from travel from one place to another, more of our land is taken, our lives are interfered with arbitrarily, our voices are prevented from reaching each other, our identity is confined to frightened little islands in an inhospitable environment of superior military force sanitized by the clinical jargon of pure administration.
>
> Thus Palestinian life is scattered, discontinuous, marked by the artificial and imposed arrangements of interrupted or confined space, by the dislocations and unsynchronized rhythms of disturbed time . . . [W]here no straight line leads from home to birthplace to school to maturity, all events are accidents, all progress is a digression, all residence is exile.

How, then, can a geography appropriate to the Palestinian condition be written? How can that 'flaw', that crack in the clinical lattices of administered space, fracture its enframing geography? In an oblique reversal of the narrativity and systematicity that he attributes to the hegemonic discourses of Orientalism, Said's response – in this essay – is to argue for a space of representation that deploys hybrid, broken, fragmentary forms to reinscribe a Palestinian presence on the map.[14]

But writing such a geography is doubly difficult. Most immediately Said's attempt to interleave Mohr's photographs with his own text is confounded by his enforced absence from Palestine. In effect, he is obliged to enframe Palestine – in the same dispiriting sense in which Heidegger used the term – and then struggle to get through the looking glass, so to speak, because he is prevented from either accompanying Mohr or following in his wake: 'I cannot reach the actual people who were photographed, except through a European photographer who saw them for me.'[15] Said's predicament is deeply personal, of course, but that is exactly the point. Other sympathetic commentators were able to visit the West Bank and the exiled communities in the camps and cities beyond its borders, and they were allowed to speak with Palestinians. In a parallel collaboration between British TV journalist Jonathan Dimbleby and photographer Donald McCullin, for example, published seven years earlier, the faces *and* voices of the dispossessed leap from the pages. The images and text coalesce into an unusually direct, dignified and passionate statement about the plight of the Palestinians, and there is none of the mediated recollection and speculation that so marks Said's anguished prose.[16] For the most part, all Said can do is provide a series of plausible annotations to Mohr's images. His inscriptions are moments in the collective recitation I wrote of earlier, but there is always a gap between these shards of memory and the particularities of the people and places captured by Mohr's lens. The effect is oddly abstracted, a disconcertingly generalized series of readings. But the poignancy of *After the Last Sky* derives much of its power precisely from this enforced absence of the subject voice.

The sadness in the sentence I cited just now – 'I cannot reach the actual people who were photographed, except through a European photographer who saw them for me' – soon spirals out into a wider set of mediations, of dislocations in time and space. Here, for example, is Said commenting on a photograph of refugee labourers packing vegetables into boxes:

When in London and Paris I see the same Jaffa oranges or Gaza vegetables grown in the *bayarat* ('orchards') and fields of my youth, but now marketed by Israeli export companies, the contrast between the rich inarticulate *thereness* of what we once knew and the systematic export of the produce into the hungry mouths of Europe strikes me with its unkind political message. The land and the peasants are bound together through work whose products seem always to have meant something to other people, to have been destined for consumption elsewhere. This observation holds force not just because the Carmel boxes and the carefully wrapped eggplants are emblems of the power that rules the sprawling fertility and enduring human labour of Palestine, but also because the discontinuity between me, out here, and the actuality there is so much more compelling now than my receding memories and experience of Palestine.[17]

The passage stages the rupture of an organic unity. In Said's careful prose, the deep imbrications of identity and rootedness represented by the image of Palestinian peasants tilling their land – in its way, a *détournement* of the timelessness of Orientalist discourse – are torn apart. But this is about more than the time–space compression of commodity capitalism – more than an invitation to think where our breakfast comes *from*, as David Harvey once urged – because Said has reversed the point of view. In effect he asks: 'Where has Palestine *gone*?'[18]

Said answers his own question in the essay in which he records his eventual visit to Palestine after an absence of forty-five years. His cousin, living in Canada, had drawn a map of Said's native village from memory, and after two hours Said found his family's house, now occupied by a Christian fundamentalist organization:

> More than anything else, it was the house I did not, could not, enter that symbolized the eerie finality of a history that looked at me from behind the shaded windows, across an immense gulf I found myself unable to cross. Palestine as I knew it was over.

This is not that commonplace of autobiography, the adult return to a childhood world made strange by the passage of years; it is, rather, the melancholy of a collective memory that is inscribed in place, in landscape and in territory. As Said and his family drove along the coast, he noticed how every open space – 'whether football field, orchard or park' – was surrounded by barbed wire, and this sense of partition and enclosure heightened his sense '. . . of a history finished, packed up, taking place elsewhere'. He drove to Gaza, entering through a gate that was locked at night, and visited the Jabalaya Camp, home to 65 000 refugees: 'The numerous children that crowd its unpaved, potholed and chaotic little streets have a spark in their eyes that is totally at odds with the expression of sadness and unending suffering frozen on adult faces.' That sadness and suffering is written across the face of the land itself: for Said, like so many others, the very heart of the Palestinian predicament is geography.[19]

I shall have occasion to underscore the stubborn materiality of all this later, but in these circumstances it's not surprising that Said should admire Gramsci so much. He explains that Gramsci

> . . . thought in geographical terms, and the *Prison Notebooks* are a kind of map of modernity. They're not a history of modernity, but his notes really try to place everything, like a military map . . . [T]here was always some struggle going on over territory.[20]

Territory is etymologically unsettled: its roots are in *terra* (earth) and *terrere* (to frighten), so that *territorium* conveys '. . . a place from which people are frightened away'.[21] Both the archaeological and the historical records provide endless instances of displacements brought about by spellbinding fear and disfiguring terror, but Said is most concerned with the distinctively modern inflection of territory. On the map of modernity, territory connotes what Foucault would call a juridicopolitical field, and it is surely no accident that Said's writings about Palestine are shot through with an imagery of partition and enclosure, 'sanitized by the clinical jargon of pure administration', that so acutely mimics Foucault. This sense of territory establishes a connective imperative among power, knowledge and geography that Said's own project seeks to disclose, call into question and, in its turn, dis-place.

2 The subversive archipelago

I invoke Gramsci and Foucault as a way of opening a second set of answers to my original question, as a way of suggesting that Said's geography is also derived from the spatialization of cultural and social theory. But I don't want this intellectual genealogy to be construed as somehow separate and distinct from the intersections between Said's biography and history. He reads and reworks the ideas of these and other thinkers in ways that are inseparable from his commitment to the struggle over Palestine, and his successive engagements with the Palestinian question have been shaped by these ideas.[22] I have located the roots of this intellectual project on the Left Bank as a toponymic shorthand – nothing more – but I want to disentangle two theoretical strands in Said's writings that bear directly on both the intellectual cultures of postwar France and the generalized political project of the left. I should say at once that Said's appropriations are, in a sense, rhizomatic rather than direct: they are reworkings and graftings, conceptual equivalents of what he describes elsewhere as 'musical elaborations'. Perhaps it is for this reason that both strands have turned out to be so contentious. Some critics have been troubled by the traces of poststructuralism they identify in Said's writings, while others have objected to his distance from historical materialism. But they all read Said in obdurately conventional (linear) ways, whereas the power of his work seems to me to derive from his deep sense of spatial figuration: the creative juxtaposition of dissonant theoretical traditions. Peter Hulme captures something of what I have in mind when he describes Said's work as a 'subversive archipelago', a series of scattered but connected interventions that simultaneously calls into question the practices of colonial discourse and fractures the plates of 'continental theory'.[23]

It does so, Hulme suggests, by conjoining Foucault and Marx, but this is not a purely theoretical project (however unlikely the conjunction may seem) and I am particularly anxious not to lose the echo in that last sentence of Palestine's shattered geographies and Said's courageous attempt to fissure the politicomilitarized surfaces – or at least the imaginative geographies and representations of space – that contain and divide its peoples. Although I want to consider the same intellectual conjunction as Hulme, then, I do not want to do so in the abstract. I make this point because one of the most common objections to Said's project is that, in his later writings concerned most directly with the canonical cultures of colonialism and imperialism, he slides into a textualism. This is put most succinctly – and most suggestively – by Neil Smith:

> There remains in much of Said's later work a significant discrepancy between the imagined geographies unearthed from his literary texts and the historical geographies with which he seeks to re-entwine them; the latter never fully crystallize out of and into the

former . . . [There is] a geographical ambivalence in Said: the invocation of geography seems to offer a vital political grounding to Said's textuality until the abstractness of that geography is realized.[24]

Much of what follows is a consideration of this claim. I will attempt to rework some of Said's thematics to reinforce the materiality so vividly present in his interventions over the Palestinian question.

I have chosen my ground carefully, however, and I need to enter two qualifications. In the first place, the vignettes that I use to illustrate my argument – the Napoleonic *Description de l'Égypte* and the Cairo première of Verdi's opera *Aida* – move Said much further from the library than he usually travels. In his discussions of these texts, Said moves deeper into the material cultures of colonialism and into their dissonant landscapes. For metropolitan French culture, the *Description* was one of the most significant legacies of the military occupation of Egypt; but, as Said emphasizes, its production also discloses the intimate connections between textualization and taking possession. The work of the scholars and scientists who accompanied the French army was illuminated not only by the torch of reason but also by the blaze of gunfire, and Said accentuates the ways in which textual violence bleeds into physical violence.[25] Equally, his essay on *Aida* turns not on disembodied score and libretto but on the physical particulars of production and performance, on culture as event, and his 'worlding' of high culture proceeds here through the clamorous entry of colonial power on to the stage of the opera house itself.[26] Yet if the materialities of these two situations are unusual in Said's work, they are hardly exceptional in the wider scheme of things. The textual practices of Orientalism were marked by corporealities and physicalities whose recovery should be a strategic moment in any critical inquiry.[27] Similarly, performance may well be the 'extreme occasion, something beyond the everyday' that Said says it is: the première of a Verdi opera, especially in Egypt, was undoubtedly out of the ordinary. But culture is itself a production and a performance, and its stubborn everydayness has to be incorporated within the critique of Orientalism. As Said notes, we need to register, as part of the 'micro-physics of imperialism', 'the daily imposition of power in the dynamics of everyday life.'[28]

In the second place, both my case studies are staged in Egypt. Unlike many critics, however, I think that one of the strengths of Said's critique of *Orientalism* was its grounding in the so-called 'middle east'. Conversely, one of the cardinal weaknesses of his magisterial account of the connective imperatives between *Culture and Imperialism* is its geographical diffuseness: it is not accidental that his essay on the Cairo première of *Aida* should be one of the most successful readings in the book. I hope this will not be misunderstood. I do not mean to imply that the discourse of Orientalism is just another local knowledge, but neither do I think that its constellations of power, knowledge and spatiality can be transferred to other colonial or neocolonial situations without (often considerable) reworking. It is of the first importance to resist that exorbitation of Orientalism through which it becomes a synonym for colonial discourse *tout court*. There are resonances, connectivities and systematicities that tie Orientalism to discourses informing the practices of other colonial powers in other places; there are also inflections, supplements and reversals that differentiate it from other colonial regimes of truth. The imaginative geographies that were used to display the middle east were different from those that displayed south Asia, sub-Saharan Africa or South America, for example, and the power of their representations – their effectivity in devising, informing and legitimating colonial practices – was guaranteed by more than metropolitan assertion.[29] As Said repeatedly emphasizes, colonial discourses were not simply airy European fantasies: they were, of necessity, grounded. I might add that I have put all this in the plural deliberately. While Said does not treat Orientalism as that contradictory discursive terrain urged upon him by Lisa Lowe – and it is the absence of contradiction rather than any presumptive totalization which is the real issue[30] – the readings he offers are by no means homogeneous: within his pages, Flaubert is not Nerval, Massignon is not Renan, Lawrence is not Burton.

But if a discriminating geography is called for, so too is a determinate one. For these reasons, like Said, I want to continue my argument through a consideration of some imaginative geographies of Egypt produced by European scholars and artists in the nineteenth century. Within the geographical imaginary of postenlightenment Europe, which is Said's primary concern, Egypt occupied a pivotal position among Europe, Asia and Africa. It was at once the cradle of ancient civilization and one of the originary landscapes of the Old Testament; it was the political and commercial gateway to India and the far east; and it was a major vein into the 'heart of Africa'. These intersections made Egypt a liminal zone, located in that 'middle east' that traced a psychogeographical area within the European imaginary from the supposed familiarity and proximity of the 'near east' to the danger and distance of the 'far east'. John Barrell suggests that the 'middle east' was thus 'a kind of itinerant barrier or buffer between what can possibly be allowed in and what must be kept out at all costs'.[31] But the membrane was never unyielding: it was always ambiguous and contradictory. For all the attempts to project a series of binary oppositions on to the screens of Europe's imaginative geographies, 'Egypt' could not be held in place 'simply as an Other'.[32]

III Imag(in)ing geography

In *Orientalism* Said treats these imaginative geographies as so many triangulations of power, knowledge and geography, and the conceptual

311

architecture of his account is derived from the spatial analytics of Michel Foucault. Said's engagements with Foucault are neither uncritical nor unchanging, but throughout his writings he retains a considerable respect for Foucault's spatial sensibility. 'Foucault's view of things,' he remarks, was intrinsically 'spatial', and, as I want to show, this 'view of things' shapes Said's geography too.[33]

In doing so, however, I will bracket two issues. First, several commentators fasten on the difficulties of yoking Said's humanism to Foucault's anti-humanism, and insist that this produces a conceptual incoherence – at best, a vacillation – at the very heart of *Orientalism*. The dilemma is largely a product of Said's ethics of critical practice, I think, and in particular his unwavering commitment to intellectual responsibility; but complaints of this sort characteristically overlook the reappearance of a parallel problem in Foucault's own later move towards an ethics of the self. In neither case can the predicament be resolved by theoretical purification.[34] Secondly, one of Said's most vituperative critics objects that his presentation of Orientalism is radically non-Foucauldian because it is suprahistorical. Aijaz Ahmad claims that Said convenes Orientalism within 'a seamless and unified history of European identity and thought' whose interpretative arch spans without interruption all the discontinuities that a Foucauldian history would place between ancient Greece and nineteenth-century Europe. I am not sure whether the objection is an empirical one – does Ahmad deny the continuities that Said posits? – or whether he is dismayed by Said's departure from intellectual fideism.[35] In any event, this is a shockingly indiscriminate reading, because the force of Said's analysis is unmistakably directed against the specifically *modern* formation of Orientalism. 'Othering the Orient' has a long history in European thought, as numerous writers have shown, and Said is no exception. He does indeed call upon Aeschylus and Euripides to demonstrate the antiquity of Europe's casting itself as puppet-master to the Orient's marionette. But I see no reason to choose between an account that charts continuities – the stagnant air of Orientalism trapped within the corridors of history – and one that throws open the shutters to admit the ill-winds that interrupt this state of affairs from time to time and place to place. Neither does Said, who argues, explicitly and unequivocally, that the French occupation of Egypt at the end of the eighteenth century inaugurated a distinctively *modern* constellation of power, knowledge and geography: that it was 'an enabling experience for modern Orientalism'.[36]

In bracketing these two issues I am not implying that they are irrelevant to Orientalism and its representations of space; but I want to describe the connections between the imaginative geographies of Foucault and Said – their 'spatial view of things' – because I think these parallels are much more significant than any disjunctures.

1 The poetics and politics of space

Said begins with a general claim. What Lévi-Strauss called 'the science of the concrete' – what Said calls 'the economy of objects and identities' – depends on the ordered, systematic and differentiated assignment of *place*. This spatial metaphoric is a vehicle for the fabrication of identity, Said argues, through the 'universal practice of designating in one's mind a familiar space which is "ours" and an unfamiliar space beyond "ours" which is "theirs"'.[37] Said means this in a literal sense. Following Bachelard, he describes the practice as a *poetics of space*:

> The objective space of a house – its corners, corridors, cellar, rooms – is far less important than what poetically it is endowed with, which is usually a quality with an imaginative or figurative value we can name and feel: thus a house may be haunted or homelike, or prisonlike or magical. So space acquires emotional and even rational sense by a kind of poetic process, whereby the vacant or anonymous reaches of distance are converted into meaning for us here.[38]

If this seems unduly abstract, think for a moment of Said's return to his family house in Talbiya: the site is imaginatively converted from 'homelike' to 'prisonlike' and one topography of identity is displaced by another. But notice that this *is* an imaginative transformation, a process of fabrication and poesis, so that in this first approximation Said effectively denaturalizes imaginative geographies.

All the same, the production of these imaginative geographies is a generalized practice. Said insists that 'the construction of identity involves establishing opposites and Said insists that 'the construction of identity involves establishing opposites and "others"' and that it 'takes place as a contest involving individuals and institutions in *all* societies'.[39] Claims of this sort can be developed in several ways. So for example, Helga Geyer-Ryan reformulates Said's argument in Lacanian terms. Her argument is that the layered doubles between body and space fashion a sense of identity – precariously constructed within the imaginary and symbolic registers – that is vulnerable to, indeed shattered by, the *dis*placements of exile and emigration.[40]

It is of course imperative to understand the ways in which anxiety, desire and fantasy enter into the production of imaginative geographies, and Said's inattention to these topographies of desire is a remarkable lacuna in his account of Orientalism.[41] But I think it is necessary to retain the tension between the transcendental claims registered by Geyer-Ryan (and others) and the historicogeographical specificity of the congruences between bodies and spaces put in place by particular constellations of Orientalism.

On the other side, as I have indicated, Said himself is less interested in transcendental than in historical arguments, and much less interested in any psychoanalytics of space than in the *politics of space*. If the construction of identity through the poetics of space is a generalized practice, he makes it perfectly clear that it is also a 'contest': that it is inseparable from determinate modalities of power. For this reason, Said argues that the most appropriate model for colonial discourse analysis is not a linguistic one – on which most psychoanalytic theory turns – but a strategic or 'geopolitical' one.[42] Hence in a second approximation he reformulates the poetics of space in Foucauldian terms, in order to draw attention to 'the force by which a signifier occupies a place': to the assignment of individuals to particular places within discursive regimes of power-knowledge. Seen like that, he insists, 'the parallel between Foucault's carceral system and Orientalism is striking'.[43]

2 Scopic regimes

I want to describe that parallel between Foucault's 'carceral system' and Said's *Orientalism* by plotting three points on their maps of power, knowledge and geography.

Division

First, both Foucault and Said describe the discursive construction of exclusionary geographies. At the heart of Foucault's work, Said remarks, 'is the variously embodied idea that conveys the sentiment of otherness', an idea that shapes not only what Foucault writes about but also the way in which he writes about it: hence 'there is no such thing as being at home in his writing'.[44] One of Foucault's central claims is that societies are discursively constituted through a series of normalizing judgements that are put into effect by a system of divisions, exclusions and oppositions. He traces this process in his histories of madness, the prison and punishment, and sexuality. Although these narratives all confine their trajectories of reason to the west, *Orientalism* can, I think, be read as Said's attempt to reconstruct the missing history of Foucault's 'great divide' between Occident and Orient. His project is thus, in part, a mapping of the cells that form the primary graticule of Orientalism's imaginative geographies:

Table 14.1 Occident and Orient

'Occident'/'same'	'Orient'/'other'
Rational	Irrational
Historical	Eternal
Masculine	Feminine

In *Orientalism* Said's treatment of these binary oppositions is highly uneven; in particular, he says remarkably little about the sexualization and eroticization of the Orient.[45] In his later writings he seeks to interrupt and displace the oppositions altogether: 'Partly because of empire,' he declares, 'all cultures are involved in one another; none is single and pure, all are hybrid, heterogeneous.[46] But setting the couplets out in this stark, schematic form shows that the discourses of Orientalism not only essentialized 'the Orient': *they also essentialised 'the Occident'*. Contrary to many of his critics, therefore, I think that the strategic essentialism which Said discloses, of both Orient and Occident, is not the pure product of his own artifice: *it is, rather, a constitutive function of Orientalism itself.*

Said wires these divisions to a grid of power that is both universalizing and differentiating, and in doing so extends Foucault's original thesis. He laments that Foucault 'does not seem interested in the fact that history is not a homogeneous French-speaking territory, but a complex interaction between uneven economies, societies and ideologies'. Some critics insist that Foucault's ethnocentrism was by no means unconsidered, and that there are difficulties in situating his work within the 'much larger picture' involving 'the relationship between Europe and the rest of the world' urged upon him by Said. But Said is determined to show not only that 'the ideas of discourse and discipline are assertively European' but also 'how, along with the idea of discipline to employ masses of detail (and human beings), discipline was used to administer, study and reconstruct – then subsequently to occupy, rule and exploit – almost the whole of the non-European world'.[47]

Detail

Second, and following closely from these observations, both Foucault and Said suggest that such a history of division, of 'discourse and discipline', is at the same time a history of detail. Foucault argues that its reconstruction brings us, at the end of the eighteenth century, to Napoleon, who dreamed of what he calls 'the world of details' and set out to organize it: 'He wished to arrange around him a mechanism of power that would enable him to see the smallest event that occurred in the state he governed.'[48] For Said, too, the power of Orientalism derived from its constitution as a discipline of detail.

> Most of all, it is as a discipline of detail, and indeed as a theory of detail by which every minute aspect of Oriental life testified to an Oriental essence it expressed, that Orientalism had the eminence, the power and the affirmative authority over the Orient that it had.

And he too attaches a particular significance to Napoleon, and most of all to the *Description de l'Égypte* carried out under his authority. Drawing on many of the same visual strategies deployed in the anatomized description of eighteenth-century France in Diderot's *Encyclopédie*, the *Description de l'Égypte* offered an unprecedented textual appropriation of one country by another. In effect, it constituted what Andrew Martin calls 'a textual empire' in which 'the subjugation of a country was to be supplemented by scriptural fortification'.[49] The withdrawal of the expeditionary force did not diminish (though it did displace) these aspirations. In their attempt to turn Egypt into 'a department of French learning', to 'render [Egypt] completely open', and 'to divide, display, schematize, tabulate, index and record everything in sight', both the surveyors and scholars on the ground and the authors and engravers in Paris put the discipline of detail into practice with minute perfection.[50] Said argues that this lineage of 'monumental description' (in more senses than one) inaugurated and continued to shape a distinctively modern Orientalism. Hence he reads Edward Lane's classic inventory of *The manners and customs of the modern Egyptians*, published in 1836, as an attempt 'to make Egypt and the Egyptians totally visible, to keep nothing hidden from his reader, to deliver the Egyptians without depth, in swollen detail'; similarly, he suggests that 'what matters' to Gustave Flaubert, busy keeping a diary of his *Voyage en Orient* in 1849–50, and fascinated by both what he sees and how he see, 'is the correct rendering of exact detail'.[51]

Visibility

Third, as the previous paragraph intimates, both Foucault and Said argue that the discipline of detail depends on 'spaces of constructed visibility'. John Rajchman, from whom I have borrowed the phrase, suggests that Foucault's histories of division and detail are also histories of 'the visual unthought' in which the production of space plays a central role: that Foucault was particularly interested in how spaces were designed to make things seeable in a specific way.[52] Said also accentuates the imbrications between power and what Foucault calls 'the empire of the gaze', but Said means it quite literally. He claims that the colonizing inscriptions of Orientalism are constituted panoramically: 'The Orientalist surveys the Orient from above, with the aim of getting hold of the whole sprawling panorama before him.' The phrasing seems to suggest what Gillian Rose perceptively identifies as 'the uneasy pleasures of power', with the Orient-as-woman reclining before the scopic virilities of the masculinist spectator.[53] But the visual repertoire of Orientalism was not confined to the panoramic, any more than its eroticization of the Orient was confined to a heterosexual imaginary: the sexual politics implicated in the empire of the gaze were more complicated and more unstable than any simple equation between Orientalism and masculinism. Joseph Boone has persuasively shown that the homoerotics of Orientalism all too often trembled on the edges of an occidental homophobia, for example, and that a careful reading of these imaginative geographies – the psychic screens on which these fantasies were projected – will have to acknowledge the ambiguities and contradictions generated by the collisions between sexual stereotypes and colonialist tropes.[54]

But this is not Said's project. Instead, in the central chapters of *Orientalism* he seeks to show how, in the course of the nineteenth century, European representations of the Orient as a sort of magic theatre, a stage 'affixed to Europe' on which were displayed the fabulations of a rich and exotic world, were overlaid (if never altogether displaced) by representations in which the Orient became a tableau, a museum and a disciplinary matrix. His chronology both repeats and interrupts Foucault's epistemological distinctions among the renaissance, the classical and the modern. There were, for example, close filiations between the languages of theatre and geography in renaissance Europe, and Orientalism mobilized these devices in its evocations of half-imagined, half-known worlds.[55] But the tableau in which the east is watched for, what the *Description de l'Égypte* calls 'bizarre jouissance', continues the theatrical imagery and, at the same time, installs a sense of exhibition that is profoundly modern. Its 'use-equivalent', so Said suggests, is to be found 'in the arcades and counters of a modern department store'. Equally, the representation of the Orient as 'an imaginary museum without walls', in which cultural fragments were reassembled and allocated among the categories of a tabular Orientalism, invokes an altogether different order of departmentalization: the textual inventory that is emblematic of Foucault's classical, eighteenth-century taxonomies. Finally, the enframing of the Orient within what Said describes as 'a sort of Benthamite Panopticon' moves the empire of the gaze beyond the tableau and the table to anticipate a system of power-knowledge in which 'things Oriental [are placed] in class, court, prison or manual for scrutiny, study, judgement, discipline or governing': it is a preliminary and a prop for the disciplinary powers inscribed within the colonizing apparatus of 'the world-as-exhibition'.[56]

3 Describing Egypt

It would no doubt be possible to extend and revise this account in several ways, but I want to underscore Said's interest in what one might call the scopic regimes of Orientalism. Indeed, his constant emphasis on the visual tropes, technologies and strategies embedded in orientalist texts is as noticeable as his inattention to the visual arts themselves.[57] To consolidate my argument about the ways in which Said imag(in)es geography, I want to provide three readings of the frontispiece to the first edition of the *Description de l'Égypte* (Figure 14.1).[58]

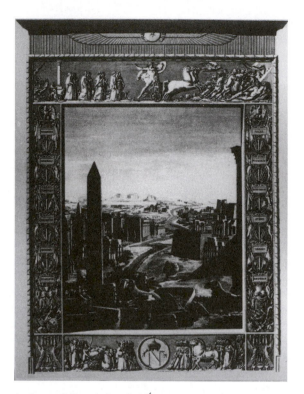

Figure 14.1 Frontispiece of *Description de l'Égypte*

The image shows a view through the portal of a stylized temple on to the monumentalized landscape of ancient Egypt, from which all signs of life – of the contemporary inhabitants of Egypt – have been erased.[59] From this position the eye commands, in a single impossible glance, a sweep of monuments from Alexandria in the foreground up the valley of the Nile to Philae in the far distance. It is a characteristic of panoramas, of what Denis Cosgrove calls 'landscape as a way of seeing', that the surveillant eye which takes in the scene is absent from the visual field.[60] But in this case the apparatus that makes such an impossible prospect *possible* is paraded in triumph across the panels enframing the panorama. The upper panel shows the French army putting the Mamelukes to flight at the pyramids through the metonymical figures of the eagle, emblem of the army, and a Roman hero, presumably Napoleon (and so, by implication, the legitimate heir of antiquity). Behind them are allegorical figures representing the scholars who accompanied the expedition and produced the surveys for the *Description*. The defeated Mamelukes reappear on the lower panel, laying down their arms and acknowledging the centrality of the Napoleonic seal which is encircled by a serpent, the symbol of immortality. The flanking panels are festooned with French battle honours in Egypt. In all, I think one can conclude power–knowledge in indissoluble union. Indeed, as Fourier noted in his preface to the *Description*:

> This great work concerns the glory of our native land; we owe it to the efforts of our warriors; it has its origins in the union between science and military might [*les sciences et les armes*]: it is both testimony to and fruit of their alliance.

But it is also surely a union of power, knowledge and *geography*. For this is a memorialized *landscape* and its surveyors are inscribed in positions of power and prominence, contemplating and conquering Egypt in the name of – and even, I think, as *part of* – France.

The originality of the scholars' contribution resided most critically in their commitment to empirical science grounded in field observation, and the *Description* was distinguished above all by the sheer detail of its representations. In the volumes devoted to ancient Egypt – the core of both the first and second editions – David Prochaska has shown that these images were organized as a hierarchical sequence of views from the panoramic down to the detailed. This is fully conformable with that 'organization of the view' implied by the modern enframing of the world-as-exhibition, but in this particular case topographies and descriptive geometries provided, in effect, a geo-graphing of Egypt whose power imposed an extraordinary unity on the dispersed volumes of the text. At each site the inventory begins with an eagle's eye view (literally so): topographic maps locate the antiquities, which are then displayed in panoramic view; these give way to perspective views, which in turn dissolve into the close-up detail of reliefs and inscriptions (Figure 14.2). The interlocking sequence is repeated at each site, and the imperial itinerary is thus organized into a sort of proto-GIS whose mobile gaze traces an arc down the Nile Valley from Philae in the south to Alexandria in the north. It is this journey, the empire of the gaze in material form, that is recapitulated in the frontispiece.[61]

This mode of detailed representation was a way of claiming not only empirical authority – that sense of 'being there' which dazzled the first European readers of the *Description* and which continues to invest contemporary ethnographies with so much of their power – but also colonial *legitimacy*: an implication that the scholars, and by extension their European audience, were *entitled* to be there and to have Egypt set out thus for their edification. I say this because it is noticeable how often the French included themselves in the plates depicting Egypt's ancient monuments, much more rarely in those of the present Egypt, and in striking contrast to the

Figure 14.2 The organization of the view in *Description de l'Égypte*

so-called orientalist picturesque which was distinguished by the studied absence of the western observer. The practice of autoinscription was not only an enduring record of the short-lived French presence in Egypt – they were forced to withdraw by the British in 1801 – but also an implicit evocation of Egypt as the cradle and mirror of French civilization, 'a sort of Eden where reason triumphed, a perfect world governed by a wise sovereign'.[62] A fantasy-Egypt, Anne Godlewska calls it, where the torch of reason illuminated an ancient Egypt somehow 'more true and more real' – *because* rational – than the dismal present 'sullied by centuries of Oriental despotism'. Hence, Godlewska argues, in a third reading, that the frontispiece shows

> all of the outstanding monuments of Egypt . . . in the foreground, out of context, as though they had all been recently collected together to be taken on board a ship about to sail into the Mediterranean . . . This is the Egypt that the writers and editor of the *Description* most wanted to capture, the Egypt that could be claimed and taken home.[63]

These three readings parallel Said's own summary account of the *Description*. 'What Napoleon and his teams found,' he writes, was 'an Egypt whose antique dimensions were screened by the Muslim, Arab and even Ottoman presence standing everywhere between the invading French army and ancient Egypt'. In order to displace that screen, to open a passage leading directly from Egyptian antiquity to European modernity, the reconstructions proceeded 'as if there were no modern Egyptians but only European spectators'. Ancient Egypt was staged 'as reflected through the imperial eye' and its material cultures were finally 'dislodged from their context and transported to Europe for use there'.[64] But I hope that my readings reveal, perhaps more clearly than Said's, the connections among power, knowledge and geography inscribed within these representations that made the imperial project possible.

This must be pressed further, because I also want to insist upon the *specificity* of the imaginative geographies of Orientalism. This matters for at least three reasons. First, it is necessary to retain the particular imbrications of place and space within colonial constellations of power–knowledge. I have some sympathy with the suggestion that Foucault's 'geometric turn', on which much of Said's spatial sensibility depends, runs the risk of 'elevating an abstract sense of space above a concrete sense of place'.[65] But, as I have indicated, Orientalism was implicated in the forced, often violent production of an abstract(ed) space and its superimposition over the particularities of different places. Those places were never the timeless, essentialized settings represented by orientalist travellers – the historicity and hybridity of Egypt's human geographies requires emphasis – but during the nineteenth century their textures were caught up in and reworked by European grids of power–knowledge. Conversely, the incorporation of those places, localized knots in wider webs of social practice, recast the spaces of Orientalism (and in so doing made the binary distinctions of its imaginative geographies highly unstable). This makes it necessary to say much more than Said usually does about the particular places brought within the imaginative geographies of Orientalism, in order to map both the violence of its projective geometries and the shifting contours of its spaces.

'Second, the production and superimposition of these abstracted spaces depended on practices and protocols that, though they were also deployed within Europe, became freighted with other modalities of power–knowledge in their extension beyond Europe. The visual practices through which nineteenth-century Paris was known by its bourgeoisie, for example, the same bourgeoisie that gazed on the plates in the *Description* early in the century, trooped through the Egyptian galleries in the Louvre at mid-century, and travelled up the Nile at the end of the century, were not radically different from those through which they knew nineteenth-century Cairo.[66] But outside Europe these visual practices were intertwined with colonialism and imperialism to produce imaginative geographies that involved not simply demarcating 'our space' from 'their space', as Said first proposed (above: 313–14), but 'our' *reaching into* 'their' space and imaginatively – and eventually materially – *appropriating* that space and claiming it *as* 'ours'. Said sees this as the singular, baleful achievement of nineteenth-century Europe. 'When it came to what lay beyond metropolitan Europe,' he argues, the arts and the disciplines of representation 'depended on the powers of Europe to bring the non-European world into representation, the better to be able to see it, to master it, and above all to hold it.[67]

Third, it is possible to resist these appropriations by turning around the practices of representation on which, historically, they have depended. Thus, for example, Said now calls for the Palestinian leadership to set in motion its own 'discipline of detail'. He notes that all the documents and maps used in the negotiations that culminated in the Declaration of Principles in Oslo in 1993 were produced by Israel, and he insists that Palestine has to devise its own maps and a systematic counterstrategy on the ground, 'in which each detail is an organic part of the whole'.[68] Said's argument turns not only on the deconstruction of the map – though he is properly sceptical of the neutrality of cartographic science – but also on the specificity of the imaginative geographies of Orientalism and, by extension, of Zionism. For if, as I have argued, the discipline of detail is implicated in the production of a specific space of constructed visibility, then effecting a *détournement* of its visual practices should enable the Palestinian people to be seen in a different way, to represent themselves, on their own ground as inhabitants of their own land.[69] There is thus a profound continuity between Said's genealogy of Orientalism and his closing observations in *After the Last Sky*:

> I would like to think that we are not just the people seen or looked
> at in these photographs: We are also looking at our observers. We

Palestinians sometimes forget that – as in country after country, the surveillance, confinement and study of Palestinians is part of the political process of reducing our status and preventing our national fulfillment except as the Other who is opposite and unequal, always on the defensive – we too are looking, we too are scrutinizing, assessing, judging. We are more than someone's object. We do more than stand passively in front of whoever, for whatever reason, has wanted to look at us. If you finally cannot see this about us, we will not allow ourselves to believe that the failure has been entirely ours. Not any more.[70]

IV Dislocating geography

In order to disclose the connections I've tried to establish in the previous paragraphs – the articulations between place and space, the modulations between Paris and Cairo – and to make possible the strategic reversals, Said suggests that it is necessary to move towards what, in *Culture and imperialism*, he calls a 'contrapuntal reading.' I want to follow in his footsteps, but I need to make two complicating observations, the first about colonialism's 'consolidated vision' and the second about its 'overlapping territories'.

1 Consolidated vision

Towards the end of the eighteenth century, the French writer Louis-Sebastien Mercier came down for breakfast in his Paris hotel. There, as Linda Colley observes,

> [He] saw imperialism displayed upon a Parisian breakfast table. The polished mahogany surfaces, like the steaming coffee, brought the colonies of the New World instantly to his imagination. The fine porcelain, he judged, had been shipped by armed merchantmen from China. The sugar spoke to him of Caribbean slavery and the scented tea of Indian plantations. The world the Europeans looted was no longer a distant enterprise. It was part of the very fabric of their lives at home. Yet, as he noted this, Mercier was clearly congratulating himself on his rare measure of perceptiveness. *He did not believe that those eating alongside him saw what he saw.*[71]

The visual image is striking, and Said makes a similar point. 'The empire functions for much of the European nineteenth century as a codified, if only marginally visible, presence in fiction,' he writes, 'very much like the servants in grand households and in novels whose work is taken for granted but scarcely ever more than named, rarely studied or given density.[72] This is a perceptive observation too, trading on another visual metaphoric, but the oversight to which it draws attention is mirrored in Said's own transition to *Culture and imperialism*, where the visual thematic that is so prominent in the central chapters of *Orientalism* is virtually eclipsed.

But this is more complicated than it seems. Said offers one other arresting visual image to conjure up the relationship between metropolitan cultures and their colonialisms, when he describes Guy de Maupassant choosing to have lunch at the Eiffel tower because it was the only place in Paris from which he couldn't see it.[73] By then, however, the world had turned. The Eiffel tower was built for the Paris exposition of 1889, which not only marked the centenary of the Revolution but also incorporated for the first time a simulacrum of a colonial city whose 'Moslem minarets, Cambodian pagodas, Algerian mosques and Tunisian casbahs' were expressly designed to display France's colonies to a metropolitan (and cosmopolitan) audience.[74] By the closing decades of the nineteenth century, the reverse projection of metropolitan cultures on to non-European landscapes was also becoming a commonplace in colonial planning discourse and in other, less instrumental but no less colonial cultural productions. At the turn of the century a guidebook published by Hachette figured Egypt thus: 'The shape of Egypt is for all the world exactly like the Eiffel Tower. The shaft is Upper Egypt, and the base is the Delta . . . All the space inside is cultivation; without is desert. At the junction of the two is Cairo.'[75] The envelope of cultivation (and, by implication, civilization) was contained within what had become an unmistakably Gallic symbol: everything beyond was aridity and sterility. Yet Said says nothing about the visual cultures and iconographies inscribed in imaginative geographies like these. For the most part his interest in the optics of colonialism and imperialism seems to be resolutely metaphorical.

Said's argument is double headed. On the one side, he agrees with Colley that there was a 'fundamental unevenness' in the receptivity of European cultures to empire. Given the scale of Britain's imperial enterprise, she insists, 'what is surely remarkable is not that this should have influenced its literary culture, but that it failed to influence it far more than it did'. Although Colley claims that Said is unwilling to confront this paradox, or at least to offer an explanation for it, he does concede that the connections between culture and imperialism are only displayed with clarity in the closing decades of the nineteenth century. 'Not until well after mid-century did the empire become a principal subject of attention in writers like Haggard, Kipling, Doyle, Conrad,' he observes, and 'when European culture finally began to take due account of imperial "delusions and discoveries"', Said argues that it did so with a characteristically modernist gesture: irony. European writers 'began to look abroad with the skepticism and confusion of people surprised, even shocked by what they saw.[76] On the other side, however, Said also suggests that colonialism and imperialism functioned as a ground for European cultural production much earlier:

> If one began to look for something like an imperial map of the world in English literature, it would turn up with amazing insistence and frequency well before the mid-nineteenth century. And turn up not only with the inert regularity suggesting something taken for granted, but – more interestingly – threaded through, forming a vital part of the texture of linguistic and cultural practice.[77]

The appeal to cartography is not accidental and is repeated at intervals throughout *Culture and imperialism*. Said argues that the 'consolidated vision of empire' (his phrase) is unlikely to be disclosed through conventional critical practices that privilege temporality. 'We have become so accustomed to thinking of the novel's plot and structure as constituted mainly by temporality,' he admonishes, 'that we have overlooked the function of space, geography and location.'[78] What is required, as a complement to the usual practices of textual criticism, is a contrapuntal reading that is specially attentive to spatial connectivity and juxtaposition. Thus in his readings of *Mansfield Park* and *Kim*, Said discloses a hierarchy of spaces that functions as a grid wiring metropolitan circuits of action to their colonial ground: What Dana Polan describes as the novel's 'projection of power across locales, its rendering of disparate situations as linked by interests and economies'. Polan's metaphor is revealing because it suggests that there is an essential – though for the most part understated – connection between the sly spatiality of colonizing cultures and the empire of the gaze. There is something phantasmagoric about this metropolitan 'projection of power'. I owe the comparison to Benjamin's critique of commodity culture. The phantasmagoria was a magic lantern which became popular in early nineteenth-century Europe through its use of back-projection to ensure that its audience remained unaware of the source of the image they were seeing. Benjamin used it to figure the ideological projections of nineteenth-century bourgeois culture and to disclose the elisions and evasions in their visual practices and 'structures of understanding'.[79] It is not, I think, unduly fanciful to glimpse something of the same in Polan's suggestion that Said reads cultural productions 'as doubly driven, pushed by the needs of imperial ideology to spatialize history, *but enabled by contrapuntal reading to reveal the traces of the very history they seek to occult in the security of spectacle*'.[80]

2 *Overlapping territories*

Closely connected to the changes in his approach to visuality, spatiality and colonialism, Said's theoretical attention switches from Foucault in *Orientalism* to Gramsci in *Culture and imperialism*. Said's relationship to historical materialism is, of course, as complex as it is contentious, not least because the very 'westernness' of western Marxism, its typical closure around the cultures of Europe and North America, makes it difficult for him to invoke it in a transcultural register.[81] It is not so much Marxism, therefore, as particular Marxists who have captured his attention: most of all, Raymond Williams and Antonio Gramsci.

Williams was one of the main inspirations for *Orientalism*. This must seem a strange role for so British a thinker, and Said admits that Williams's work is limited by his 'stubborn Anglocentrism' and its implication that 'English literature is mainly about England.'[82] But he still has the greatest admiration for him, and says that his own project was particularly affected by the dialectic between acquisition and representation that animates *The country and the city*. He cites Williams's readings of seventeenth-century English country-house poems as exemplary instances of a critical strategy that interprets cultural productions not so much for what they represent as 'what they *are* as the result of contested social and political relationships'. Of all Williams's writings it is probably *The country and the city* that has had the greatest impact on geography, but it will be clear from what I have thus far said that Said's geographical sensibility is radically different from Williams's deep love of landscape and what he usually called 'working country'. Perhaps for this reason his influence on Said's early writings is at once pervasive yet oblique. His theoretical formulations are rarely invoked, but the shape and form of his critical practice animates Said's work.

In *Culture and imperialism*, however, it is Gramsci who supports the main architecture of the text, and in similarly pervasive but none the less oblique fashion. His main function, I suggest, is to provide Said with another way of mapping the intersections among power, knowledge and geography. His specific contribution seems to be twofold. In the first place, Said is attracted by Gramsci's emphasis on the productivities and positivities – what he called the work of 'elaboration' – through which power and culture are conjoined. Gramsci, he says,

> loses sight neither of the great central facts of power, and how they flow through a whole network of agencies operating by rational consent, nor of the detail – diffuse, quotidian, unsystematic, thick – from which inevitably power draws its sustenance, on which power depends for its daily bread. Well before Foucault, Gramsci had grasped the idea that culture severs authority and ultimately

the national state, not because it represses and coerces but because it is affirmative, positive and persuasive.[83]

Accordingly, when Said draws attention to a geographical notation, to the 'imperial map' that licensed the 'cultural vision', and suggests that 'common to both is an *elaboration* of power', this should, I think, be understood in exactly Gramsci's sense.[84]

In the second place, the mainstream of western Marxism was thoroughly Hegelian and, as Said recognizes, attached a special importance to History and historicity, whereas Gramsci's writings display a contrary emphasis on space (without that imperial capital) and spatiality. Said finds this 'explicitly geographical model' in Gramsci's essay on 'Some aspects of the southern question', which functions as a prelude to his *Prison notebooks* 'in which he gave, as his towering counterpart Lukàcs did not, paramount focus to the territorial, spatial, geographical foundations of social life'. More specifically, Said argues that Gramsci was not interested in some transcendent logic by means of which antinomies are resolved within the telos of History, but rather 'in working them out as discrepant realities, physically, on the ground'.[85] This sense of what, following Benjamin, might be seen as a sort of geographical constellation, a configuration formed by the forceful conjunction of distanciated geographies in a particular place, intersects with Said's objection to treating colonialism as a one-way street:

> In one instance, we assume that the better part of history in colonial territories was a function of the imperial intervention; in the other, there is the equally obstinate assumption that the colonial undertakings were marginal and perhaps even eccentric to the central activities of the great metropolitan cultures.[86]

Said believes that by mapping the interpenetrations of culture and imperialism as 'overlapping territories, intertwined histories', it is possible 'to reinterpret the Western cultural archive as if fractured geographically by the activated imperial divide.'[87]

And yet Gramsci remains a spectral figure in *Culture and imperialism*, always in the margins and shadows of the text, haunting the interline, so to speak, yet repeatedly invoked by Said in his subsequent interviews and commentaries. Strangely, Said neither develops nor even fully deploys Gramsci's conceptual apparatus. Had he done so, he would, I think, have been obliged to treat the colonial 'elaboration of power' in less cohesive terms; that is to say, he would have been able to recognize those 'textual gaps, indeterminacies and contradictions' that fissured colonial discourse and, in consequence, would have found it easier to map those spaces of resistance, that 'mutual siege' as Gramsci called it, which Parry suggests splintered the 'consolidated vision' of colonialism and imperialism.[88] One of the main differences between *Orientalism* and *Culture and imperialism* is in fact Said's determination to open a space for resistance, and he now suggests that one of the major limitations of Foucault's work is its portrayal of 'an irresistible colonizing movement': 'The individual [is] dissolved in an ineluctably advancing "microphysics of power" that it is hopeless to resist.' Yet Gramsci's passionate sense of the power of collective agency – of subaltern resistance and displacement – is passed over and his relation to the project of subaltern studies, surely of cardinal significance to any re-visioning of colonialism, is barely noted.[89]

Said elects to explain his way of working not through any discussion of critical theory – and I understand his wariness about its institutionalization, neutralization and, indeed, trivialization – but by invoking a metaphor derived from music. He proposes to model his work not on a symphony, 'as earlier notions of comparative literature were', but on an 'atonal ensemble'. In effect, he transposes Gramsci's mapping of a complex and uneven cultural topography into his own practice of contrapuntal reading:

> As we look back at the cultural archive, we begin to reread it not univocally but *contrapuntally*, with a simultaneous awareness both of the metropolitan history that is narrated and of those other histories against which (and together with which), the dominating discourse acts. In the counterpoint of Western classical music, various themes play off one another, with only a provisional privilege being given to any particular one; yet in the resulting polyphony there is concert and order, an organized interplay that derives from the themes, not from a rigorous melodic or formal principle outside the work. In the same way, I believe, we can read and interpret English novels, for example, whose engagement with the West Indies or India, say, is shaped and perhaps even determined by the specific history of colonization, resistance and finally native nationalism.[90]

Most of Said's writing about music has focused on contrapuntal work. Forms like opera interest him for that very reason, 'forms in which many things go on simultaneously', and he says that he made a considered decision to organize the essays that compose *Culture and imperialism* around the same musical form: his intention was thereby to exhibit 'a kind of exfoliating structure of variation'.[91] It is entirely appropriate, therefore, that the clearest illustration of Said's contrapuntal method should be his account of the Cairo première of Verdi's opera *Aida*.[92] But I want to consider this essay for reasons other than the purely methodological. Its substantive significance rests, in part, on the place classical opera occupied within the bourgeois cultures of late nineteenth-century Europe and, by extension, on the intersections between the cultural formations of the bourgeoisie and the cultures of Orientalism.[93] But I think it is also possible to connect Said's treatment of this production to the scopic regimes of colonialism and imperialism and the geographies of truth inscribed within them and so reactivate my previous discussion.

3 'Aida' and the geography of truth

Aida was commissioned by the Khedive of Egypt after a series of protracted negotiations during the first six months of 1870. The Superintendent of the Khedival Theatres, Paul Draneht, had originally tried to persuade Verdi to compose a celebratory hymn to mark the opening of the Suez Canal, but Verdi's polite refusal only heightened the Khedive's ambitions, and now he envisioned 'a purely ancient and Egyptian opera'.[94] The story-line was drafted by Auguste Mariette, a distinguished French Egyptologist whom the Khedive had placed in charge of archaeological excavations throughout Egypt. It traced the tragic love-affair between Radames, a captain in the Egyptian army, and Aida, the daughter of the king of Ethiopia; she had been captured and made to work as a slave in the household of the pharaoh's daughter. The story revolved around jealousy and betrayal, played out against the background of military conflict and aggression.[95]

The interpretative politics of *Aida* are extremely complicated. Anthony Arblaster agrees that, from some perspectives, it might appear to be 'a triumphalist work from the high noon of European imperialism'; but, as he says, it can also be seen as a transposed commentary on European geopolitics. Thus Said argues that the British tacitly encouraged Egyptian expansionism in east Africa in order to frustrate French and Italian ambitions in the region, so that 'from the French point of view, incorporated by Mariette, *Aida* dramatized the dangers of a successful Egyptian policy of force in Ethiopia'.[96] 'From the French point of view' it may have done exactly that; but Said fails to note that Verdi himself was no defender of imperialism and that he was frankly appalled by the territorial ambitions of the European powers. He took an active part in developing Mariette's original outline and shaping the libretto, and this allows Arblaster to argue that *Aida*'s indictment of the bellicosity and cruelty of ancient Egypt and its priesthood was intended by Verdi as a pointed comparison with the Prussians, 'whose success in the Franco-Prussian war he regretted and whose growing power and ambition he (rightly) feared'.[97] Whatever the merits of these readings, however, Said's central claim is that the production 'is not so much *about* but *of* imperial domination', and I want to explain what I think he means by sketching the geography of truth that underwrote the Cairo production.[98]

This was double-edged. One the one side, the Khedive placed an unprecedented premium on accuracy and authenticity. Mariette was no stranger to the meticulous staging of Egypt. He was closely involved in the Egyptian installation at the 1867 exposition in Paris, which he described as 'a living lesson in archaeology'. At its centre was a copy of the temple at Philae, and Mariette had its architect work from precise measurements and photographs taken at Philae. Although a number of compromises had to be made, Mariette insisted on 'the greatest authenticity in the ensemble and in the minutest detail'.[99]

And yet: although the opera was *about* Egypt and was to have its world première *in* Egypt, it had behind it the most displaced of geographies that tied Egypt umbilically to Europe. On the other side, therefore, *Aida* was written in Italy (and in Italian: there was never any question of its being sung in Arabic); Mariette was sent to Paris to oversee the preparation of sets and costumes by French craftsmen and costumiers; the company was cast in Italy and the Khedive was prepared to allow rehearsals to be held in Paris, Milan or Genoa. This was time–space compression on a grand scale, and Hans Busch's wonderful documentary history of *Aida* – on which Said relies too – shows that the postbags bulged and the telegraph wires hummed between Genoa, Paris and Cairo. By July 1870 Mariette was in the thick of things in Paris. 'In order to follow the instructions the [Khedive] have given me,' he wrote, 'to make a scholarly as well as a picturesque *mise-en-scène*, a whole world must be set in motion.'[100] A few weeks later it was. The Prussian army invaded France and laid seige to Paris; the only communication with the outside world was by pigeon or balloon, and by November all work on the sets and costumes had been suspended and the première postponed. The contractors were unable to resume work until the following summer, but Mariette was inordinately satisfied with the result. 'The view of the pyramids is completed and crated,' he wrote to Draneht. 'It is very lovely, and I am pleased with it. *At the raising of the curtain one will truly believe oneself in Egypt.*'[101]

But then, of course, one would *be* in Egypt . . . Think about that for a moment. The audience in the Opera House in Cairo will 'truly believe' itself transported to Egypt, not because it already *is* in Egypt, not through any theatrical suspension of belief – always awkward in opera[102] – but because Egypt would be presented as a spectacle 'more true and more real' (above: 321) than the streets and bazaars outside the theatre. This Egypt would be more true and more real precisely because the view would be *organized*: the set would be *framed*; it would have depth, perspective, coherence – in a word, *meaning* – that the 'other Egypt', the Egypt checked at the doors of the Opera House, was supposed to lack.[103]

Mariette's sentiment was, I think, sparked by what Jean-Louis Comoli calls 'the frenzy of the visible' that exploded in the second half of the nineteenth century, a moment in which, so he suggests, 'the whole world becomes visible at the same time that it becomes appropriatable'.[104] The interconnections of vision and appropriation within the cultures of colonialism and imperialism are of the first importance, because what lay behind Mariette's proud boast, what gave 'his' Egypt its verisimilitude, was a regime of truth imposed through a sort of archaeology-in-reverse. The sets for *Aida* were explicitly based on the *Description de l'Égypte*, and just as the scholars had sketched plans and views and, on occasion, disassembled and crated artifacts to be shipped to Paris, so now those same plates were to be used to reconstruct an 'authentic Egypt' in Paris which was then to be disassembled, crated and returned to Egypt. I have called all this a geography of truth because it was clear to all the European principals that accuracy and authenticity could not be found in Egypt. Mariette was only the most recent in a long line of French intellectuals who claimed to be unable to find the 'real' Orient there. Gérard de Nerval despaired of ever being able to furnish Théophile Gautier with descriptions of Cairo that could be reproduced as sets for the Paris Opéra: 'I will find at the Opéra the real Cairo, the Orient that escapes me'. In the end, as Mitchell remarks, 'only the Orient that one finds in Paris, the simulation of what is itself a series of representations to begin with, can offer a satisfying spectacle'.[105] But this time Mariette was also aware that authenticity would not be found in the customary assumptions of the Paris *ateliers* either. He was determined to eschew 'imaginary Egyptians as they are usually seen on the stage', and although he knew French designers who could provide him with 'Egyptian architecture of great fantasy', he was adamant 'that is not what is needed'. Authenticity could only be found in the pages of the *Description*.[106]

The result certainly impressed the first-night audience. According to one of its members,

> *Aida* was accepted, generally, as an opera *faithful* to its historic import; as one which is, beyond question, among the most *conscientious* works of the century; as a spectacle with splendid and *truthful* scenery, princely costumes and massive music; as history written on the scale, tradition glowing on the canvas. Viewed in this light by the Egyptologist it is *utilitarian* and *instructive*, and it is the first example where poetic license has not been freely indulged by the composer ... [B]ut to Verdi does not belong all the credit of this success. It must be shared with Mariette Bey, a most eminent Egyptologist, who went to Paris by special order of the [Khedive] to oversee the preparation of the costumes. To the *minutest degree* they reproduced the *acknowledged* dress of the ancients. The stage scenery too was prepared with like *fidelity*.[107]

And the Khedive was also impressed, so much so that, in a final truly spectacular twist, he bought the Villa La Spina on the shores of Lake

Maggiore – not far from Verdi's home at Sant' Agata – and had its gardens landscaped into a fantastic version of the *Aida* sets. The lakeside village of Oggebbio eventually became a popular tourist destination known as 'Little Cairo'.[108]

And yet, as I showed earlier, the *Description's* Egypt was a fantasy-Egypt too. Said captures something of this when he argues that the 'projective grandeur' of its plates produced not so much description as ascription:

> As you leaf through the *Description* you know that what you are looking at are drawings, diagrams, paintings of dusty, decrepit and neglected pharaonic sites looking ideal and splendid as if there were no modern Egyptians but only European spectators . . . The most striking pages of the *Description* seem to beseech some very grand actions or personages to fill them, and their emptiness and scale look like opera sets waiting to be populated. Their implied European context is a theater of power and knowledge, while their actual Egyptian setting in the nineteenth century has simply dropped away.[109]

The reverse-archaeology of *Aida* was also directed at a European audience. Said argues that Verdi recognized that 'the opera was first composed and designed for a place that was decidedly not Paris, Milan or Vienna', and suggests that this accounts for some of its incongruities and irresolutions.[110] Against this reading, however, I suggest that the opera was conceived and presented for a place that decidedly *was* Europe. This was clearly true of Verdi's involvement. He was never greatly interested in the Cairo production and did not bother to attend the première on 24 December 1871; he was always much more exercised by its première at La Scala in Milan, and arrived there in early January to begin rehearsals.[111] But it was also true of the Cairo production itself. Not only did it present, as Said says, 'an Orientalized Egypt', but its audience was also largely European. Special steamers ran from the main Mediterranean ports bringing 'amateurs and artists anxious to see the operative sensation of the day', and on the first night, according to one critic,

> The curiosity, the frenzy of the Egyptian public to attend the premier of *Aida* were such that, for a fortnight, all the seats had been brought up, and at the last moment the speculators sold boxes and stalls for their weight in gold. When I say the Egyptian public, I speak especially of the Europeans; for the Arabs, even the rich, do not care for our kind of theatre; they prefer the miaouing of their own chants, the monotonous beatings of their tambourines . . . [and] it is a perfect miracle to see a fez in the theatres of Cairo.[112]

I don't know what the ordinary inhabitants of Cairo made of the production, if they did so at all; but the series of displaced and dispersed European geographies to which it was tied also bound *Aida* to the city in which it was staged. According to one member of the audience,

> The drop curtain was a work of art, representing old Egypt on the right, with decayed temples, pyramids, obelisks and mausoleums, and on the left its new green fields, railroads, telegraphs and modern agriculture. This alone expresses the purpose of *Aida* – to advertise the progressive works of the Khedive.[113]

Those 'progressive works' were inscribed on the landscapes of both the country and the city. The editor of Murray's revised *Handbook for Egypt*, published in 1873, justified his new edition by the changes that had taken place over the previous decade:

> Since the accession of the Khedive, Ismail Pasha, the work of change has been carried on in Egypt at an almost feverish rate of speed. Several hundred miles of railway have been completed and are in full operation. The telegraph wires intersect every part of the country. Many parts of Alexandria and Cairo are so changed that those who saw them only a few years ago would hardly recognise them.[114]

And in the closing passages of his essay, Said opens the door of the Opera House to confront the shimmering, teeming, buzzing city that lay outside. It was a city caught up in the desperate toils of capitalist modernity, and as Said describes the congeries of European merchant bankers, loan corporations and commercial adventures who were involved in the runaway transformation of the Egyptian economy, he also makes it clear that Cairo was indeed in the eye of the storm. Unlike Alexandria, he writes,

> Cairo was an Arab and Islamic city ... Cairo's past did not communicate easily or well with Europe; there were no Hellenistic or Levantine associations, no gentle sea-breezes, no bustling Mediterranean port life. Cairo's massive centrality to Africa, to Islam, to the Arab and Ottoman worlds seemed like an intransigent barrier to European investors, and the hope of making it more accessible and attractive to them surely prompted Ismail to support the city's modernization. This he did essentially by dividing Cairo.[115]

I suspect, too, that the symbolic importance of both the opera and the Opera House derived from the Paris of the second empire. After all, why did the Khedive of Egypt place such a premium on this particular European cultural form? Said does not say, but when Ismail visited the French capital, grand opera was an established imperial institution and the extravagance and exuberance of its productions reflected both the splendour of the imperial court and the sophistication of bourgeois culture. Work on Garnier's new opera house had started in 1862, five years earlier, and Penelope Woolf argues that the new building was intended '. . . to establish an historical orthodoxy the radical modernity of its age'. Thus '. . . the Opera House joined banks, market halls and currency and commodity exchanges as an indicator of opulence and prosperity'. It lay at the centre of – and in a sense was the crowning glory of – the Hausmannization of Paris.[116] Ismail surely could not have overlooked the intense public interest in the Opéra nor its symbolic importance in the new urban landscape.

Transposing these iconographies to Cairo, then in much the same way that the production of *Aida* was intended to mark the threshold of a radically modern Egypt, so the Opera House marked the boundary of the new city: and both were supposed to mirror the mastery of the Khedive. It was fortunate that the Cairo Opera House was not modelled on Garnier's Opéra, which did not open until 1875. Instead it was modelled on La Scala, designed by two Italian architects and completed in a mere five months, just in time for the performance of *Rigoletto* that celebrated the opening of the Suez Canal.[117] As Said remarks, it turned its back on the traditional eastern city to face the modern western city: 'Behind the Opera House lay the teeming quarters of Muski, Sayida Zeinab, 'Ataba al Khadra, held back the Opera House's imposing size and European authority.' In Said's view, clearly, what counted was not so much the reflected glory of the Khedive as the refracted power of Europe. Hence he concludes that

> *Aida*'s Egyptican identity was part of the city's European façade, its simplicity and rigor inscribed on those imaginary walls dividing the colonial city's native from its imperial quarters. *Aida* is an aesthetic of separation . . . [and] for most of Egypt was an imperial *article de luxe* purchased by credit for a tiny clientele whose entertainment was incidental to their real purposes . . . [It was] an imperial spectacle designed to alienate and impress an almost exclusively European audience.[118]

But it was, I think, more than a façade; it was also part of a much deeper process of cultural appropriation. Just as the Opera House, the conventions of operatic form and the reverse-archaeology of *Aida* staged a spectacular appropriation of Egyptian history, so too the familiar sites of the modern city – the western hotels, banks, booksellers, telegraph offices and, from 1873, the office of Thomas Cook in the grounds of Shepheard's

Hotel – were platforms from which western visitors could issue out to inspect the exotic sights of the traditional city.[119]

V Learning from Luxor

I began by recalling Said's attempt to 'rethink geography', and I hope to have shown that his imaginative geographies are indeed different from the mental maps and images recovered by our own disciplinary traditions that have been concerned with behavioural geographies and environmental perceptions. His are profoundly ideological landscapes whose representations of space are entangled with relations of power. They cannot be counterposed to a 'more true and more real' geography whose objective fixity is disclosed through the technologies of science – for example – because those technologies are always and everywhere techno*cultures*: they are embedded in distinctive regimes (and geographies) of truth too, and their representations are also partial and situated. As Donna Haraway has reminded us, however, situated knowledge is not a barrier to understanding but rather its very condition. In much the same way, and for much the same reason, mapping imaginative geographies can be said to constitute a 'cartography of identities' (above: 302), provided it is conducted as a process of negotiated understanding and not an exercise in surveillance and confinement, because there is a sense in which 'knowing oneself' is, in part, a matter of 'mapping where one stands'. Certainly there is in Said's work, from both the West Bank and the Left Bank, an intimate connection between the spatialities of these imaginative geographies and the precarious and partial formation of identity.[120]

But I do have reservations about the way in which that connection is usually construed. In the first place, neither 'knowing oneself' nor 'mapping where one stands' imply that space is rendered transparent. Imaginative geographies cannot be understood as the free and fully coherent projections of all-knowing subjects. It is necessary to find ways to interrogate the unconscious and to explore the multiple spatialities inscribed within the geographical imaginary; these inclusions create analytical openings for the contradictions that are contained within (often contained by) dominant constellations of power, knowledge and geography. Said's contrapuntal reading needs to register these dissonances and 'atonalities' more explicitly, but there is nothing in his critical practice that excludes them. In so far as such a project will have to pay particular attention to the ways in which imaginative geographies congeal into a socially constituted geographical imaginary, it will require a careful working out of the tensions between psychoanalytic theory on the one side and social theory on the other. As I have indicated, however, Said's interest in psychoanalytic theory is strangely attentuated: the allusions to a 'manifest' and 'latent' Orientalism, the appeals to Deleuze and Guattari in *Culture and imperialism*, are wonderfully suggestive but radically undeveloped. It is symptomatic, I think, that when Fanon is invoked in *Culture and imperialism*, it is always his celebrated account of *The Wretched of the Earth* that occupies centre-stage, written in ten short weeks after he learnt he had leukaemia, while *Black Sin, White Masks* – of which Bhabha makes so much – slips into the wings and endnotes where Said mutters darkly about his 'early psychologizing style'.[121] And so I start to wonder about Said's reservations . . .

Clip one: on the desk, surrounded by other antiquities and figurines, is a statue of the Egyptian god Amon-Re from the city of Thebes, the site of Luxor and Karnak; hanging over the couch in the consulting-room is a colour print of the temples at Abu-Simbel, dedicated to Ramses and associated with Amon-Re. The apartment belongs to Sigmund Freud. He often referred to his classic *The interpretation of dreams*, which was first published in 1900, as his 'Egyptian dream-book', and he was clearly fascinated by the art and archaeology of ancient Egypt. What are we to make of his obsession? Most obviously, archaeology provided Freud with a linguistic model for psychoanalytic practice. 'Amon-Re' means 'the hidden one', for example, and – although he expressed reservations about the analogy from time to time – Freud seems to have thought of psycho-analysis as a process of quasi-archaeological excavation and disclosure. He insisted on the continued presence of the past in the present, at once unex-pected and unacknowledged, and drew upon archaeology for a stratified and spatial figure of the psyche in which those things hidden and concealed from 'surface consciousness' can be brought into the light.[122] But the appeal to archaeology also allowed Freud to invoke a remarkably successful and highly popular science as a cover, or at any rate a theatrical guise, for the otherwise suspect and even discreditable science of psychoanalysis.[123] There is no doubting the spectacular public success of archaeology in the early decades of the twentieth century. One only has to think of the hoopla surrounding Carter, Caernarvon and the discovery of Tutankhamun's tomb in 1922. But this was also the moment at which archaeology reached its imperialist climax, when western archaeologists and adventurers fought over the spoils of the valley of the Nile in the full glare of publicity.[124] What, then, was the impact of all this on Freud's archaeological metaphor and, by extension, on his thought and practice? He once confided to a close friend that he was '. . . not at all a man of science, not an observer, not an experimenter, not a thinker'. 'I am by temperament,' he wrote, 'nothing but a conquistador . . .'[125]

It would, of course, be absurd to use one casual remark to claim that psychoanalytic theory is indelibly and inescapably marked with colonialist trappings. In its various post-Freudian forms it can, I think, help to elucidate the connections between imaginative geographies and the formation of identities and, as I have implied, the work of Fanon and Bhabha shows that these ideas can be enlisted in struggles *against* colonialism and imperialism.

Perhaps, too, psychoanalytic theory should be examined for its colonial signs, because there is at least the possibility that its repressed past might have entered, unexpected and unacknowledged, into its critical present.

In the second place, and my second set of reservations, the production – the inscription and contestation – of imaginative geographies cannot be confined to the realm of high culture. It is of course important to demonstrate that high culture is not immune from the corruptions of colonialism, and Said does this with exemplary tact and patience, but the connections between spatiality and identity are continuous with the production of everyday life in *all* its particulars. To insist on this is not to agree with those critics who charge Said with abstraction or textualism. As I have tried to show, the spatialities of Orientalism were – *are* – abstractions, and the canonical texts in which they are articulated are marked by corporealities and physicalities. In short, there *is* a materialism in Said's work; he himself notes that we live in a world not only of commodities but also of representations, and these are at once abstractions *and* densely concrete fabrications.[126] In the late twentieth century, however, commodities and representations have become interlaced in ever more complex forms, and what Said makes much less of is the way in which these connective tissues have challenged – if not altogether dissolved – the (other) 'great divide' between high and popular culture.

Clip two: the pyramid of Luxor rises 350 feet from the desert floor into the shimmering blue sky; its entrance is guarded by a great Sphinx; beyond, an obelisk from the Temple of Karnak towers into space; inside the tomb of Tutankhamun a golden sarcophagus, amulets, masks and scarabs gleam in the darkness; outside, boats full of tourists cruise along the River Nile – making their way from the lobby to the elevators and the casino. The desert is in fact the Mojave, the obelisk and the tomb are replicas, the Nile is artificial, and the Luxor is one of the newest resort hotels to open on the Strip in Las Vegas. This 'entertainment megastore' is owned by Circus Circus Enterprises, and one of the press releases for its opening was headlined 'Ancient civilization discovered in Las Vegas'. The Luxor was conceived 'as a vast archaeological dig, where the mysteries of ancient Egypt are revealed as though in a state of 'excavation' throughout the interior'. The tomb of Tutankhamun is a reproduction of the site 'as it was found' by Carter and Caernarvon; the measurements of the rooms are 'exact', the artifacts have been reproduced using 'the same materials and methods' as the original artisans, and each is 'meticulously positioned according to the records maintained by Carter'. The intention of its promoters, so they insist, 'is a homage – not an exploitation'. In the casino, where homage of a rather different kind is the order of the day, ancient Egypt is 'brought to life' with reproductions from the temples of Luxor and Karnak. Advertised as 'the next wonder of the world', 'the copywriter had in mind, a pace where 'history is about to be rewritten', the resort boasts 'inclinators' that travel up the inside slopes of the pyramid to the oversized guest rooms; adventure entertainments conceived by the special-effects designer for *Blade Runner* that promise to 'race you through time and astound your sense of reality'; galleries and boutiques with authenticated Egyptian antiquities; theme restaurants that offer not only 'feasts fit for a pharaoh' but also 'a Kosher-style deli located on the River Nile', presumably to tempt the Israelites to return; and, finally, a floor show devised by Peter Jackson, including a team of acrobats wrapped in linen shrouds called 'The Flying Mummies', 'extravagant dance numbers, belly dancing, nail-biting stunts and original special effects'. The production tells the story of 'a long-lost pharoah whose resting place is desecrated by a band of thieves'.[127]

I don't intend to mimic Frederic Jameson's odyssey in the Bonaventure Hotel in Los Angeles, but situating the Luxor Las Vegas within an itinerary inaugurated by the *Description de l'Égypte* and continued through the production of Verdi's *Aida* shows how the Luxor's imaginative geography stages the interpenetration of a colonial past and a neocolonial present. There is, perhaps, a knowing irony in the promotional copy, a parodic re-presentation of the colonial connections installed within the world-as-exhibition between visualization and appropriation, but these fantasy-architectures none the less provide a physical site at which particular spatialities are captured, displaced and hollowed out, and by means of which identities are fashioned, negotiated and contested.

It should be clear from my two clips – I use the filmic phrase deliberately – that the historical geographies I have described in the preceding pages open passages into our own present. The critical reading of late twentieth-century cultural geographies cannot turn its back on the past. Two hundred years ago, before the French army engaged the Mamelukes at the battle of the pyramids, Napoleon dismissed his immediate entourage with the instruction to 'think that from the heights of these monuments, forty centuries are watching us'. Robert Young's remarkable account of *White mythologies: writing history and the west* has on its cover a photograph of the former Egyptian President, Anwar Sadat, gazing back up at the pyramids. There are lessons there for the writing of geography too, and for our own productions of imaginative geographies.

Acknowledgements

The *Progress in Human Geography* lecture, delivered at the Annual Meeting of the Association of American Geographers, Chicago, March 1995. I am very grateful to Alison Blunt, Noel Castree, Dan Clayton, Felix Driver, Cole Harris, Jennifer Hyndman, David Ley, Lynn Stewart, Joan Schwartz, Joanne Sharp and Bruce Willems-Braun for their helpful comments. This paper forms part of a research project funded by the Social Science and Humanities Research Council of Canada.

Notes

1 Said, E. 1994: Edward Said's *Culture and imperialism: A symposium. Social Text* 40, 21.
2 Fox, R. 1992: East of Said. In Sprinkler, M., editor, *Edward Said: a Critical Reader*. Oxford: Blackwell, 144–56 the quotation is from p. 144; Gilroy, P. 1993: Travelling theorist. *New Statesman and Society* 12 February, 46–47. Gilroy's implicit assimilation of Said's work to postmodernism, which he shares with several other commentators, is misleading. While Said makes no secret of the crisis of modernism – and in particular the dislocations and displacements brought about by its imbrications with imperialism: see Said, E. 1993: *Culture and imperialism*. New York: Alfred Knopf, 186–90 – this does not make him a post-modernist. In fact, Said reserves some of his most trenchant criticism for the postmodern assault on metanarrative. 'The purpose of the intellectual's activity is to advance human freedom and knowledge', he insists, and when Lyotard and his followers dismiss 'grand narratives of emancipation and enlighten-ment' then, so Said claims, they are 'admitting their own lazy incapacities' rather than acknowledging the politico-intellectual challenges and opportuni-ties that remain 'despite postmodernism': Said, E. 1994: *Representations of the Intellectual*. London: Vintage, 13–14.
3 Said, E. 1995: *Orientalism*. Harmondsworth: Penguin Books (1st edn, 1978), 215.
4 But see Driver, F. 1992: Geography's empire: histories of geographical knowl-edge. *Environment and Planning D: Society and Space* 10, 23–40; Rogers, A. 1992: The boundaries of reason: the world, the homeland and Edward Said. *Environment and Planning D: Society and Space* 10, 511–26.
5 Said, *Orientalism*, 215–18; Driver, Geography's empire; Livingstone, D. 1993: A 'sternly practical' pursuit: geography, race and empire. In *The Geographical Tradition: Episodes in the History of a Contested Enterprise*. Oxford: Blackwell, 216–59; Godlewska, A. and Smith, N. editors, 1994: *Geography and Empire*. Oxford and Cambridge, MA: Blackwell.
6 Smith, N. 1992: Real wars, theory wars. *Progress in Human Geography* 16, 257–71.
7 These estimates are inevitably contentious: United Nations, Palestinian and Israeli sources all differ. See Hadawi, S. 1989: *Bitter Harvest: a Modern History of Palestine*. New York: Olive Branch Press; Tessler, M. 1994: *A History of the Israeli-Palestinian Conflict*. Bloomington IN: Indiana University Press.
8 Said, E. 1966: *Joseph Conrad and the Fiction of Autobiography*. Cambridge, MA: Harvard University Press.
9 Said, *Orientalism*, 25.
10 Said describes the productive force of 'peripheral, off-center work that grad-ually enters the West and then requires acknowledgement' as 'the voyage in' (*Culture and imperialism*, 216, 239). Cf. Robbins, B. 1994: Secularism, elitism, progress and other transgressions: on Edward Said's 'voyage in'. *Social Text* 40, 25–37.
11 Said, E. 1986: *After the Last Sky: Palestinian Lives*. New York: Pantheon Books, 115.
12 Said, *After the Last Sky*, 14; Wicke, J. and Sprinkler, M. 1992: Interview with Edward Said. In Sprinker, editor, *Said*, 221–64; the quotation is from p. 222, where Said sketches his own 'imaginative geography' of Cairo. There are important bonds between imaginative geographies and imagined communities, as Anderson now recognizes, but his discussions needs to be reworked to take into account the particular predicaments of dispersed and displaced national communities: Anderson, B. 1991: *Imagined Communities: Reflections on the Origin and Spread of Nationalism*. London: Verso (2nd edn).

13 For a larger discussion of the culture of dispossession and exile and its imbrications with geography – though one that, unaccountably, makes no mention of Said – see Parmenter, B. M. 1994: *Giving Voice to Stones: Place and Identity in Palestinian Literature.* Austin, TX: University of Texas Press. The poems I cite here are 'We travel like other people' and 'The earth is closing in on us', reprinted in Darwish, M., 1984: *Victims of a Map.* London: Al Saqi Books. The other quotations are from Said, *After the Last Sky*, 6, 17; see also Rushdie, S. 1993: On Palestinian identity: an interview with Edward Said, In *Imaginary Homelands: Essays and Criticism 1981–1991.* Harmondsworth: Penguin Books, 166–84.

14 Said, *After the Last Sky*, 19–21, 38. Rogers, *Boundaries*, 521, suggests that Said's language is redolent of the bleakness of the postmodern condition, but he also urges that the experience is particular not universal – '"We" are not all Palestinians' – and argues that, for Said, these dislocations and dispersions are 'to be fought against rather than embraced'. Exactly so.

15 Said, *After the Last Sky*, 12.

16 Dimbleby, J. 1979: *The Palestinians.* London: Quartet Books.

17 Said, *After the Last Sky*, 28.

18 In Chicago, Edward Soja urged me to reflect on this question as an antidote to the 'historicism' he saw underwriting my presentation. Historicism usually means a view of history as a telos, however, and I don't think this is a proper characterization of my argument (or Said's project). What Soja was cautioning against was the privileging of history over geography, but I believe that the power of Said's work derives from his remarkable ability to think the two *together.* He certainly does not diminish the importance of historical inquiry. When he concluded that 'Palestine does not exist, except as a memory or, more importantly as an idea, a political and human experience, and an act of sustained popular will', for example, he was surely drawing attention to the continued salience of the past – imaginatively, culturally, critically – for contemporary politics. See Said, E. 1979: *The Question of Palestine.* New York: Times Books, 5. Elsewhere Said does accentuate the importance of spatial counterpoint rather than temporal sequence, but this is an argument about affinity, influence and connection that scarcely diminishes the significance of historicity: see *Culture and imperialism*, 81.

19 Said, E. 1994: Return to Palestine – Israel. In *The Politics of Dispossession: the Struggle for Palestinian Self-determination 1969–1994.* New York: Pantheon, 175–99.

20 Said, E. 1993: An interview with Edward Said. *Boundary* 2, 1–25; the quotation is from pp. 12–13.

21 Bhabha, H. 1994: *The Location of Culture.* London: Routledge, 99–100.

22 Said provides his own narrative of these engagements in *Politics of dispossession*, i–xxxiv; he was a member of the Palestine National Council from 1977 to 1991. For a review and commentary, see Hovsepian, N. 1992: Connections with Palestine. In Sprinker, editor, *Said*, 5–18.

23 Hulme, P. 1990: Subversive archipelagos: colonial discourse and the break-up of continental theory. *Dispositio* 15, 1–23. By 'continental theory' Hulme means a postwar discursive terrain triangulated by Marx, Nietzsche and Freud that is distinctly European in both its origins and its presumptive universality.

24 Smith, N. 1994: Geography, empire and social theory. *Progress in Human Geography* 18, 491–500; the quotation is from pp. 492–93.

25 Said, *Orientalism*, 80–87; *Culture and imperialism*, 33–35.

26 Said, E. 1993: The empire at work: Verdi's *Aida.* In *Culture and imperialism*, 111–32.

27 See Gregory, D. 1995: Between the book and the lamp: imaginative geographies of Egypt, 1849–50. *Transactions, Institute of British Geographers* 20, 29–57.

28 Said, E. 1991: *Musical Elaboration*, New York: Columbia University Press, 17; *Culture and imperialism*, 109.

29 On south Asia, see Breckenridge, C. and van der Veer, P., editors, 1993: *Orientalism and the Postcolonial Predicament: Perspectives on South Asia*. Philadelphia, PA: University of Pennsylvania Press; on 'Orientalism and South America, see the cautions of Mary-Louise Pratt in 'Symposium', 4–6.

30 Lowe, L. 1991: Discourse and heterogeneity: situating orientalism. In *Critical Terrains: French and British Orientalisms*. Ithaca, NY: Cornell University Press, 1–29. There are multiple ways in which 'contradiction' can be conceptualized, as the vocabularies of social theory and psychoanalytic theory suggest, and it is not my intention to privilege any one of them here.

31 Barrell, J. 1991: *The Infection of Thomas de Quincey: a Psychopathology of Imperialism*. New Haven, CT: Yale University Press, 16.

32 Lant, A. 1992: The curse of the pharoah, or how cinema contracted Egyptomania. *October* 59, 86–112; the quotation is from p. 98.

33 Said, E. 1986: Foucault and the imagination of power. In, Hoy, D. C., editor, *Foucault: a Critical Reader*. Oxford: Blackwell, 149–55. I do not, of course, claim that Foucault is the only source but, for reasons that will become clear, I disagree with Brennan's otherwise engaging appreciation when he treats Foucault as 'a minor player' in *Orientalism*: Brennan, T. 1992: places of mind, occupied lands: Edward Said and philology. In Sprinker, editor, *Said*, 74–95.

34 For the criticisms of Said, see Clifford, J. 1988: On *Orientalism*. In *The Predicament of Culture: Twentieth-century Ethnography, Literature and Art*. Cambridge, MA: Harvard University Press, 255–76; Young, R. 1991: Disorienting orientalism. In *White Mythologies: Writing History and the West*. London: Routledge, 119–40. For a critical discussion of Foucault's ethics of the self, see McNay, L. 1992: *Foucault and Feminism: Power, Gender and the Self*. Cambridge: Polity Press.

35 Ahmad, A. 1994: *Orientalism* and after: ambivalence and metropolitan location in the work of Edward Said. In *In Theory: Classes, Nations, Literatures*. London: Verso, 159–219; see especially pp. 165–66. On Ahmad's fideism – his 'rage against critics who do not declare themselves Marxists' – see Parry, B. 1993: A critique mishandled. *Social Text* 35, 121–33; Levinson, M. 1993: News from nowhere: the discontents of Aijaz Ahmad. *Public Culture* 6, 97–131.

36 Said, *Orientalism*, 57, 122; see also pp. 42–43, 76, 87, 120, 201. For premodern genealogies of Orientalism, see Hentsch, T. 1992: *Imagining the Middle East*. Montreal: Black Rose Books.

37 Said, *Orientalism*, 54.

38 Said, *Orientalism*, 55; Bachelard, G. 1969: *The Poetics of Space*. Boston, MA: Beacon Press.

39 Said, E. 1995: East isn't east: the impending end of the age of Orientalism. *The Times Literary Supplement* 3 February, emphasis added.

40 Geyer-Ryan, H. 1994: Space, gender and national identity. In *Fables of Desire: Studies in the Ethics of Art and Gender*. Cambridge: Polity Press, 155–63. I have shifted the weight of Geyer-Ryna's argument, which is concerned less with the exiled – 'though they bear the cruellest burden' – and more with those 'who are confronted with immigration and experience the arbitrariness and relativity of the symbolic order through their encounter with the other'.

41 Said's distinction between 'manifest' and 'latent' Orientalism trembles on the edge of psychoanalytic theory, but this remains undeveloped in his work: for an elaboration, see Bhabha, *Location of Culture*, 71–75.

42 The opposition is not an inevitable one, of course, and the work of Gerard O'Tuathail in particular suggests several ways in which a critical geopolitics might be informed by psychoanalytic theory.

43   Said, E. 1984: Criticism between culture and system. In *The World, the Text and the Critic*. London: Faber & Faber, 178–225; the quotations are from pp. 219–22. There are important filiations between Bachelard and Foucault, but these reside in Foucault's archaeology rather than the genealogy that Said is attempting here.

44   Said, E. 1988: Michel Foucault, 1926–1984. In Arac, J., editor, *After Foucault*. New Brunswick, NJ: Rutgers University Press, 1–11; the quotations are from p. 5.

45   Said, *Orientalism*, 186–88, 207–208.

46   Said, *Culture and imperialism*, xxv.

47   Said, Criticism, 222.

48   Foucault, M. 1979: *Discipline and Punish*. New York: Vintage Books, 140–41.

49   Martin, A. 1988: *The Knowledge of Ignorance*. Cambridge: Cambridge University Press, 81.

50   Said, *Orientalism*, 80–95. On the surveys in Egypt, see Laurens, H., Gillispie, C., Golvin, J.-C. and Traunecker, C. 1989; *L'Expédition d'Égypte 1798–1801*. Paris: Armand Colin; on the publishing project in Paris, see Albin, M. 1980; Napoleon's *Description de l'Égypte*: problems of corporate authorship. *Publishing History* 8, 65–85.

51   Said, Criticism, 223; *Orientalism*, 162,-85.

52   Rajchman, J. 1991: Foucault's art of seeing. In *Philosophical Events: Essays of the 80s*. New York: Columbia University Press, 68–102; see also Flynn, T. 1993: Foucault and the eclipse of vision. In Levin, D. M., editor *Modernity and the Hegemony of Vision*. Berkeley, CA: University of California Press, 273–86.

53   Rose, G. 1993: Looking at landscape: the uneasy pleasures of power. In *Feminism and Geography: the Limits of Geographical Knowledge*. Cambridge: Polity Press, 86–112; for a discussion of masculinism and Orientalism, see Kabbani, R. 1986: *Imperial Fictions: Europe's Myths of Orient*. London: Macmillan.

54   Boone, J. 1995: Vacation cruises, or the homoerotics of Orientalism, homophobia, masochism. *Diacritics* 24, 151–68.

55   See, for example, Gillies, J. 1994: Theatres of the world. In *Shakespeare and the Geography of Difference*. Cambridge: Cambridge University Press, 70–98; Lestringant, F. 1994: Ancient lessons; a bookish Orient. In *Mapping the Renaissance World: the Geographical Imagination in the Age of Discovery*. Cambridge: Polity Press, 37–52.

56   Said, *Orientalism*, 63, 103, 127, 166. On the world-as-exhibition, see Mitchell, T. 1988: *Colonising Egypt*. Cambridge: Cambridge University Press; Mitchell, T. 1992; Orientalism and the exhibitionary order. In Dirks, N., editor, *Colonialism and Culture*. Ann Arbor, MI: University of Michigan Press, 289–317.

57   There is now a rich literature on Orientalism in art history: see, for example, Nochlin, L. 1989: The imaginary Orient. In *The Politics of Vision: Essays on Nineteenth-century Art and Society*. New York: Harper & Row, 33–59. But I suspect that nineteenth-century photography is even more important to Said's argument for at least two reasons. First, photography encapsulated the ideology of definitive truth in ways that drew together imperialism and the discipline of detail. As Solomon-Godeau notes, 'the mid-nineteenth century was the great period of taxonomies, inventories and physiologies, and photography was understood to be the agent par excellence for listing, knowing and possessing, as it were, the things of the world'. Indeed, one of the best-known calotypists of Egypt, Félix Teynard, advertised his systematic survey of Egypt's monuments as a complement to the *Description de l'Égypte*. Second, for purely technical reasons, calotypists focused on immobile scenes – especially the exterior of buildings – and so depicted them as 'essentially vacant spaces': It is reasonable to assume that such photographic documentation showing so much of

the world to be empty was unconsciously assimilated to the justifications for an expanding empire.' See Solomon Godeau, A. 1991: A photographer in Jersusalem, 1855: Auguste Salzmann and his times. In *Photography at the Dock: Essays on Photographic History, Institutions and Practices*. Minneapolis, MN: University of Minnesota Press, 150–68; the quotations are from pp. 155, 159; see also Howe, K. 1994: *Excursions along the Nile: the Photographic Discovery of Ancient Egypt*. Santa Barbara, CA: Santa Barbara Museum of Art.

58   The frontispiece and its accompanying explication are reprinted in Gillispie, C. C. and Dewachter, M., editors 1987: *Monuments of Egypt: the Napoleonic Edition*. Princeton, NJ: Princeton Architectural Press.

59   This is a commonplace of colonial discourse: see Pratt, M. L. 1992: *Imperial Eyes: Transculturation and Travel Writing*. London: Routledge. But this discourse of negotiation required special strategies in Egypt, since it plainly could not proceed through the usual evacuation of the inhabitants as a 'people without history': the whole purpose was to reclaim and recover that (ancient) history as both prolegomenon to and *part of* European history. This eventually licensed a racialization of the past, in which the people of ancient Egypt were assumed to be white – and hence proto-European – unlike most of the contemporary inhabitants of Egypt.

60   Cosgrove, D. 1985: Prospect, perspective and the evolution of the landscape idea. *Transactions, Institute of British Geographers* 10, 45–62.

61   Prochaska, D. 1994: Art of colonialism, colonialism of art: the *Description de l'Égypte* (1809–1828). *L'Esprit créateur* 34, 69–912; on 'the organization of the view', see Mitchell, *Colonising Egypt*, 12. For a wonderfully detailed exposition of the geographical apparatus involved in the production of the *Description*, see Godlewska, A. 1988: The Napoleonic survey of Egypt. *Cartographica* 25, monograph 38–39; Godlewska, A. 1994: Napoleon's geographers (1797–1815): imperialist soldiers of modernity. In Godlewska and Smith, editors, *Geography and Empire*.

62   Laurens *et al.*, *L'expédition*, 352–53; Said, *Orientalism*, 88, speaks to much the same point when he identifies 'highly stylized simulacra, elaborately wrought imitations' produced by subsequent European writers who conceived of the Orient as 'a kind of womb out of which they were brought forth'.

63   Godlewska, A. 1995: Map, text and image: representing the mentality of enlightened conquerors. *Transactions, Institute of British Geographers* 20, 5–28.

64   Said, *Culture and imperialism*, 118.

65   Philo, C. 1992: Foucault's geography. *Environment and Planning D: Society and Space* 10, 137–61.

66   On envisioning nineteenth-century Paris, see Asendorf, C. 1993: *Batteries of Life: on the History of Things and their Perception in Modernity*. Berkeley, CA: University of California Press, 46–47; Clark, T.J. 1984: The view from Notre Dame. In *The Painting of Modern Life: Paris in the Art of Manet and his Followers*. Princeton, NJ: Princeton University Press; Green, N. 1990: *The Spectacle of Nature: Landscape and Bourgeois Culture in Nineteenth-century France*. Manchester: Manchester University Press, 29–31; Prendergast, C. 1992: *Paris and the Nineteenth Century*. Oxford: Blackwell. The parallels I have in mind are touched upon in Behdad, A. 1994: Notes on notes, or with Flaubert in Paris, Egypt. In *Belated travelers: Orientalism in the age of colonial dissolution*. Durham, NC: Duke University Press, 53–72 and Shields, R. 1994: Fancy footwork: Walter Benjamin's notes on *flânerie*. In Tester, K., *The Flâneur*. London: Routledge, 61–80.

67   Said, *Culture and imperialism*, 99–100.

68   Said, *Politics of dispossession*, 416–17; Said, 1995: Symbols versus substance a year after the Declaration of Principles: an interview with Edward Said. *Journal of Palestine Studies* 24, 60–72.

69 Cf. the epigraph from Marx's *Eighteenth Brumaire of Louis Bonaparte* that Said uses to head *Orientalism*: 'They cannot represent themselves; they must be represented.'

70 Said, *After the Last Sky*, 116.

71 Colley, L. 1993: The imperial embrace. *Yale Review* 81, 92–98; the quotation is from p. 92, emphasis added.

72 Said, *Culture and imperialism*, 63.

73 Said, *Culture and imperialism*, 239.

74 Silverman, D. 1977: The 1889 exhibition: the crisis of bourgeois individualism. *Oppositions* 8, 70–91. The exhibition also included the first Rue du Caire: see Mitchell, *Colonising Egypt*, 1–4; Çelik, Z. 1992: *Displaying the Orient: Architecture of Islam at Nineteenth-century World's Fairs*. Berkeley, CA: University of California Press. In fact, Maupassant entertained in his apartment 'a troop of Arab dancers, acrobats and musicians' who were in Paris for the exhibition: Steegmuller, F. 1950: *Maupassant*. London: Collins, 279.

75 *Egypt and How to See it*. Paris: Hachette, London: Ballantyne, 131.

76 Colley, Imperial embrace, 94; Said, *Culture and imperialism*, 74, 189. Said is, of course, talking about his culture in general and the novel in particular. Two other cultural productions are of special importance for the colonial construction and circulation of imaginative geographies. That there are intimate connection between travel-writing and colonialism is shown by a stream of illuminating studies: see, for example, Blunt, A. 1994: *Travel, Gender and Imperialism: Mary Kingsley and West Africa*. New York: Guilford Press; Mills, S. 1991: *Discourses of Difference: an Analysis of Women's Travel Writing and Colonialism*. London: Routledge; Pratt, *Imperial Eyes*; Spurr, D. 1993: *The Rhetoric of Empire: Colonial Discourse in Journalism, Travel Writing and Imperial Administration*. Durham, NC: Duke University Press. The connections between the rise of cinema and the age of imperialism have attracted less attention, but see Browne, N. 1989: Orientalism as an ideological form: American film theory in the silent period. *Wide Angle* 11, 23–31; Lant, Curse of the pharoah; Shohat, E. and Stam, R. 1994: *Unthinking Eurocentrism: Multi-culturalism and the Media*. London: Routledge.

77 Said, *Culture and imperialism*, 82–83.

78 Said, *Culture and imperialism*, 84.

79 Cohen, M. 1993: *Le diable à Paris*: Benjamin's phantasmagoria. In *Profane Illumination: Walter Benjamin and the Paris of Surrealist Revolution*. Berkeley, CA: University of California Press, 217–59.

80 Polan, D. 1994: Art, society and 'contrapuntal criticism': a review of Edward Said's *Culture and imperialism*. *Clio* 24, 69–79; the quotations are from: 73, 75. Polan's appeal to 'the needs of imperial ideology' is more functionalist than Said's argument warrants, but the essential point stands. In counterpoint to Said, Behdad has argued for what he calls an 'anamnesiac reading' of colonial discourse that 'unmasks what the object holds back and exposes the violence it represses in its consciousness'; but in so far as this critical strategy privileges historicity it obscures the *geopolitical* violence that is Said's main concern. See Behdad, *Belated Orientalism*, 8.

81 Cf. Said, *Culture and imperialism*, 278; see also Young, *White mythologies*. Nevertheless, Benita Parry describes Said as 'a secret sharer in that socialist project which nourishes hope in the possibility of human agency effecting a transfigured secular future from which exploitation and coercion have been erased'. I am not sure he would see it quite like that, but Said's recent writings have been distinguished by a (qualified) recuperation of the enlightenment project and an affirmation of its values of truth, reason and emancipation that distances him from postmodernism, post-Marxism and probably most of what

passes for 'postcolonialism' in the metropolitan academy. See Parry, B. 1993: Imagining empire: from *Mansfield Park* to Antigua. *New Formations* 20, 181–88; Norris, C. 1994: *Truth and the Ethics of Criticism*. Manchester: Manchester University Press, 67–69, 110–12.

82   Said, *Culture and imperialism*, 14. Said's criticism is itself oddly displaced, given the importance of Williams's Welsh roots, but the wider claim surely stands. See also Viswanathan, G. 1993: Raymond Williams and British colonialism: the limits of metropolitan theory. In Dworkin, D. and Roman L., editors, *Views Beyond the Border Country: Raymond Williams and Cultural Politics*. London: Routledge, 217–30; Radhakrishnan, R., 1993; Cultural theory and the politics of location. In Dworkin and Roman, editors, *Views*, 275–94.

83   Said, E. 1984: Reflections on American 'left' literary criticism. In *The World, the Text and the Critic*. London: Faber & Faber, 158–77; the quotation is from p. 171.

84   Said, *Culture and imperialism*, 48, 59.

85   Said, *Culture and imperialism*, 49. Edward Soja makes the same point about 'Gramsci's geography': 'Gramsci is so much more spatial than the other founding fathers of Western Marxism'. Soja, E. 1989: *Postmodern Geographies: The Reassertion of Space in Critical and Social Theory*. London: Verso, 46n.

86   Said, *Culture and imperialism*, 35. Said subsequently argues that the importance of these conjunctions and displacements has been heightened at the end of the twentieth century. In doing so, he develops a contrast between the 'unhoused and decentred' politico-intellectual formulations of Virilio, Delleuze and Guattari and others – which propose a radical break between past and present – and the condition of countless refugees, migrants and exiles whose predicament continues to articulate 'the tensions, irresolutions and contradictions in the overlapping territories shown on the cultural map of imperialism' (p. 332).

87   Said, *Culture and imperialism*, 50.

88   Said, *Culture and imperialism*, 195; Parry's reservations will be found in her 'Imagining empire', 182–83 and in her 'Overlapping territories, intertwined histories: Edward Said's postcolonial cosmopolitanism', in Sprinker, editor, *Said*, 19–45.

89   Said, *Culture and imperialism*, 266, 278. Said makes much more of the debt to Gramsci in his 'Foreword' to Guha R. and Spivak G. C., editors, 1988: *Selected Subaltern Studies*. New York: Oxford University Press, v–x. The project of subaltern studies is of particular relevance to any consideration of Said's work because it has become a touchstone in debates surrounding the relations among western humanism, poststructuralism and postcolonialism: I have provided an outline of that discussion in Gregory, D. 1994: *Geographical Imaginations*. Oxford and Cambridge, MA: Blackwell, 183–93.

90   Said, *Culture and imperialism*, 51, 318.

91   Said, Interview, 2–3.

92   Most commentators agree with me. There are dissenters, however, most notably the historian John Mackenzie. But I think he comprehensively misrepresents Said's argument: see Mackenzie, J. 1993: Occidentalism, counterpoint and counter-polemic. *Journal of Historical Geography* 19, 339–44; and Mackenzie, J. 1994: Edward Said and the historians. *Nineteenth-century Contexts* 18, 9–25.

93   See Adorno, T. 1993: Bourgeois opera. In Levin, D. editor, *Opera Through Other Eyes*. Stanford, CA: Stanford University Press, 25–43. Opera has found a wider audience in late twentieth-century Europe and North America, as much through recordings as productions; and on my way to Chicago I learnt that Elton John and Tim Rice are currently collaborating on a version of *Aida* that Disney intends to bring to Broadway: see *Time*, 13 March 1995.

94 Mariette to Du Locle, in Busch, H. 1978: *Verdi's* Aida: *the History of an Opera in Letters and Documents*. Minneapolis, MN: University of Minnesota Press, 11.

95 Mariette's part in the storyline is a matter of some controversy. He claimed to have collected material for the short story on which *Aida* was based during an archaeological trip through Upper Egypt in 1866 (whose main purpose was to collect artifacts for the Paris exposition). His son Edouard later claimed that he had drafted the first version of the story, but most critics think this improbable. Mariette was quite explicit: '*Aida* is, in effect, a product of my work,' he declared. 'I am the one who convinced the Viceroy to order its presentation; *Aida* is, in a word, a creation of my brain': Busch, *Verdi's* Aida, 186. Matters are further complicated because, much later, Verdi made light of Mariette's contribution, but that can probably be explained as Verdi's reaction to a threatened copyright action. More problematic is the suggestion that the outline was the work of Temistocle Solera, who had provided the libretto for *Nabucco* – there are similarities between the two operas – and who had organized the celebrations for the opening of the Suez Canal: 'Verdi's refusal to be reconciled with Solera, over a period for more than twenty years, would likely have caused problems had he been told that the author of *Aida* was Solera. As for Solera's failure to claim authorship, he could not have done so without ruining Mariette's reputation and defaming the Khedive.' See Phillips-Matz, M. J. 1993: *Verdi: a Biography*. Oxford: Oxford University Press, 570–72. Whatever one makes of all this, my own argument depends much less on Mariette's involvement in the storyline and much more on his part in the *mise-en-scène*, about which there is no doubt.

96 Said, Empire at work, 126.

97 Arblaster, A. 1992: *Viva la Libertà! Politics in opera*. London: Verso, 141–44; see also Mackenzie, Said and the historians.

98 Said, Empire at work, 114.

99 Çelik, *Displaying the Orient*, 115–16.

100 Mariette to Draneht, in Busch, *Verdi's* Aida, 33–34.

101 Mariette to Draneht, in Busch, *Verdi's*Aida, 209, emphasis added.

102 I have in mind the outrage that greeted the Frankfurt Opera's production of *Aida* in 1981, when the curtain rose on the second act to confront the audience 'with something like its mirror image: the original first-night audience of the opera's European premiere at La Scala in 1872': Weber, S. 1993: Taking place: toward a theatre of dislocation. In Levin, editor, *Opera Through Other Eyes*, 107–46. As Weber notes, a production that calls attention to its own staging in this way also calls into question the 'individualist attitude' to opera (p, 113).

103 Mitchell, *Colonising Egypt*, 12.

104 Comoli, J.-L. 1980: Machines of the visible. In de Lauretis, T. and Heath, S., editors, *The Cinematic Apparatus*. New York: St Martin's Press, 122–23.

105 Mitchell, *Colonising Egypt*, 29–30; de Nerval, G. *Oeuvres* I, 787–79, 882, 883. The Paris Opéra staged a series of orientalist operas, including in 1827 Rossini's *Moïse* (whose sets were based, in part, on the *Description*) and in 1850 d'Aubert's *L'enfant prodigue* (whose sets and costumes were drawn from Champollion's *Monuments de l'Égypte et de la Nubie*). See Humbert, J.-M., Pantazzi, M. and Ziegler, C. 1994: *Egyptomania: L'Égypte dans l'art occidental 1730–1930*. Paris: Louvre, 395.

106 Mariette to Draneht, in Busch, *Verdi's* Aida, 33, 44; Humbert *et al.*, *Egyptomania*, 423–28.

107 Southworth, A. 1875: *Four Thousand Miles of African Travel*. New York: Baker, Pratt; London: Sampson & Low, 45–47, emphases added.

108 Phillips-Matz, *Verdi*, 570.

109 Said, Empire at work, 118, 120.

110 Said, Empire at work, 124–25.

111 Budden, J. 1992: *The operas of Verdi. Vol. 3: from* Don Carlos *to* Falstaff. Oxford: Clarendon Press, 183.

112 The opera critic Filippo Filippi writing in the Milanese newspaper *La Perseveranza*, in Osborne, C. 1987: *Verdi: a life in the theatre*. London: Weidenfeld & Nicholson, 223.

113 Southworth, *Four thousand miles*, 45.

114 *A handbook for the traveller in Egypt*. London: John Murray, 1873, v; this revised edition was based on a series of visits made between 1863 and 1871.

115 Said, Empire at work, 128; Said draws in particular on Landes, D. 1958; *Bankers and Pashas*. Cambridge, MA: Harvard University Press.

116 Woolf, P. 1988: Symbol of Second Empire cultural politics and the Paris Opera House. In Cosgrove, D. and Daniels, S., editors *The Iconography of Landscape: Essays on the Symbolic Representation, Design and Use of Past Environments*. Cambridge: Cambridge University Press, 214–35; the quotations are from p. 219.

117 Mostyn, T. 1989: The finest opera house in the world. In *Egypt's Belle Époque: Cairo 1869–1952*. London: Quartet, 72–82.

118 Said, Empire at work, 129–30. Garnier's Opéra also staged an aesthetic of separation at the heart of Second Empire Paris, dramatizing the division between the affluence of the western quarters and the poverty of the east. According to one critic, it too was a façade, 'the showground of the epoch's poverty, masked by wealth': Woolf, Symbol, 229.

119 I discuss this in detail in Gregory, D. in preparation: *Describing Egypt*. Minneapolis, MN: University of Minnesota Press, London: Routledge.

120 For the purposes of this discussion I have focused on the ways in which representations of space are implicated in the sort of cultural distinctions that Said describes because this is his main concern, and one that has come to be shared by some of the most innovative work in our contemporary discipline: see Keith, M. and Pile, S., editors, *Place and the Politics of Identity*. London: Routledge. But a more comprehensive discussion of the imaginative geographies of colonialism and imperialism would also have to consider the ways in which representations of 'nature' entered into these cultural discriminations. I am thinking in particular of the connections among landscape, nature and colonial identity described so brilliantly for the rain forest of South America by Michael Taussig in his *Shamanism, Colonialism and the Wild Man* (Chicago: University of Chicago Press, 1987), 74–92. I think it would be possible to show, for example, that in the course of the nineteenth century, Europe's imaginative geographies of Egypt assimilated its native inhabitants to the desert while, both poetically and physically, the west took possession of the Nile, and that these discursive strategies played through colonial and imperial constructions of identity.

121 Fanon, F. 1986: *Black Skin, White Masks*. London: Pluto Press (originally published in Paris, 1952); Fanon, F. 1967; *The Wretched of the Earth*. Harmondsworth: Penguin Books (originally published in Paris, 1961); Said, *Culture and imperialism*, 267–68, 351n. On the contrasting appropriations of Fanon by Bhabha and Said (and others), see Gates, H. L., Jr 1991: Critical Fanonism. *Critical inquiry* 17, 457–70.

122 This paragraph relies on Spitz, E. H. 1989: Psychoanalysis and the legacies of antiquity. In Gamwell L. and Wells, R., editors, *Sigmund Freud and Art: his Personal Collection of Antiquities*. New York: Harry Abrams, 153–71; Torgovnick, M. 1990: Entering Freud's study. In *Gone Primitive: Savage Intellects, Modern Lives*. Chicago, IL: University of Chicago Press, 194–209; Forrester, J. 1994: 'Mille

e tre': Freud and collecting. In Elsner, J. and Cardinal, R., editors, *The Cultures of Collecting*. Cambridge, MA: Harvard University Press, 224–51.

123 I owe this suggestion to Kuspit, D. 1989: The analogy of archaeology and psychoanalysis. In Gamwell and Wells, editors, *Freud and Art*, 133–51.

124 See Frayling, C. 1992: *The Face of Tutankhamun*. London: Faber & Faber. When Howard Carter first entered the tomb, incidentally, he recorded that 'the first impressions suggested the property room of an opera of a vanished civilisation' (p. 4). The connections between archaeology and empire in the previous century are described in Fagan, B. 1992: *The Rape of the Nile: Tomb Robbers, Tourists and Archaeologists in Egypt*. Wakefield, RI: Moyer Bell.

125 Freud to Fliess, 1 February 1900, in Masson, J. M., editor, *The Complete Letters of Sigmund Freud to Wilhelm Fliess 1887–1904*. Cambridge, MA: Harvard University Press, 398.

126 Said, *Culture and imperialism*, 56.

127 The descriptions are all taken from promotional materials produced by Luxor Vegas for its opening in October 1993. See also Chabon, M. 1994: Las Vegas: glitz and dust. *New York Times Magazine* 13 November.

15

'ALTERNATIVE' FILM OR 'OTHER' FILM? IN AND AGAINST THE WEST WITH TRINH MINH-HA

Alastair Bonnett

Introduction

'Non-West' is one of the most elusive of the categories deployed within the contemporary geographical and political imagination. It is the place where the West is not. Not that it is, of course, just a place. It is also the site of 'a (non-Western) perspective', 'a (non-Western) point of view'. Through non-imperial eyes the world, its history, its people, look different. Such sentiments have the somewhat eerie quality of appearing simultaneously obvious and meaningless. They are constantly employed but their reliability and veracity is, just as regularly, placed in doubt.

The way we position ourselves in relation to the West, the way we embrace it or refuse it, provides one of the most complex and pressing of modern dilemmas. This chapter addresses this fraught scene through the work of the film-maker and writer Trinh Minh-ha. More specifically, I will be drawing on two of Trinh's most influential films – 'Reassemblage' and 'Naked Spaces' – in order to explore how the geographical constructs, 'Western' and 'non-Western', operate in her work. Both films are critiques of Western ideas about Africans. Both are in English and distributed principally within the Western 'alternative', 'art house', cinema mileux.[1] I will be arguing that Trinh's attempts to film 'from the margins' and to employ a discourse of otherness to interpret her films, have had the ironic effect of placing her work, at least in part, within a familiar history of Western avant-garde cultural production.

Perhaps, I should admit straight away that the argument I will be advancing here arose serendipitously. The arrangement of my bookshelf played a key role. At one end I have carefully stacked what I once regarded as a discrete area of interest, namely the theories and practices of the avant-garde: surrealist wanderings, situationist slogans as well as any number of beautifully produced, highly aesthetised, avant-garde classics and compendiums. This section also contains a fair number of works on avant-garde film; texts and videos that subversively cut and paste images of commodity capitalism (such as Debord's 'Society of the Spectacle', 1973; see Debord, 1992), or that provide quirky mediations on the everyday, employing the recognisably avant-garde 'look' of rough and idiosyncratic cutting and sound editing. Originally placed far away from these cultural explorations are a whole bunch of books on another area that has always fascinated me, theories of race. Both collections have grown over the years and now rub shoulders. However, sometime in the summer of 1997, after a series of book-buying binges, I began to have the unnerving experience of not being able to tell where one collection began and the other ended. More specifically, many of my new race books had all the hallmarks of avant-garde ones. A not untypical example is the volume on Frantz Fanon published by the Institute of Contemporary Arts in London (Farr, 1995; see also Read, 1996). This particular work is replete with cryptic photographs and pages entirely blank apart from single pithy quotes from Fanon, set out, 'avant-garde style', as incendiary epithets. One page (p. 10) is devoted to the words: 'O my body, make of me always a man who questions!'. Another example is a text co-edited by Trinh (Ferguson *et al.*, 1990), *Out There*. Again it is published through an art gallery (The New Museum of Contemporary Art in New York). Again it contains many pages of art photography. And again it is a collection whose tone – aphoristic, aesthetically laden, self-consciously marginal – confused my attempts at the compartmentalisation of knowledge. This anecdote may be taken to betray a certain, typically academic, impulse on my part to control and regulate the world into discreet fields on inquiry. But the collapsing of these intellectual cages encouraged me to begin thinking about the possibility that the 'post-colonial other' may be being framed by and, perhaps, be inhabiting, the same space of aestheticised radicalism and oppositionality once occupied by the white middle-class males of the historical avant-garde. In other words, that the non-Western and non-white 'other' is being constructed as a recognisable location of critique, an avant-garde, with all the implications of self-conscious marginality that that role implies. Is this 'other', I wondered, merely 'alternative'? I recognise that these formulations are pretty clumsy – too neat and too Eurocentric – to adequately address the diverse ways post-colonial criticism is being developed. However, I would also suggest that they provide a potentially revealing starting point from which to begin looking critically at post-colonial cultural production. More specifically, they have provided me with a point of departure from which to explore 'Naked Spaces' and 'Reassemblage', two films that, as I shall be explaining, are simultaneously in and against Western traditions of representation.

Lacerating narrativity: 'Naked Spaces' and 'Reassemblage'

Both 'Naked Spaces: Living Is Round' (1985; 135 minutes), and 'Reassemblage: From the Firelight to the Screen' (1982, 40 minutes) offer critiques of the Western anthropological gaze. Both films comprise a series of disjointed images of what appear to be traditional rural communities in West Africa. Each is overlain by an unexpressive, almost dead-pan, commentary. Neither film offers a conventional narrative but rather a series of images of everyday life, particular attention being paid to the tactile and formal qualities of houses, pots, mats, skin. This aesthetic quality is heightening by both films' insistent attention to rhythm, especially the rhythms of dancing, music and food preparation. Both movies also have an innovative approach to sound. Indeed, sometimes the soundtrack is 'switched off', leaving only images. At no time does the commentary in either film attempt to directly address or explain the actions or scenes appearing in it.

The voice-over in 'Reassemblage' comprises a set of short statements that collide local voices, anthropological discourse and the film-maker's own account of the making of the film. Although often appearing randomly organised, these statements are positioned in such a way as to engender a kind of persistent melancholy; a sense, not that the people represented in the film 'cannot speak', but that their voices can only exist in the context of a dominant Western mode of (mis)representation. Thus, for example, as the camera jumps between scenes of village huts, village women's breasts, and traditional village activities (first we see weaving, then thatching, then cooking upon an open fire) the commentary intones:

A film about what? my friends ask.
A film about Senegal; but what in Senegal?
I feel less and less the need to express myself
Is that something else I've lost?
Something else I've lost?

(Voices: same conversation in Sereer language)

Filming in Africa means for many of us
Colorful images, naked breast women, exotic dances and fearful
     rites.
The unusual

First create needs, then, help
Ethnologists handle the camera the way they handle words
Recuperated collected preserved

351

The Bauman the Bassari the Bobo
What are *your* people called again? an ethnologist asks a fellow
    of his.

(Trinh, 1992: 98)

The irreducible otherness of Trinh's subjects is most strongly conveyed by
the use of untranslated Sereer. The same concern emerges in 'Naked
Spaces', another film that critically deconstructs the representation of
everyday rural life in West Africa. It is a more ambitious work than
'Reassemblage', presenting a concerted attempt, not simply to critique
anthropology, but to develop a vision of the interactivity, the interpenetration, of 'Western' and 'African' discourses. As explained by Trinh (1985;
also Trinh, 1989a) the film engages in the act of 'lacerating narrativity
without resulting in a state of meaninglessness'. A concern to make acts
of interpretation and representation transparent, whilst confusing and
colliding Western knowledges, can be found throughout the film, most
clearly in the soundtrack. As explained by Trinh in her script note

> Text written for three women's voices, represented here by three
> types of printed letters. The low voice [bold], the only one that
> can sound assertive, quotes from the villagers' sayings and state-
> ments, as well as African writers' words. The high-range voice
> [plain] informs according to Western logic and mainly cites
> Western thinkers. The medium-range voice [italics] speaks in the
> first person and relates personal feelings and observations.

(Trinh, 1992: 3)

The three voices are spoken by three different women, the personal voice
being Trinh's own. As with 'Reassemblage', 'Naked Spaces' is marked by
'rough', 'avant-garde-style' camera work. However, the film is consider-
ably longer, with lengthy periods where nothing is spoken and the camera
simply rests or roves amongst its subject matter. The polemical dynamic
of 'Reassemblage' is thinned into a more contemplative and stiller ambience. With the camera dwelling for minutes at a time on particular village
situations the viewer is invited to enter the rhythms of rural life, to think
about how the spoken commentaries are all re-presentations, overlaying
the physical realities, and beauties, of the everyday. Indeed, the emphasis
in 'Naked Spaces' on the rhythmic and aesthetic qualities of the quotidian
provides its presiding sensibility. The notion of roundness, in lives, in things
(huts, bowls, bodies, etc.), is repeatedly alluded to visually and in the script;
one telling 'villager's saying' being 'Everything round invites touch and
caress'. In her critique of 'Naked Spaces' Henrietta Moore (1994) argues
that it seeks to establish an equivalence between the exterior and interior,
the public and the private, spaces of village life. As this suggests, the film

352

offers a vision of a holistic, organic geographical imagination. The buildings people live in, the artefacts they use, the thoughts they have, all are merged into a portrait of a unified and indivisible culture. Moore suggests that

> An explicit homology is established [in 'Naked Spaces'] between architecture or physical space and cosmological beliefs or people's 'inner lives'. This homology is reinforced technically through the use of shots through spaces (i.e., from the inside of the house, through a framed doorway or pool of light, to the outside) and through sequences of images which establish visual connections between spheres and patterns in architectural forms, deconstructive motifs, items of material culture and parts of human bodies.
>
> (Moore, 1994: 119)

The opening sequence of the film, lasting nearly four minutes, commences with a ceremonial procession and dancing and moves on to images of food preparation, a ceremonial display of gun firing, then some fixed camera shots from inside dark dwelling places, out into the bright sunlight. The commentary during these scenes draws on each of the three 'voices' described earlier and runs as follows:

People of the earth

Not descriptive, not informative, not interesting
Sounds are bubbles on the surface of silence

Untrue, superstitious, supernatural. The civilised mind qualifies many of the realities it does not understand untrue, superstitious, supernatural

Truth and fact
Naked and plain
A wise Dogon man used to say

'to be naked is speechless'

Since the impression of thematic disjuncture resides largely within the spoken soundtrack, with the visual imagery remaining focused and relatively unchanging, it is the latter, the scenes of village life, that come to seem like the 'real stuff', the raw essence, both of the film and of the lives it portrays. Indeed, whilst the commentaries twist and turn, delivering doubt and disorientation, the imagery we see conveys a romantic sense of rootedness, of attachment. This effect is abetted by two facts. First, that Trinh has chosen to film amongst traditional rural communities: 'modern' incursions, whether in the form of villagers' 'sayings', clothing, communications, or machinery, are scarce. Second, the three voices offered provide as many moments of conservative stereotype as subversive play. The villagers' voice in particular, confined to the sphere of 'sayings and statements' (as well as, every so often, an 'African writer'), is heavy with timeless sagacity. Indeed, whilst the other two voices – the voice of 'Western logic' and the voice of 'personal feelings' – appear engaged with each other, as the romantic and rationalist side of Western modernity, the 'villagers' sayings and statements' often seem anachronistic and artful. Explaining the film, Trinh has noted that the three voices represent positions within one subjectivity, 'In Naked Spaces ... the viewer hears [the voices] not so much as contradictions or as separate entities, but as differences within the same subjectivity' (1992: 184). This explanation certainly adds to the interest of the commentary but it does little to address the problems identified above. After all, the three voices are still presented as drawing from distinct traditions, albeit traditions warring within one consciousness. Moreover, once one understands that the commentary is coming from the 'same subjectivity' the question arises, whose? Certainly not the Africans represented, whose voice is so insistently 'traditional', and not Western anthropologists, who are portrayed as oppressors. Rather, once we begin to hear the three voices as expressing one subjectivity, the voices of the villagers and 'Western logic' become subservient to the central, authoritative, personal voice of Trinh; mere stereotypes engineered to highlight the individual, unique qualities of her own interior musings.

(An)other kind of avant-garde: (an)other kind of primitivism?

Trinh's films are watched, for the most part, in Western 'art house' cinemas; they are consumed as challenging and authentic cinema about the representation of the other. Reviews of the two movies under discussion have tended to focus on their 'tactile beauty'. 'The images', notes Armatage (1985), 'seem to be edited by an almost intuitively associational process, unified by geographical force and by the rhythms of their repetition'. The reviews also invariably mention that Trinh was born and brought up in Vietnam (sealing her work's non-Western status).

However, the style of Trinh's films, and the way they treat their audience, are evocative of familiar forms of Western, avant-garde, filmmaking. As befits her position as Professor of Cinema at San Francisco State University, Trinh is highly literate in the genres and vocabulary of film. However, she has repeatedly refused to locate 'Naked Spaces' and 'Reassemblage' as part of, or engaged with, any particular filmic tradition; claiming instead that their form and style arose organically, developing spontaneously from the act of filming and personal response (Trinh, 1992).

This position may be contrasted with the stance Trinh adopted in her first book (1981) *Un Art sans Oeuvre*, which extolled the virtues of those modernist cultural workers, such as Cage, Artuad and the Dadaists, who worked against artistic conventions of authorial and fixed identity. Trinh has interpreted her move away from this modernist tradition in terms of a shift away from an academic to a personal approach (see also Trinh, 1989b; 1991): 'The approach I adopted earlier differs from the one I have now in that, in the former case there is no "I" – I alternatively and anonymously speak through the voices of those whose works I discuss (1992: 237).'

Trinh's move towards personal account is, in part, enabled by a simultaneous emphasis on the autonomous, self-activating, nature of her film work. 'I am always working at the borderlines of several shifting categories', she asserts, 'stretching me out to the limits of things, learning about my own limits and how to modify them' (1992: 137). As the following revealing dialogue with Scott MacDonald indicates, Trinh does not wish to be seen as part of an established tradition.

SCOTT MACDONALD: Often in Reassemblage there'll be an abrupt movement of the camera or a sudden cut in the middle of a motion that in a normal film would be allowed to have a sense of completion. Coming to the films from the area of experimental moviemaking, I felt familiar with those kinds of tactics. Had you seen much of what in this country is called 'avant-garde film' or 'experimental film?'. I'm sorry to be so persistent in trying to relate you to film! I can see it troubles you.

TRINH MINH-HA: [Laughter] I think it's an interesting problem because your attempt is to situate me somewhere in relation to a film tradition, whereas I feel the experimentation is an attitude that develops with the making process when one is plunged into a film. As one advances, one explores the different ways that one can do things without having to lug about heavy belongings. The term 'experimental' becomes questionable when it refers to techniques and vocabularies that allow one to classify a film as 'belonging' to the 'avant-garde' category. ... So, while the techniques are not surprising to avant-garde film makers, the film still does not quite belong to that world of filmmaking. It differs perhaps because it exposes its politics of representation instead of seeking to transcend representation in favour of visionary presence and spontaneity which often constitute the prime criteria for what the avant-garde considers to be Art. But it also differs because all the strategies I came up with in Reassemblage were directly generated by the material and the context that define the work.

(Trinh, 1992: 113–114)

In this exchange Trinh's insistence that the material itself guided her hand is offered in contrast to the artificial strategies of the avant-garde, more specifically their spontaneism and aestheticism. However, this attempt to distance her work from a recognized film tradition, and by implication, to claim it as original, is not entirely effective. For allowing oneself to be led by one's material (which entails giving oneself up to chance) is itself a variety of spontaneism. Moreover, the evidently political function of much (indeed, I would suggest, most) avant-garde film-making (for example, Godard and Debord), means that it cannot be categorised as merely aesthetic, no more than Trinh's own work. Trinh's refusal to be contaminated by Western tradition demands that she employ the language of originality, refusal and exploration to interpret her own work. This is, ironically, the very language of the historical avant-garde, that rag-bag of mostly male, mostly white, cultural workers whose enterprises relied on claims of otherness and marginality.

Perhaps the most damaging result of Trinh's lack of reflexivity in this area is the operation of primitivism within the two films under discussion. Trinh locates primitivism and other colonial discourses as Western constructs, more specifically associating them with Western anthropology. However, since colonialism and primitivism have been institutionalised as items of debate within anthropology for many decades, her portrayal of the discipline appears oddly dated. The image of old-style colonial anthropology Trinh uses seems designed to establish the authentically non-Western nature of the other voices (both her own and those of Africans) in her films. This procedure also acts to obscure the presence of primitivism in her narrative. Thus Trinh's focus on a dated image of colonial anthropologists conceals the way primitivism also animated (and continues to animate) the avant-garde. 'The other', for the latter, was valued because it could be used to connote and develop their own image as cultural subversives and transgressors; 'the other' was 'alternative'. In other words, the modernist avant-garde adopted, and identified itself with, the 'tribal' and non-Western in order to position itself as outside Western civilized society, to locate itself as a challenging force representing raw nature and real art. Thus, to mention just one example, Tzara, a founder of Dada, proclaimed, 'We want to continue the tradition of the Negro, Egyptian, Byzantine and gothic art and destroy in ourselves the atavistic sensitivity bequeathed to us by the detestable era that followed the quattrocento' (1992: 63; first published 1919). 'The primitive' is created within the West as a critique of the West. It is a critique that has insistently compared the naturalness, and the everyday rhythmic simplicity, of 'tribal' life with the civilised, bureaucratic and mechanical West (Jordan and Weedon, 1995). To cite Tzara again, in his 'Note on Negro art' (1992; first published 1917), 'My other brother is naive and good, and laughs. He eats in Africa or along the South Sea Islands . . . From blackness, let us extract light. Simple, rich luminous naiveté'. Lapsing into primitivist aphorism, Tzara continues 'Art, in the infancy of time, was prayer, wood and stone were truth. In man I see the moon, plants, blackness, metal, stars, fish' (p. 58).

Although Trinh wishes 'Naked Spaces' and 'Reassemblage' to be understood as outside of any recognisable tradition, the thematic and formal qualities of the two films echo many of the central conceits of the historical avant-garde. Unlike the latter, Trinh's work foregrounds issues of cultural bias and the alterity of communities, or whole nations, of people. However, her insistent representation of the timeless, natural qualities of African village life, her romantic invocations of the rhythmic, round, qualities of villagers objects, houses, 'sayings' and actions, her positioning of these representations in opposition to a 'West' defined by its logic and rationalism, as well as her repertoire of filmic techniques, all suggest that Trinh is as much inside as outside the 'avant-garde tradition' and indeed, 'Western representation'. Trinh has described as 'perceptive' the view of 'Reassemblage' as 'an amorous invasion' (1992: 182). Yet hers is not the first such incursion into Africa; it has been loved in this way before.

Conclusions

Trinh is aware of the attraction of her films to Western 'art-house' audiences. 'The margins,' she notes 'our site of survival, become our fighting grounds and their site of pilgrimage' (1990: 330). Yet this formulation betrays a certain self-romanticism: 'they', the Western audience, must leave home, and journey to visit 'our', non-Western, sites of resistance. The fact that these 'sites' are mutually constituted, that this marginal, non-Western, area is as much constructed within as outside the West, is filtered out of Trinh's account. Indeed, her self-positioning as marginal makes it difficult to theorise the presence or use of the West in her work at all. When the West is 'acknowledged' its presence must necessarily be displaced. Thus, for example, considering the recognisably post-modern interest in fractured and fluid identities that permeates her work, she notes 'if I am interested in Barthes, in Western contemporary music, in feminism, in post-structuralism, it is mainly because, in my view, these ways of thinking do not exclude and therefore appeal more to non-Western thinking' (Trinh, 1992: 233).

The notion of 'non-Western thinking', of a non-Western perspective, animates and structures the two films of Trinh I have discussed in this chapter. Trinh does not offer this mode of thought, this perspective, as readily available, as compliant, in any traditional anthropological sense. She portrays it rather as something that is irretrievable, as the 'not to be grasped', the other. Yet this latter conceit encourages and enables an aesthetic packaging and a series of historical resonances: the irretrievable other which can only be defined negatively (as the non-West); a sphere of marginality and, by association, mysterious beauty and exotic difference;

an always-distant site whose raw nature is designed to remind the civilised world what it has lost.

Trinh often appears to want to escape Western modes of representation. In 'Naked Spaces' and 'Reassemblage' she is not driven by a desire to speak for the other, but rather not to speak from within the familiar repertoire of Western interpretation. However, the romantic dynamic behind such refusal, such voyaging away, is, at least in part, built on and ideologically driven by past attempts to leave the West behind. Trinh's work is not necessarily defined by or bemired in 'Western traditions'. It does exist, however, within and against Western modes of representation, and of the Western avant-garde; its real 'site', I would submit, is not some aloof 'margin' but that of messy and contradictory struggle.

Acknowledgements

Thanks to Mike Crang for his encouragement and helpful comments.

Notes

1 'Naked Spaces: Living is Round' (1985, 135-minute colour film), distributed by: Woman make Movies (New York); The Museum of Modern Art (New York); Idera (Vancouver); Cinenova (London); The National Library of Australia (Canberra). 'Reassemblage' (1982, 40-minute colour film, distributed by Woman make Movies (New York); The Museum of Modern Art (New York); Idera (Vancouver); Cinenova (London); The National Library of Australia (Canberra); Third World Newsreel (New York); Lightcone (Paris); Image Forum (Tokyo).

References

Armatage, K. (1985) 'Naked Spaces', review in 'Festival of Festivals' programme, September 5–14th, 1985 (Festival of Festivals, Toronto).

Debord, G. (1992) *Society of the Spectacle and Other Films* (Rebel Press, London).

Farr, R. (ed.) (1995) *Mirage: Enigmas of Race, Difference and Desire* (Institute of Contemporary Arts/Institute of International Visual Arts, London).

Ferguson, R. *et al.* (eds) (1990) *Out There: Marginalization and Contemporary Cultures* (The New Museum of Contemporary Art, New York/The MIT Press, Cambridge).

Jordan, G. and Weedon, C. (1995) *Cultural Politics: Class, Gender, Race and the Postmodern World* (Blackwell, Oxford).

Read, A. (ed.) (1996) *The Fact of Blackness: Frantz Fanon and Visual Representation* (Institute of Contemporary Arts/Institute of International Visual Arts, London).

Moore, H. (1994) 'Trinh T. Minh-ha observed: anthropology and others', in Taylor, L. (ed.) *Visualizing Theory: Selected Essays from V.A.R. 1990–1994* (Routledge, New York) 115–125.

Trinh, M. (1981) *Un Art sans Oeuvre* (International Book Publishers, Troy, Michigan).

Trinh, M. (1985) 'Naked Spaces', paper presented at the San Francisco Cinematheque, October 27th.

Trinh, M. (1989a) 'Outside in inside out', in Pines, J. and Willman, P. (eds) *Questions of Third Cinema* (BFI, London): 133–149.

Trinh, M. (1989b) *Woman, Native, Other: Writing Postcoloniality and Feminism* (Indiana University Press, Bloomington).

Trinh. M. (1990) 'Cotton and iron', in Ferguson, R. *et al.* (eds) (1990) *Out There: Marginalization and Contemporary Cultures* (The New Museum of Contemporary Art, New York/The MIT Press, Cambridge): 327–336.

Trinh, M. (1991) *When the Moon Waxes Red: Representation, Gender and Cultural Politics* (Routledge, New York).

Trinh, M. (1992) *Framer Framed* (Routledge, New York).

Tzara, T. (1992) *Seven Dada Manifestos and Lampisteries* (Calder Press, London).

16

THINKING GEOPOLITICAL SPACE

The spatiality of war, speed and vision in the work of Paul Virilio

Tim Luke and Gearóid Ó Tuathail

Born in Paris in 1932, Paul Virilio is a child of the Third Republic. His intellectual project, in many ways, centres upon the tremendous military, economic, and cultural forces that blitzed the republic of his birth in less than a month. Virilio describes his childhood as one wracked by warfare, recalling the destruction of Nantes in 1942 as a traumatic event (1983, 2, 24). In the preface to *The Insecurity of Territory*, Virilio describes war as his father and his mother. After the trauma of World War II, Virilio's intimate relationship with war continued as he was drafted into the French army to fight in the Algerian War. 'War,' Virilio once claimed, 'was my University' (1983, 24).

Trained as a city planner and architect, Virilio's experience of, and indeed fascination with, military affairs and weapon technologies shaped his approach to the intellectual questions of landscape morphology and urban design. In 1958, he began researching and photographing the fortified emplacements of Hitler's Atlantic Wall. The result was an exhibition organized by the Centre for Industrial Creation and presented at the Museum of Decorative Arts in Paris from December 1975 to February 1976. Out of this show came *Bunker Archaeology*, a collection of the exhibition's photographs together with a brooding exegesis by Virilio on military space and the historical tendencies, institutions, personalities and aesthetics conditioning the spatiality of war.

By this time, Virilio was already a well established figure within the French architectural world. In 1963 with Claude Parent, Virilio founded the 'Architecture Principle' group, and oversaw the construction of two important structures: the Sainte Bernadette de Nevers parochial centre in 1966 and the aerospace research centre of Thomson-Houston in

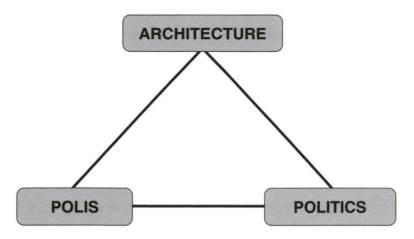

Figure 16.1 Military power/knowledge/technology

Villacoublay during 1969. Named professor and workshop director of the Ecole Spéciale d'Architecture in Paris during 1969, he was promoted to director of studies in 1973 and president in 1990.

Beginning with his pathbreaking *Bunker Archaeology*, Virilio has published a series of innovative and suggestive 'think pieces' on transhistorical tendencies in warfare, technology, human settlement forms, communications, media and cinema, many but not all of which have been translated into English and other languages (see References). The wide-ranging scope and eclectic nature of these writings have made Virilio a difficult intellectual to categorize. He is, at one and the same time, a historian of warfare, technology and photography, a philosopher of architecture, military strategy and cinema, and a politically engaged provocative commentator on history, terrorism, mass media and human–machine relations.

Nevertheless, it is not difficult to identify the problematic that Virilio has been addressing since the 1970s. This problematic can be conceived of in terms of two triangles, the first disciplinary and categorical in a conventional sense (see Figure 16.1) while the second (see Figure 16.2) is more fully conceptual and idiosyncratic, revealing the three overarching themes that preoccupy or, perhaps as some might argue, obsess Virilio in his writings.

The first triangle maps out the linkage between Virilio's different professional identities as an architect, an analyst of military strategy and an engaged political figure. As an architect, Virilio is deeply concerned with the nature of urban form. It is reflection on the urban that leads him directly to politics via the polis. '[T]he relation to the city, for me, is immediately a relation to politics. Furthermore, urbanist and politician,

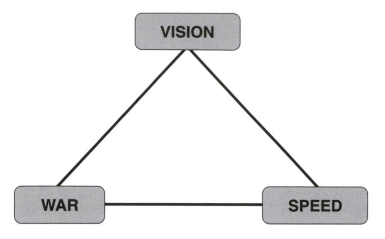

Figure 16.2 Unifying concern: human–machine interfaces

etymologically speaking, are the same thing. Modern political ideologies have obscured the fact that politics is first and foremost about the *polis* (1983, 2–3). Playing the Latin roots of 'urbanist' against the Greek derivations of 'politician,' Virilio turns this linguistic collision into an important observation. If the urbanist, as a student of the *polis*, and the politician, as an actor within a *polis*, are one, then reflections on urbanism are inevitably also reflections on politics. Etymology also leads him to reflection on warfare, the primordial human activity that has always shaped the very form of human settlements and the possibilities of the city. As he points out in his writings, etymologically an urbanist is one who builds cities in order to defend them (1983, 86). Like Louis Mumford, with whom his writings share certain key elements, Virilio reimagines the city, in part, as the material anticipation and outcome of war-making, and urbanistic reasoning becomes, in part, the constant preparation for it (Mumford, 1963, 1970).

Uniting all three corners of this first triangle is the problematic of military power, knowledge and technology. With qualitative changes in the combined functioning of the latter during the twentieth century, the very nature of the former is transformed. Virilio's single-minded pursuit of this problematic of military power/knowledge/technology pushes him into a wide-ranging reconsideration of some of the most complex and challenging questions of our time, problematics that register as concerns with war, speed and vision but which are unified at a deeper level by Virilio's ongoing fixation upon the contraction of human control over the machines that frame, condition and threaten life at the end of the twentieth century. This concern with the human–machine equation is not, of course, particular to Virilio but a dominant concern of late twentieth-century French thought, finding expression in the work of Baudrillard, Deleuze, Derrida, Foucault, Guattari, Latour and others.

It is the elemental qualities of this second triangle that concern us in this paper. More specifically, we propose to examine the nature and implications of Virilio's thinking on the machinic qualities of war, speed and vision for the thinking space of geopolitics. Geopolitics, it is worth noting, is not a fixed and unified body of work or field of endeavour. Rather, it is a constellation of concerns with logistical technology, territorial space, global vision, imperial strategy and power projection, concerns that have historically come together in different ways in different places in the writings of certain canonical intellectuals and the practices of powerful and hegemonic states (Matellart, 1994, 1996; Ó Tuathail, 1996). Virilio's writings can certainly be considered as operating within the constellation of geopolitics; it is how they help us re-map the elemental forces of this constellation that is of greatest interest to us in this chapter.

Before considering the machinic qualities of war, vision and speed, it is worth commenting a little on Virilio's method and writing style. Virilio himself has noted that he is primarily interested in 'tendencies,' not 'episodes,' quoting Winston Churchill to the effect that in ancient warfare 'the episodes were more important than the tendencies' whereas 'in modern warfare, the tendencies are more important than the episodes' (1983, 11). This separation is important to understand Virilio's work. It marks an important distinction between his own resolutely trans-historical gaze, which is centered upon identifying essential tendencies, trajectories and trends, and the more historically embedded vision of other scholars, who are interested in messy empirical reality. In deploying such a gaze to generate 'insights,' Virilio is working within the tradition of many grand strategists and geopoliticians who are also interested in decontextualized tendencies. All are inclined to generate 'timeless truths' about strategy as they sweep across the record of human history, from ancient warfare to its most contemporary forms (Agnew, 1998). Tendencies are essentially naturalized, transcultural constants, while episodes are culturally contingent and historically grounded.

In Virilio's case, this restless trans-historical gaze is combined with a particular elliptical style which values suggestion more than explanation. Virilio himself has endorsed such a position declaring that he does not believe in explanations:

> Being an urbanist and architect, I am too used to constructing clear systems, machines that work well. I don't believe it's writing's job to do the same thing. I don't like two-and-two-is-four writing . . . I work in staircases . . . I begin a sentence, I work out an idea and when I consider it suggestive enough, I jump a step to another idea without bothering with the development. Developments are

GEARÓID Ó TUATHAIL AND TIM LUKE

the episodes. I try to reach the tendency. Tendency is the level of change.

(1983, 38–9)

Lotringer, in her conversations with Virilio, somewhat charitably described his as a 'writing in a state of emergency', a writing on war that is at war in order to draw attention to the nuclear terror that is warfare in the late twentieth century (or during the era of Cold War nuclear deterrence at least).

This ignores, however, the somewhat serious deficiencies of Virilio's suggestive method. First, Virilio's method is inclined to launch rhetorical bombs: clean, little declarative statements about urbanism, warfare, states, speed and technology that are clearer and cleaner than the messy explosions of history warrant. Like Baudrillard, Virilio's rhetoric tends towards overstatement and hyperbole as he spins out observations on speed and violence. He can be a quipmeister, turning out sound-bite theory for sound-bite times. His writings are often no more than journalistic musings which leap dizzyingly from one historical age to a different one in the space of paragraph. At times, his writing is sloganistic, displaying an obsessive fascination with essential mantras and timeless truths, like Sun Tzu's 'speed is the essence of war' or William Perry's (a former U.S. Secretary of Defense) 'once you can see the target, you can expect to destroy it'. At other times, Virilio's writing stumbles off the staircase altogether, descending into absurdity and mysticism, with elements and echoes of Christian themes and apocalypticism (see, for example, his reading of death, Vietnam and Nixon (1983, 160–1), women, families and war (1990, 81) and the condemnation of sexual perversion and diversion in cyberspace (1995, 103–18)).

Second, Virilio's writings are infused by almost paranoid fantasies, which bring the tendencies he identifies into their purest form, the pure war of totally automated battlefields or the purifying Doomsday Machine of *Dr Strangelove* fame with its automated declaration of war for example. It is important to note that this paranoid style can reveal much to us and has been used to good effect by other theorists like Donna Haraway (1997). This apocalyptic style is a Cold War artifact, which was not unjustified during the so-called 'Second Cold War' of renewed American–Soviet conflict from 1979 to 1989. Nevertheless, it can lead to sweeping declarations that sound unsubstantiated or, even worse, insubstantial. Another postmodern French huckster to some, yet a prescient techno-savvy strategist to others, Virilio's writings always provoke his readers to reason beyond their inherited and conventional ontologies.

I. Colonizing war machines: the spatiality of war

To Paul Virilio, all human geography is ultimately a product of warfare, because space is always imagined as the zones of defensive barriers and/or offensive operations. The requirements of military geography establish the possibilities and parameters for human geography. At root, war and the preparation for it produces the space–time of the human experience as a function of projectile speeds, logistical rates of transport, or intelligence insight gathering. The territorial organization of space into human settlements and political units of authority, from the earliest human village settlements to medieval city-states, modern nation-states and world-wide empires, reveals a constant tendency: they express different orders of military power, knowledge and technological organization.

For Virilio, there are three distinct orders of military knowledge: tactics, strategy and logistics. Virilio imputes tactics to 'the art of the hunt' in early human civilizations. These civilizations exist without wars in the modern sense, the clashing of different tribes generating mere 'tumults' (1983, 4). Virilio associates strategy with the emergence of the Greek city-states through to the development of the commercial city-states of feudal Europe. It is the organization of space as a theatre in preparation for war, with a city-state fixed at its centre fortified and capable of defending itself and its supporting military–political system should war break out. Tactics do not disappear as an order of military knowledge but are merely subordinate to strategy.

Beginning in the late nineteenth century, however, both tactics and strategy slowly are displaced by logistics as a new order of military power/knowledge/technology associated with modern mechanized war economies realizes the emergent possibilities of vast destruction in the horrific actuality of waging total war. By the time of Hiroshima in 1945, logistics has become the dominant order of military power for Virilio. It is, he suggests, quoting from a Pentagon statement at this time, 'the procedure following which a nation's potential is transferred to its armed forces, in times of peace as in times of war' (1983, 16). With logistics, the distinction between times of peace and times of war disappears; there is only the perpetual preparation for war.

Like many scholars, Virilio reads early modern states as little more than war machines (see Mann, 1986; McNeill, 1982; Mumford, 1970). They are predatory organizations that colonize both space and human populations, organizing space into a military system of segmented and striated, parcelled and protected territory, and human populations into temporal relationships which support the functioning and perpetuation of military machines. The semi-colonial economy of feudalism, Virilio suggests, 'this military protection racket, forms the constitutional basis of the great modern States' (1990, 46). The French Revolution marks a qualitative change in the ordering of space and time by the state as a military war machine, for it unleashes the political idea of 'nations on the move' (1986, 34). The idea of democratic revolution, according to Virilio, realizes itself as 'dromocratic revolution', a revolution of acceleration and speed. Gathering momentum, the state-as-war-machine spreads 'the state of siege of the communal city-machine, immobile in the middle of its logistic glaces and domestic lodgings, over the totality of the national territory' (1986, 14). A new order of time, space and state-as-war-machine is consolidated, specified synecdochically by Virilio as polis, police and highway surveillance (1986, 14).

Overstated and underspecified, Virilio's remarks on speed and politics in the modern era can be read as a different version of David Harvey's (1989, 1996) well known interpretation of the logic of time–space compression. Virilio's argument, however, is crucially different from Harvey's for its central motor is not the dynamics of economic capital accumulation, but the dynamics of military weapons accumulation. It is not the means of production that interests Virilio; it is instead the evolution of the means of destruction. Virilio's claims about the military power/knowledge/technology nexus are not modest. The very aim of strategic action, as Sun Tzu (1971) would agree, is to 'redefine the space' one's enemy 'must cross or the time he has to live.' This makes the practicability of war, 'the coherent plan devised in time and space that can, through repetition, be imposed upon the enemy,' 'not the instrument but the origin of a totalitarian language of History.' The dynamics of military accumulation consumes European states and then the world 'thus giving it the stature of an absolute takeover of world history by Western military intelligence' (1990, 17).

This paranoid vision of the state as war machine realizing 'an absolute takeover of world history' is expressed for Virilio in the concepts of 'total war' and 'pure war'. He traces total war to the rise of logistics as the significant dimension of military activity, finding that it begins first in the great naval powers of the early twentieth century. It also marks a new order of space–time where speed and manoeuvrability are highly valued: 'it is first waged on the sea because the naval glacis naturally presents no permanent obstacle to a vehicular movement of planetary dimensions' (1986, 50). The introduction of the tank by British forces on the Somme marked a revolution in speed and manoeuvrability on land, it being both an automotive fort and a terrestrial battleship (1986, 56). Another historic date in Virilio's schema is Joseph Goebbels's declaration at the fortified Sports Palace in Berlin on 18 February 1943 that 'total war' as a radical intensification of existing warfare will be unleashed like a 'storm' (1994a, 58). A key historic figure Virilio sees representing all of the tendencies he deems significant, from military technology to logistics to architecture, is Alfred Speer (1994a, 55–61).

The drift towards total war in the twentieth century has multiple consequences for twentieth-century civilization. It deepens the colonization of
the social by the military so that distinctions between the 'civilian' and the
'military' become blurred. To be a citizen means acquiring 'a right to die'
(1990: 79) as one one-millionth of a megadeath. It also intensifies warfare against the environment. Total war quickly leads to the ultimate dimensions of technologically feasible ecological warfare, wars against the built
and natural environmental ecosystems that support one's enemy (1986,
75). Furthermore, it brings a new absolutism to political life and the
dynamics of warfare. Because it mobilizes the whole of society in a gigantic
logistically driven war effort, its goal becomes not simply to defeat one's
enemy but to destroy his very identity and soul.

Warfare develops a qualitatively new character at the end of World War
II. Hiroshima inaugurates a new era of nuclear warfare. And the earlier
launchings by Germany of the V-1 cruise missile and V-2 rockets against
London initiate the epoch of inter-continental strategic missiles. As the
logistics of both these technological innovations became further refined,
they helped constitute the Cold War system of global nuclear deterrence.
To Virilio, the era of global nuclear deterrence is not 'total' or 'absolute',
but 'pure war':

> Deterrence is the development of an arms capacity that assures
> total peace. The fact of having increasingly sophisticated weaponry
> deters the enemy more and more. At that point, war is no longer
> in its execution, but in its preparation. The perpetuation of war
> is what I call Pure War, war which isn't acted out in repetition
> but in infinite preparation. Only this infinite preparation, the
> advent of logistics, also entails the non-development of society in
> the sense of civilian consumption.
>
> (1983, 92)

Pure war challenges the very distinctions that have made warfare meaningful historically. It is neither peace nor war but permanent logistic struggle
in which warfare preparations reorganize social and economic relations in
order to secure 'peace' (Luke, 1989). As the Strategic Air Command said
amidst the Cold War, 'Peace is Our Profession.' The distinction between
offensive and defensive is no longer relevant (1991a, 131). 'War is no longer
directly identifiable with declared conflict, with battles' (1990, 36) but with
the speed logistics of nuclear vehicles:

> The will-to-defense and the will-to-power are indifferently blended
> into a single amalgam ... The speed of violence becomes the
> violence of an unsurpassed speed, and the speed of light becomes
> the standard measure for war, in its context, its essence and its

nature. Pure war contributes to the inversion of all terms of power, as it leads each antagonist to the immediate reversibility of the conditions of the possibility of confrontation.

(1991a, 138)

Pure war is such because the logic of logistics in the age of deterrence has reached such a machinic level that humans are becoming less and less significant elements of the war machine.

Here, Virilio recognizes how thoroughly semantic the inter-operation of nuclear tactics, strategies, and logistics becomes within the world's mass media markets. Indeed, television and film prove to be the most pervasive mode of delivering nuclear payloads as their photographic effects are extremely fast, virtually unstoppable, and infinitely relaunchable. Even though most nuclear strategists admit that the heat/blast/radioactive yield of nuclear weapons cannot be used rationally in the post-Hiroshima world system, the delivery vehicles with payloads are operated every day in such a way as to give credibility to the photo-realistic powers of their deterrence yield (Luke, 1991). In typical overstatement, Virilio declares that 'pure war no longer needs men, and that's why it's pure' (1983, 171). In his paranoid vision, pure war marks a new level of the endo-colonization of populations by the logic, technology and time–space requirements of the nuclear war machine. 'The Russian–American realization of global nuclear deterrence is,' Virilio concludes in part of his work, 'a catastrophic process of total colonization' (1990, 34). This colonization of society and economy by military–industrial complexes – Eisenhower, credited with coining the term, is another historic figure in Virilio's schema (1983, 14, 93) – leads Virilio, as noted above, to claim that these tendencies will lead to economic stagnation ('non-development') and zero growth. While a somewhat glib prediction at the time, Virilio was not entirely wrong in suggesting that permanent war economies would stagnate certain states in Europe and elsewhere (1983, 93). The acute difficulties of the Latin American and southern European military–bureaucratic dictatorships in the seventies and early eighties and the Soviet Union and its allies in the late eighties can in large part be attributed to the economic, political and social contradictions induced by endo-colonizing militarism.

More provocative is a second consequence Virilio extrapolates from global nuclear deterrence as pure war: the disappearance of politics. As global nuclear war machines have elaborated an increasingly technological and machinic system of mutual deterrence, the space–time of politics has been radically reduced and compressed. As nuclear war becomes an increasingly electronic decision, there has been a loss in the duration of politics. Politics is reduced to the instance of launch code authentication in an era of attack on alert deterrence (1991a, 129–30). The time for debate and diplomacy, reflection and rethinking disappears (1983, 58). Like

Baudrillard, Virilio speaks of this as a condition of 'trans-politics' though he strongly states that he considers such a situation totally negative. 'It's the contamination of traditional political thought by military thought, period! . . . It's not post-politics, it's not the end of politics, it is its cont-amination. It's completely negative. Trans-politics means no more politics at all' (1983, 144). Similarly, war becomes a transbellicose game as nuclear operations 'have also gradually taken on the aspect of large-scale elec-tronic games, a *Kriegspiel* requiring whole territories over which the various procedures and materials of modern war are reconstituted' (1989: 86).

Virilio's arguments in the 1970s and early 1980s about the system of Cold War deterrence are neither exceptional nor unique (see Sherry, 1996). In Great Britain, E. P. Thompson elaborated in a richer, materialist and more finely contextualized manner similar arguments about what he termed 'the logic of exterminism' found in the Cold War nuclear strategy of both power blocs. The Cold War, Thompson argued, had developed an exterminist logic of its own that was divorced from its origins and rational political decision-making. It had become a self-perpetuating system dominated by two mutually dependent military–industrial complexes. Weapons innovation within these blocs was self-generating, the impulse to 'modernize' and to experiment continuing independently 'of the ebb and flow of international diplomacy' (Thompson, 1982b, 5). The result was an exterminist culture, logic and momentum that threatened to push geopolitical antagonism 'in a direction whose outcome must be the extermination of multitudes' (Thompson 1982b, 20). Thompson's argu-ments were justly critiqued for technological determinism but the debate they provoked is much more conceptually nuanced than that found in Virilio (see New Left Review, 1982; Thompson, 1982a; Kaldor and Falk, 1987).

II. Territories warped by transportation technologies: the spatiality of speed

A provocative consequence of pure war that is more particular to Virilio is his argument about the eclipse of geopolitics by chronopolitics or the politics of time. Virilio equates geopolitics with the strategic value of terri-tory whereas chronopolitics is associated with the emergent strategic value of telemetricality. The former's strategic value, he argues, has been declining while the significance of technological systems has increased. Space, he suggests, 'is no longer in geography – it's in electronics':

> Politics is less in physical space than in the time systems admin-istered by various technologies, from telecommunications to airplanes, passing by the TGV, etc. There is a movement from geo- to chrono-politics: the distribution of territory becomes the

distribution of time. The distribution of territory is outmoded, minimal.

(1983, 115)

At other points, he reads this tendency as the discrediting of '*geopolitical extensivity* in favor of a *transpolitical intensivity* of exchange and communication' which has declinist implications for states as territorial entities (1991a, 92, emphasis his). The 'war of real time has clearly supplanted the war in real space of geographical territories that long ago conditioned the history of nations and peoples' (1994a, 206). 'Territory has lost its significance in favor of the projectile. *In fact, the strategic value of the non-place of speed has definitely supplanted that of place*, and the question of possession of Time has revived that of territorial appropriation' (1986, 133, emphasis his). Places disappear in a world delimited by the 'vehicular extermination' of the global nuclear war by virtue of deterrence machines (1986, 134).

These polemical claims by Virilio are certainly overstated, but they should not be underestimated. Virilio's opposition of geopolitics to chronopolitics is a crude and misleading one inasmuch as questions of technology, transportation and speed have always been central to geopolitical theorizing. The pivot in Halford Mackinder's famous 1904 'geographical pivot of history' paper is the relationship between physical geography and transportation technology or what he called 'mobilities of power' (Mackinder, 1904). The dominant mobility of power of Mackinder's pre-Columbian epoch was the horse and camel, the dominant drama the horseback Asiatic invasions of Europe and the ascendant region the land-power of the Asian steppes. In the Columbian epoch, the dominant mobilities of power lay with the most advanced seapower states which were able to construct vast overseas empires for themselves. In the post-Columbian epoch Mackinder envisioned, beginning with the disappearance of the last open spaces for colonial conquest, land-based mobilities of power, particularly railways, would supposedly be dominant.

Mackinder's schema was, of course, crude, sketchy and seriously flawed but it does illustrate how technologies of movement and speed have always been important in geopolitical theorizing. Virilio's equally sweeping speculations take Mackinder's mode of reasoning a step further when he questions the displacement of place by twentieth-century logistics:

What seems central to me is the question of place. In some way, place is challenged. Ancient societies were built by distributing territory. Whether on a family scale, the group scale, the tribal scale or the national scale, memory was the earth; inheritance was the earth. The foundation of politics was the inscription of laws, not only on tables, but in the formation of a region, nation, or

VIRILIO

city. And I believe this is what is now challenged, contradicted by
technology ... Now, technology – Gilles Deleuze said it – is deterritorialization ... Deterritorialization is the question for the end
of this century

(1983, 142)

Just as total war inspired militarist dreams of a perfect arrangement of
territory, and partly though unevenly realized these dreams in its fortress
and bunker landscapes, so also has pure war incited visions of new strategic
order and landscapes appropriate to it. The space–time of pure war is a
strategic order where 'the violence of speed has become both the location and the law, the world's destiny and its destination' (1986, 151). As
the name for terracentric orders of strategic knowledge, geopolitics has
not disappeared but it is no longer at the heart of the war machine.
As the name for the space–time problematic of war more generally, geopolitics is becoming intensively dromological. In the era of pure war,
geopolitical space begins to warp under the gun of speed, for we inhabit
accelerating times and spaces. 'We no longer populate stationariness; we
populate the time spent changing place' (1983, 60). Yet, territory remains
a unit of power's measure as weapons and ideologies mark their ranges
in terms of distances travelled in time (1983, 116). So, we still have not
yet reached his state of chrono-political nirvana, because there is still functional space somewhere, and this space still imposes a few constraints (1983,
166).

The speed-body of dromological societies reconstitutes the time/space
of society's structuration and acculturation around the conditions of
permanent mobilization. Their imbrication with living beings running at
metabolic speed forces humans to accept automated perception, robotic
reasoning, networked community, and computerized communication as
part and parcel of any effective collaboration with other and non-living
beings running at technological speeds (Castells, 1996). This techno-logistical supra-nationalism is totalitarian, and essentially irresistible. To be
borne by these techno-logistics, all are reborn continuously and painfully
with each new generation of techno-logistical complex which now hosts
almost all human life.

Inhabiting chronopolitical acceleration rather than geopolitical space is
not a liberation of movement but a tyranny of speed: 'The blindness of
the speed of means of communicating destruction is not a liberation from
geopolitical servitude, but the extermination of space as the field of political freedom ... the more speed increases, the faster freedom decreases'
(1986, 142).

In everyday life this tyranny of speed provokes a plethora of new social
and political ills: overwork, burn-out, motion sickness, information overload, xenophobic nationalist *resistance* against the speeding flows of

globalization (Barber, 1996; Brook and Boal, 1995; Schor, 1992; Luke and Ó Tuathail, 1998).

III. Virtual insights and geographies: the spatiality of vision

One domain where the (con)fusion of war and speed must be fixed is intelligence, where visual rhetorics of command/control/communication detect and discriminate between fast threats and slow problems. Virilio argues that the vision machines of cinema, television and intelligence satellites often pre-map the spaces that war and speed will occupy, confirming Baudrillard's (1994) beliefs that models precede territory in our age of simulation. Ultimately, the media for Virilio operate as speed and war vehicles. Today, he suggests, 'Blitzkrieg' is more often fought as 'Fernsehenkrieg' in the total warfare of global media markets:

> today, in order to create a *totalitarian Lebensraum*, it is no longer necessary to resort to extraordinary invasions with the motorized vehicles, tanks and stukas of lightning warfare, since one can use the *ordinary penetration* of the new media, the information blitz (1990, 70).

Much of Virilio's work explores the implications of mechanizing, automating, and virtualizing perception, particularly vision. In a world where videocameras coupled with digital scanners in networks of computers are empowered to verify human identities by sweeping their sightless vision over a person's eyeballs to authenticate subjectivity from retinal variations with digital heuristics, this project is quite significant. Virilio's insights, then, flow from 'the philosophical question of the *splitting of viewpoint*, the sharing of perception of the environment between the animate (the living subject) and the inanimate (the object, the seeing machine)', which leads, in turn, to the (con)fusion of 'the factual (or operational, if you prefer) and the virtual; the ascendancy of the "reality effect" over a reality principle already largely contested elsewhere' (Virilio, 1994b, 60).

Splitting sight, then, can paradoxically also split sites, creating reality effects of new spaces beyond, behind, between or beneath those ordinarily accorded to the principal geophysical/sociocultural spaces disclosed by the living subject's reality principle. Motorization and computerization by means of accelerating and virtualizing perception are generating their own hyperchronic or hypertopic properties, which, in the same way as nuclear deterrence has done with war, are transferring human activities '*from the actual to the virtual*' (Virilio, 1994b: 67). Images of the real spaces of objects, data about the real properties of subjects, telemetry on the real-time behaviours of objects interacting with subjects now (dis)place/(re)place actual observables with virtual non-observables whose reality effects are more real than the actual events experienced by those living subjects left out of the data streams or image flows. Such synthetic illusions, however, cannot be easily dismissed, because these virtual environments increasingly are where motorized and computerized subjectivities most materially now dwell.

The real space of the Iranian Airbus was neutral, the real properties of its passengers were peaceful, and the real behaviours of their flight were nonthreatening, but the Aegis-class battle-command centre aboard the U.S.S. *Vincennes* sensed non-observable menace in its battle-management datascapes whose real effects necessitated the tragic shootdown (Der Derian, 1990). On one level, this event perhaps was merely a lethal accident, but on another level it marks a foreseeable collision of the actual and the virtual in the acceleration lanes of infobahn traffic. Speed rules, but speed also kills. Hypermotorization through actual space and/or hypermediatization through virtual space, as Virilio asserts, put reality effects on speed. It perverts 'the illusory order of normal perception, the order of arrival of information. What could have seemed simultaneous is diversified and decomposes . . . it is this intervention that destroys the world as we know it' (Virilio, 1991a: 100–101). Still, speed also recreates the world as we have not known it, but now these effects 'are preparing the way for the *automation of perceptions*, for the innovation of artificial vision, delegating the analysis of objective reality to a machine' (Virilio, 1995: 59) to explore its diverse and decomposed dimensions.

The media thrive on packaging and promoting not the war of all against all, but rather the wars of some against all and all against some (Cumings, 1992). When the world becomes one media market, as it is now, the cameras extend '*multiple solitude* to billions of individuals, the counter-culture of the (postindustrial, postnational, posturban) ghetto now spreading over the whole of the planet that cannot shake off its status as ghetto of the cosmos' (Virilio, 1995, 11). Mediatized by the dromologies of fast capitalism, the ghetto dwellers are 'the chaos that destabilizes mass media caught in the trap of the internal act of war, violation of human rights – the fascinating spectacle, endlessly replayed, of immolation and long, slow death' (Virilio, 1995, 11). The real-time fire fight, for example, of the North Hollywood bank robbery in February 1997 typifies the chaotic televisual consciousness of the spectacular fascia-nation eager to watch an assault upon itself in real-time on live helicam TV. Unable to work, two unemployed 'losers' in full body Kevlar attacked a bank in broad daylight. Botching the robbery, they brazenly remade *Dog Day Afternoon* in the street with AKs mounted with drum magazines. Spraying hundreds of rounds on police and by-standers, they wounded eight police and twenty ordinary citizens before being taken out by high-powered weapons borrowed from a local gun shop. For days, their long, slow, videotaped death displayed how internal war flares up in big-time media markets, as these bad guys, like O.J. in his white Bronco, got their proverbial fifteen minutes of fame in 'live action cam' real-time video. After seeing their regular bullets bounce off criminals on live TV, the Los Angeles police subsequently acquired 600 M-16 army assault rifles to restore their televisual credibility by blurring the coercive lines between police and military force. On the surveillance screens of both institutions, territory has become a dramatic battlespace where 'operations other than war' nevertheless require techno-military techniques, methods and firepower.

The density of dromological systems, then, acquire their own quiddity, becoming in the last analysis features on the mediascapes of third nature (Wark, 1994). At this juncture, an entirely new virtual geography is needed to map their material infrastructures and effective ranges. For Virilio, the built environments of second nature – cities and towns – have not expanded as profusely as the conduits of motorization and mediatization:

> If you want proof, you need only look at a map of the physical geography of France ... this one showing the totality – visible and invisible – of communication networks: canals, railways, airways, highways and, from the visual path of Claude Chappe's ocular telegraph to the electronic age, radar. We immediately realize that during the last two centuries of our history, the physical geography of France has completely disappeared under the inextricable tangle of different media systems; that *not only does delocalization occupy more territory than does* localization, *but it occupies it in totalitarian* fashion ... if, as NATO wishes, we strip every communications systems of [the] neutrality conferred on it by the notion of public service and make the whole thing entirely techno-logistical; then you will have before your eyes *the true physical body of the modern totalitarian state, its speed-body.*
>
> (Virilio, 1990: 91–92, his emphasis)

Dromological existence is delocalized, mobilized, and instrumentalized living within the hyperchronic flow and hypertopic domain of speed. The totalizing reach of the media – electronic and machinic – represent for Virilio the inversion of Clausewitzian war reasoning, because the speed-body of the State must endocolonize its actual territoriality with virtual telemetricalities. Politics now is war carried on by other means, and the doctrine of security founded upon this recognition leads to 'the saturation of time and space by speed, making daily life the last theater of operations, the ultimate scene of strategic foresight' (Virilio, 1990: 92). And, victory in these internal wars comes in fully mediatized forms; indeed, *'beating an enemy involves not so much capturing as captivating them'* (Virilio, 1995: 14, his emphasis). So the heavy artillery of the modern totalitarian regime fires advertorial pitches and infomercial rhetorics out all of its tubes in commodified imageries of communion, desire, and power (for a battlefield conceptualization of this as 'shock and awe' see Ullman *et al.*, 1996).

Thinking geopolitical space with Virilio is a re-thinking of the modern geopolitical gaze under erasure by technoscience and its speeding vehicular technologies (Ó Tuathail, 1997). For Virilio, 'speed is less useful in terms of getting around than in terms of seeing and conceiving more or less clearly' (1994b: 71). The split viewpoint of actual materiality and real virtuality turns all of lived/embodied space–time into evasive manoeuvres or decoy effects, causing the principle of relative illumination (biophysical sight in optical range or radioelectric images looking over horizons/through matter/back in time) to shift. Consequently,

> *the time frequency of light* has become a determining factor in the apperception of phenomena, leaving *the spatial frequency of matter* for dead. . . . Today 'extensive' time, which worked at deepening the wholeness of infinitely great time, has given way to 'intensive' time . . . this *relative difference* between them reconstitutes a new real generation, a degenerate reality in which speed prevails over time and space, just as light already prevails over matter, or energy over the inanimate.
>
> (Virilio, 1994b, 71–72, his emphasis)

Hence, vision must be supplanted by the coming 'vision machine', whose characteristic qualities surpass the sighting of observables or non-observables with a sightless vision that senses stealthier image energies or digital effects as instrumental cities. Such active machinic optics 'will become the latest and last form of industrialization: *the industrialization of the non-gaze*' (Virilio, 1994b, 73, his emphasis) as machinic sensors generate perceptual feeds of observed energy, image space or figurative matter to represent sights and sites. Thus, in worlds of speed, 'we urgently need to evaluate light signals of perceptual reality in terms of intensity, that is "speed," rather than in terms of "light and dark" or reflection or any of the other now dated shorthand' (Virilio, 1990, 74).

Realities of space and time for Virilio, therefore, become relativities between phenomena illuminated or not by transparent lighting effects. Time warps and space distorts, leaving zones of communicating space for light to traverse marking duration absolutized. That is, photo-graphs, or light writing, now describes/enscribes geo-graphs, or space writing. To Virilio, 'if the path of light is absolute, as its zero sign indicates, this is because the principle of instantaneous emission and reception *change-over* has already superseded the principle of *communication* which still required a certain delay,' and so these new forms of constant light energy 'help modify the very definition of the real and the figurative, since the question of REALITY would become the PATH of the light interval, rather than a matter of the OBJECT and space–time intervals' (1994b, 74). So 'chrono-politics', the powers of time, apprehended as speed effects, sublate 'geo-politics', the powers of space, understood as spatial extension.

These interpretations of the vision machine are fascinating, but the fixation on tendencies – in light, speed, war – can be seen as almost fetishistic. Virilio's photofetishism, at times, bleeds off into wild hyperbole. To underscore what he sees as the remarkable changes of speed, for example, he asks us to forsake our cosmological principles, and embrace *illumination* as the force that creates everything. So, 'the center of the universe is no longer the geocentric Earth or anthropocentric human. It is now the luminous point of a helio-centrism, or, better yet, of a lumino-centrism, one that special relativity helped install, whose uncontrolled ambitions derive from the purposes of general relativity' (1991a: 43). Therefore, true consciousness of what is to be done follows from 'subliminal light, the light of the velocity of light that illuminates the world, in the instant in which it offers up its representation' (1991a: 62), and, thus, 'this matter–light – the energetic perception of the contemporary cosmos – replaces the *ether* of earlier physicians and metaphysicians' (1991a: 64). Not everyone, of course, can accept Virilio's revelations that 'In the Beginning, there was the Flash.' This fetishized photo-dictive dimension does not alter as many material realities as Virilio imagines, because at the end of the day there are still very real material machineries, discursive exchanges, and living populations coping with the messy realities of what he dismisses as 'the de-realization of the world' (1991a: 42).

The significant point resting within Virilio's exaggerations is that speed subliminalizes much of human vision, rational reflection, and normal consciousness. Future shock mostly is a motion sickness stemming from 'the rapidity of images and signs in the mirror of the journey, windshield, television or computer screen,' which simplify and distort 'the dromoscopic vision of the world' (Virilio, 1991a: 86) accelerating ahead towards hyper-modernization. Power, then, can no longer simply see panoptically, and thereby enforce its disciplinary designs; it must, instead, more than ever 'fore-see, in other words to go faster, *to see before*' (Virilio, 1990: 87). Risk assessment, game theorizing, operational simulation all are dromoscopic experiments, seeking to reposition state agency systematically in a partially anticipated future so that it might enact its designs as it tried to foresee them. Such chronoptometric manoeuvres usually fail, but risk analysts do everything in their power to transform the positive probabilities of their simulated scenarios into self-fulfilling prophecies.

Conclusion: the end of geopolitics as we know it

For students of geopolitics still clinging to the notion that territoriality is power and hegemonies are built upon its resources, Virilio's recent variations on his ascendant chronopolitics theme identify a networked condition of 'omnipolitanism' as the successor to states, citizenship and territoriality (1997, 75). Reprising familiar themes Virilio argues that the face of society is becoming teleface, the settled history of nations a flux of transitory media representations, while citizenship is overshadowed and overcome by contemporaneity (1997, 74). Politics is eclipsed by technology as citizens separate out into either caches of netizens networking in the fast lanes of the global economy or the trashbins of lumpen techno-proletarians stuck at the dead ends of networks.

The real space of national geography and the world space of geopolitics gradually are giving way to the real time of international communications and the world time of chronostrategic proximity (1997, 69). Old military and industrial complexes will be superseded by informational metropolitan complexes 'associated with the omnipotence of the absolute speed of the waves conveying the various signals' (1997, 83). Instead of the cosmopolis modelled on ancient Rome, a new world-city will surge forth, a hyperconnected omnipolis whose major defining characteristic is the interconnected global stock exchange. Typically, Virilio's argument is a more extreme technological vision of the literature identifying the emergence of an interlinked system of global cities (Sassen, 1991; Taylor, 1996, 186–88). Urban areas are becoming delocalized 'cities of bits' while the architecture that counts is increasingly the architecture within computers, information systems and networks (Mitchell, 1995). Concrete presence is fading in the face of the telepresence offered by information superhighways, real-time video transmissions and planetary networks of perpetual communication. The 'metropolization that we should fear for the coming century involves not so much concentration of populations in this or that 'city network,' as the hyperconcentration of the *world-city*, the city to end all cities, a virtual city of which every real city will ultimately be a suburb, a sort of *omnipolitan* periphery whose *center will be nowhere and circumference everywhere*' (1997, 74, emphasis in original).

This tendency is extremely dangerous from Virilio's point of view for it makes more likely the possibility of a general accident, a delocalized global event of irresistible force, like a stock market crash, which he compares to an informational Chernobyl. A disturbance or failure in one part of the omnipolitan network has implications for all, bringing with it the possibility of a generalized technological and therefore social crash almost immediately. The post-geopolitical world of the hyperconnected global cities blending into one invests power in networks of computers which can break the central bank of any state and wreck its best laid defences. Precariousness is the new law of an international politics under the rule of real-time networks.

Virilio's recent writings continue to develop themes in his work established decades ago. Undoubtedly a creative theorist of what can be described as the *postmodernization of geopolitics* (Ó Tuathail, 1998), his analyses have an intriguing and seductive appeal. Yet, in analysing tendencies, disturbing and otherwise, in the technoculture of postmodernity, Virilio is also deeply complicitous with the tropes of digital culture, with its apocalyptic visions, its sound-bite futurism, and normalization of hyperbole. His analysis is often as unrestrained as the tendencies he describes and condemns. The significance of geopolitics may appear to be fading for some; yet, as Bosnia, Rwanda, Taiwan, Kashmir and numerous other places remind us, its heavy hand still shapes life and death across the planet.

Acknowledgements

This paper was originally presented at the annual meeting of the Association of American Geographers, Fort Worth, Texas, April 2–6, 1997.

References

Agnew, J. 1998. *Geopolitics: Revisioning World Politics*, London: Routledge.
Barber, B. 1996. *Jihad vs McWorld*, New York: Ballantine.
Baudrillard, J. [1981] 1994. *Simulacra and Simulation*, Ann Arbor: University of Michigan Press.
Brook, J. and Boal, I. 1995. *Resisting the Virtual Life*, San Francisco: City Lights.
Castells, M. 1996. *The Rise of the Network Society*, Oxford: Blackwell.
Cumings, B. 1992. *War and Television*, London: Verso.
Der Derian, J. 1990. The (s)pace of international relations: simulation, surveillance and speed. *International Studies Quarterly* 34: 295–310.
Haraway D. 1997. *Modest Witness @ Second Millennium. Female Man© Meets Onco Mouse™: Feminism and Technoscience*, New York: Routledge.
Harvey, D. 1989. *The Condition of Postmodernity*, Oxford: Blackwell.
Harvey, D. 1996. *Justice, Nature and the Geography of Difference*, Oxford: Blackwell.
Kaldor, M. and R. Falk. 1987. *Dealignment*, Oxford: Blackwell.
Luke T. 1989. 'What's Wrong with Deterrence?' A Semiotic Interpretation of National Security Policy, in J. Der Derian and M. Shapiro (eds). *Intertextual/International Relations: Postmodern and Poststructural Readings of World Politics*, Lexington, MA: Lexington Books.
Luke, T. 1991. The Discourse of Deterrence: National Security as Communicative Interaction. *Journal of Social Philosophy*, XXII: 30–44.
Luke, T. and Ó Tuathail, G. 1998. Global flowmations, local fundamentalism, and fast geopolitics: 'America' in an accelerating world order, in A. Herod, G. Ó Tuathail, and S. Roberts, (eds). *An Unruly World? Globalization, Governance and Geography*, London: Routledge.
Mackinder, H. 1904. The Geographical Pivot of History, *Geographical Journal*, 23: 421–44.
Mann, M. 1986. *The Sources of Social Power. Volume 1*, Cambridge: Cambridge University Press.
Matellart, A. [1991] 1994. *Mapping World Communication*, Minneapolis: University of Minnesota Press.
Matellart, A. [1994] 1996. *The Invention of Communication*, Minneapolis: University of Minnesota Press.

McNeill, W. 1982. *The Pursuit of Power*, Chicago: University of Chicago Press.
Mitchell, W. 1995. *City of Bits: Space, Place and the Infobahn*, Cambridge, MA: MIT Press.
Mumford, L. 1963. *Technics and Civilization*, San Diego: Harcourt Brace Jovanovich.
Mumford, L. 1970. *The Pentagon of Power*, San Diego: Harcourt Brace Jovanovich.
New Left Review, 1982. *Exterminism and Cold War*, London: Verso.
Ó Tuathail, G. 1996. *Critical Geopolitics*, Minneapolis: University of Minnesota Press.
Ó Tuathail, G. 1997. At the end of geopolitics? Reflections on a plural problematic at the century's end. *Alternatives*, 22: 35–56.
Ó Tuathail, G. 1998. Postmodern geopolitics? The modern geopolitical imagination and beyond, in G. Ó Tuathail and Simon Dalby (eds). *Re-Thinking Geopolitics*, London: Routledge.
Sassen, S. 1991 *The Global City*, Princeton: Princeton University Press.
Schor, J. 1992. *The Overworked American*, New York: Basic Books.
Sherry, M. 1996. *Under the Shadow of War*, New Haven: Yale University Press.
Taylor, P. 1996. *The Way the Modern World Works*, Chicester: John Wiley.
Thompson, E.P. 1982a. *Beyond the Cold War*, New York: Pantheon.
Thompson, E.P. 1982b. 'Notes on Exterminism, the last stage of civilization,' in New Left Review (ed.) *Exterminism and Cold War*, London: Verso.
Thompson, E.P. 1985. *The Heavy Dancers*, New York: Pantheon Books.
Tzu, Sun. 1971. *The Art of War*, Oxford: Oxford University Press.
Ullman, H., Wade, P., Edney, L., Franks, F., Horner, C., Howe, J. Brendley, K., 1996. *Shock and Awe: Achieving Rapid Dominance*, Washington D.C: Institute for National Strategic Studies.
Virilio, P. and Lotringer, S. 1983. *Pure War*, New York: Semiotexte.
Virilio, P. [1977] 1986. *Speed and Politics*, New York: Semiotext(e).
Virilio, P. 1989. *War and Cinema: The Logistics of Perception*, London: Verso.
Virilio, P. [1978] 1990. *Popular Defense and Ecological Struggles*, New York: Semiotext(e).
Virilio, P. [1980] 1991a. *The Aesthetics of Disappearance*, New York: Semiotext(e).
Virilio, P. [1984] 1991b. *The Lost Dimension*, New York: Semiotext(e).
Virilio, P. [1976] 1993. *L'Insécurité au Territoire*, Paris: Galilee.
Virilio, P. [1975, 1991, 1994] 1994a. *Bunker Archeology*, New York: Princeton Architectural Press.
Virilio, P. [1988] 1994b. *The Vision Machine*, Bloomington: Indiana University Press.
Virilio, P. [1993] 1995. *The Art of the Motor*, Minneapolis: University of Minnesota Press.
Virilio, P. [1995] 1997. *Open Sky*, London: Verso.
Wark, M. 1994. *Virtual Geography*, Bloomington: Indiana University Press.

INDEX

Note: **emboldened** page references refer to chapters